Y0-BXH-101

Organization of Prokaryotic Cell Membranes

Volume I

Editor

Bijan K. Ghosh, D.Sc.
Professor
Department of Physiology and
Biophysics
CMDJ-Rutgers Medical School
Piscataway, New Jersey

CRC Press, Inc.
Boca Raton, Florida

Library of Congress Cataloging in Publication Data

Main entry under title:

Organization of prokaryotic cell membranes.

Bibliography: p.
Includes index.
1. Bacterial cell walls. 2. Cell membranes.
I. Ghosh, Bijan K., 1935- [DNLM:
1. Bacteria—Cytology. 2. Cell membrane—Ultrastructure. 3. Cells. QW 51 068]
QR77.073 589.9'0875 80-25355
ISBN 0-8493-5653-9 (v. 1)

This book represents information obtained from authentic and highly regarded sources. Reprinted material is quoted with permission, and sources are indicated. A wide variety of references are listed. Every reasonable effort has been made to give reliable data and information, but the author and the publisher cannot assume responsibility for the validity of all materials or for the consequences of their use.

Direct all inquiries to CRC Press, Inc., 2000 N.W. 24th Street, Boca Raton, Florida 33431.

International Standard Book Number 0-8493-5653-9 (Volume I)
International Standard Book Number 0-8493-5654-7 (Volume II)
International Standard Book Number 0-8493-5659-8 (Volume III)

Library of Congress Card Number 80-25355
Printed in the United States

FOREWORD

The pioneering investigations of Gorter and Grendel (1925) and Danielli and Davson (1935) in which cell membranes were visualized as bimolecular leaflets ushered in what might be termed the modern era of biomembrane research. Electron microscopy played a key role in establishing the universal existence and anatomical features of biological membranes in cells of animal, plant, and microbial origin, and indeed provided the necessary methodology for the isolation and characterization of biomembranes. Moreover, the ultrastructural studies pointed to the essential differences in surface organization and membranous organelles of eukaryotic and prokaryotic cells. The robust bacterial cell walls of Gram-positive organisms and the envelopes of Gram-negative bacteria became amenable to isolation in the 1950s, and soon after the pioneering work of Weibull (1953) paved the way for the study of prokaryotic cytoplasmic membranes. Three decades of interest in the physiological and biochemical properties of bacterial plasma membranes have witnessed great advances in the state of our knowledge of their structure and functions. Dr. Bijan Ghosh is to be congratulated in bringing together so many distinguished leaders in the field of prokaryotic membrane research in three Volumes devoted to the "Organization of Prokaryotic Cell Membranes." The collection of authoritative articles covering the most active areas of prokaryotic biomembrane investigations into the several volumes has provided a great service not only to those interested in the field but also to microbiologists in general. We are deeply indebted to Dr. Bijan Ghosh for his considerable editorial efforts in assembling truly valuable contributions to our understanding of such basic aspects of bacterial membrane studies as transport functions, energizing membranes, the biochemistry and immunochemistry of membranes, and the structure-function relationships of photosynthetic membranes, gas vacuoles, and the more controversial mesosomes. The extensive reference lists will be invaluable for students and research workers in the various fields of prokaryotic membrane research especially in the "exploding" segments of the molecular and genetic aspects of Gram-negative cell membranes. These monographs will also serve to focus attention on prokaryotic membranes that are so often ignored by eukaryotic "membraneologists" and will provide an excellent reference source for many years to come.

Milton R. J. Salton
Department of Microbiology
New York University
School of Medicine
New York, New York

PREFACE

The hallmarks of bacterial physiology are the fast growth rate of these prokaryotes, the minuteness of their size, and a high degree of adaptability to differing growth conditions. Their fast rate of growth and colonization enable the bacterial cells to flourish on limited and transient resources. As a result of their smallness these cells have a high membrane-to-cytoplasm ratio. It is possible that this high ratio is important for the establishment of greater contact between the biochemical machinery of the cells and the biosphere.

It has been frequently suggested in the literature that eukaryotic cells have evolved from prokaryotic ancestors. These forerunners of the eukaryotes might have been basically prokaryotic, but it cannot be denied that contemporary prokaryotic cells have evolved equally as much as the eukaryotic cells. Therefore, a dual direction of cell evolution can be suggested.

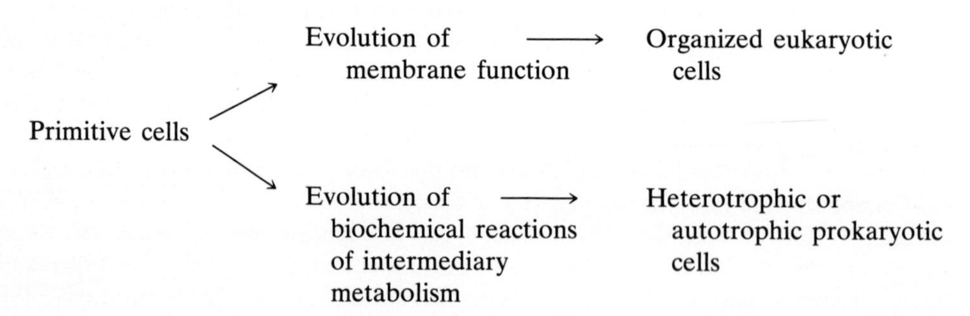

In general, the prokaryotic cell membrane lacks plasticity and ability to differentiate; however, these properties are inherent in the eukaryotic cell membranes. It is possible that the physiological functions related to membrane plasticity and differentiation, e.g., phagocytosis, pinocytosis, organelle formation, etc., did not develop in the prokaryotic cells. Evolution in prokaryotic cells may have progressed towards the diversification of the biochemical reactions for intermediary metabolism. Because of this metabolic diversity the habitat for prokaryotic cells varies widely. The following concepts are consistent with the properties of the prokaryotic membranes: (1) because of the lack of differentiation, the characteristics of the primitive ancestral cell membrane might have been conserved through evolution in prokaryotic cells; (2) the diversity of the intermediary metabolism in different groups of prokaryotic cells is likely to be accompanied by biochemical differences of the membranes. Due to this extreme variability, it is difficult to form a general concept of the structure and function of the prokaryotic cell membrane.

One group of eukaryotic cells (i.e., fungi) does not fit in this simplistic model of cell evolution. Wide diversity of intermediary metabolism is well known in fungal organisms. As regards subcellular morphology, some fungal cells may be richly endowed with a variety of organelles commonly found in plant or animal cells, whereas the others may show such paucity of organelles that hardly any subcellular body could be demonstrated besides nucleus. This notable variation in subcellular organelle content of fungal cells may be frequently correlated with subtle differences in their growth conditions. One may speculate from these observations that the organelle formation, presumably dependent upon membrane differentiation, is at an intermediate stage of development in fungal cells. This possibility, when taken together with the present view that fungal cells are highly evolved, suggests that membrane differentiation may have failed to progress enough in the ancestral cells of fungal organisms. Hence, both differentiability and primitive property of the membrane has been conserved in fungi. In other words, the membrane system in fungal cells may have retained the full potential to revert back to its primitive undifferentiated character. However, this idea (which is conjectural) is presented to correlate current views on evolution of membrane and intermediary metabolism in one general model. A critical study on membrane evolution may be a rewarding area of research in cell biology.

The above discussion shows the difficulty in organizing a comprehensive text on the prokaryotic cell membrane. I have attempted to obviate these problems by combining discussions on both the general and specialized properties of prokaryotic membranes. A general feeling has been expressed by colleagues in the field of membrane research that the work in this area is progressing at a very fast rate. Therefore, the scientific material in a manuscript becomes largely out of date because of the delay inherent in the publication of a book. Hence, the aim was to organize a general text written by experts in the field. I requested that the authors present a thorough review of the available scientific material within a general conceptual framework and to indicate the future direction. In addition, an attempt has been made to provide an extensive bibliography.

The subjects covered may be divided into two parts: (A) general considerations of structure and physiological functions, and (B) specialized membranes of different organisms. Chapters in category (A) include: The Role of Membranes in the Transport of Small Molecules; the Role of the Membrane in the Bioenergetics of Bacterial Cells; Immunology of the Bacterial Membrane; Bacterial Cell Surface Receptors; Biosynthesis of Bacterial Membrane Proteins; Submicroscopic Morphology of Bacterial Membranes; the Role of the Membrane in the Transport of Macromolecules in Bacterial Cells; and, Some Evolutionary Considerations of Prokaryotic Cell Membranes. Category (B) includes: The Mycoplasma Membrane; The Mesosome; The Gas Vesicle: A Rigid Membrane Enclosing a Hollow Space; Membranes of Phototropic Bacteria; and Membranes of Hydrocarbon Utilizing, Nitrifying and Sulfur Bacteria. The chapters of groups A and B have been distributed within three volumes.

Frequently, investigators working with a specific organism lose appreciation for the diversity of the prokaryotic membrane material. However, it is obvious that the flow of information from research on the different types of membranes will be helpful in formulating a unified approach in prokaryotic membrane studies. Therefore, chapters from both group A and group B have been incorporated in each of the individual volumes. It is hoped that this unified approach will be helpful to students and research workers in the field of prokaryotic membranes.

It was mentioned earlier that the bacterial cytoplasm directly interacts with the external environment. Membrane is the interface of this interaction. Therefore, the information exchange between the cytoplasmic material (enzymes and other factors) and the extracellular environment is mediated through the membrane. There is a strong possibility that the prokaryotic cell membranes receive the information input from the biosphere and regulate physiological activity accordingly. Vigorous research activity will develop in this area of coupled receptor regulator activity of membranes. A thorough understanding of the regulatory role of membranes will stimulate the development of technology for programming the bacterial cells for the production and secretion of industrially important substances in the growth (fermentation) medium.

In fact, industrial uses of bacterial cells are steadily increasing in the field of bioorganic industry. The future holds the possibility of extensive use of bacterial cells in the area of biomass utilization. The production of enzymes by microorganisms is already a $100 million industry and has the potential to increase further. Thus, study of the prokaryotic cell membrane is rewarding, both for the understanding of basic biological phenomena and for the development of a technology for the greater industrial use of prokaryotic cells. I hope the materials in these three volumes will stimulate further research and will help the students in the field of prokaryotic cell physiology.

I must thank all the contributors for their valuable articles. In spite of extremely busy schedules, they sympathetically considered my proposal and gave their time. With their help, I must say that we may not have touched the success mark, but we have reached the area of near success in our project.

In the initial planning, bibliography collection, and other matters I am indebted to Benita Budd, Rita Deb, and Ranjan Ghosh. Finally, I thank Helen Sedlowski for all the rush typing work.

B. K. G.

THE EDITOR

Bijan K. Ghosh, D.Sc., is Professor in Physiology and Biophysics in the Department of Physiology and Biophysics at the College of Medicine and Dentistry of New Jersey—Rutgers Medical School, Piscataway, and Honorary Professor in Microbiology at the Waksman Institute of Microbiology, Rutgers State University, N.J.

Dr. Ghosh received the B.Sc. and the M.Sc. degrees in physiology from Presidency College, Calcutta University. His doctorate was awarded by Calcutta University while he was working in the Indian Institute of Experimental Medicine in 1963. He engaged in postgraduate study at various institutions including the Woods Hole Oceanographic Institute, Woods Hole, Massachusetts, and the Anatomy Institute of the University of Bern, Switzerland.

Dr. Ghosh was an instructor and subsequently, a junior research fellow, at Presidency College during 1958 and 1959. At the Indian Institute of Experimental Medicine Dr. Ghosh was a Junior Research Fellow from 1959 to 1961, and a Senior Scientific Assistant until 1964. From July 1965 until November 1966 he was a Medical Research Council Canada Postdoctoral Fellow at the Department of Bacteriology and Immunology of the University of Western Ontario. He became associated with Rutgers University as Waksman-Merck Postdoctoral Fellow in November 1966, and served there as an assistant professor from 1967 to 1973. He was a visiting Professor at the University of Amsterdam, Netherlands in 1973. He moved to the Rutgers Medical School at the end of 1973.

Dr. Ghosh is a member of the Editorial Board of Journal of Bacteriology, and he is very active in the Morphology and Ultrastructure Division of the American Society for Microbiology. He organized several symposia in the general area of structure/function interrelationship in microorganisms.

Dr. Ghosh is a member of the Canadian Society of Biochemistry, the American Society for Microbiology, the Electron Microscopic Society of America, and the American Association for the Advancement of Science. He is a fellow of the American Institute of Chemists and a member of the New York Academy of Sciences. He is author or co-author of 65 original papers including some reviews and chapters on Bacterial and Fungal Ultrastructure in the CRC Handbook of Microbiology.

Among Dr. Ghosh's awards are a University Gold Medal from the Calcutta University, the Medical Research Council of Canada Postdoctoral Fellowship, a Waksman-Merck Postdoctoral Fellowship at the Waksman Institute of Microbiology of Rutgers University, and the Research Career Development Award from the National Institute of General Medical Sciences of the National Institutes of Health.

Dr. Ghosh has done extensive research on bacterial fine structure, particularly on the microbial membranes and the bacterial mesosomes, and on the evolution of subcellular organelles membrane phenomena of enzyme secretion in microorganisms.

CONTRIBUTORS

Volume I

A. J. Apperson
Biology Department
University of California,
San Diego
La Jolla, California

Steven S. Dills
Research Associate
Warner-Lambert Company
Morris Plains, New Jersey

Arthur A. Guffanti, Ph.D.
Department of Biochemistry
Mt. Sinai School of Medicine
New York, New York

Terry Ann Krulwich, Ph.D.
Associate Professor of Biochemistry
Department of Biochemistry
Mt. Sinai School of Medicine
City University of New York
New York, New York

C. A. Lee
Biology Department
University of California,
San Diego
La Jolla, California

John E. Leonard
Department of Biology
University of California,
San Diego
La Jolla, California

Kenneth G. Mandel
Department of Biochemistry
Mount Sinai School of Medicine
New York, New York

Peter Owen, Ph.D.
Lecturer in Microbiology
Dept. of Microbiology
The Moyne Institute
Trinity College
University of Dublin
Dublin, Ireland

Shmuel Razin, Ph.D.
Professor of Microbiology
Department of Membrane and
Ultrastructure Research
The Hebrew University-Hadassah
Medical School
Jerusalem, Israel

Milton H. Saier, Ph.D.
Associate Professor
Department of Biology
University of California
at San Diego
La Jolla, California

ORGANIZATION OF PROKARYOTIC CELL MEMBRANES

Volume I

Volume II

Chapter 1

THE ROLE OF MEMBRANES IN THE TRANSPORT OF SMALL MOLECULES

J. E. Leonard, C. A. Lee, A. J. Apperson, S. S. Dills, and M. H. Saier, Jr.

TABLE OF CONTENTS

Surprised by Joy—impatient as the Wind,
I turned to share the transport.

W. Wordsworth

When one sees eternity in things that pass away and infinity in finite things, then one has pure knowledge. But if one merely sees the diversity of things with their divisions and limitations, then one has impure knowledge.

From the Bhagavad Gita

I. INTRODUCTION

The plasma membrane of any living cell must provide two primary functions: (1) it must serve as a permeability barrier, preventing entry of unwanted substances into the cytoplasm as well as leakage of essential cytoplasmic constituents from the cell interior and (2) it must allow selective permeability of substances which either may be of utility to the cell and are therefore accumulated within the cytoplasm or are deleterious and must be actively extruded. The Gram-negative bacterium possesses two permeability barriers, the inner and the outer membranes of the cell envelope (Section II.) Both of these membranes consist of phospholipid bilayers into which integral membrane proteins are inserted and with which peripheral membrane proteins may be associated. There are several classes of phospholipids, and these contain characteristic fatty acyl moieties.

Fatty acid synthesis and phospholipid assembly are accomplished by enzymes which, for the most part, are integral constituents of the cytoplasmic membrane (Section III). Genetic alteration of the phospholipid biosynthetic apparatus allows the investigator to modify the cellular lipid content within certain limits and thereby to ascertain the physiological significance of a particular lipid constituent to the fluidity and integrity of the membrane. Such manipulations also permit evaluation of the requirements of integral membrane enzymes and transport proteins for specific classes of phospholipids (Section IV).

An understanding of the dependencies of transmembrane solute transport systems on the phospholipid matrix ultimately depends on a thorough knowledge of the detailed structural and functional properties of the proteins which catalyze vectorial reactions. Acquisition of this knowledge will require that these proteins be isolated and their modes of action analyzed. Only in recent years have the protein constituents of bacterial permeases become available in homogeneous form. This accomplishment has led to the realization that many small hydrophilic solutes permeate the outer membrane through proteinaceous aqueous pores (Section V). Solute-specific permeases are generally absent from this membrane. By contrast, nonspecific pores are not found in the cytoplasmic membrane. Instead, solute-specific energy-linked transport systems confer upon this membrane the ability to generate electrochemical gradients of electrolytes and concentration gradients of uncharged solutes. While many of these transport systems function with inwardly directed polarity, others pump ions and metabolites out of the cell. Some of the best characterized bacterial permeases and their possible translocation mechanisms are discussed in Section VI.

Finally, the extent to which a nutrient present in the external medium is accumulated in the cytoplasm is dependent not only on the activities of the permease proteins but also on their rates of synthesis. Numerous regulatory interactions control the syntheses of permeases and enzymes, and representative examples are discussed in Section VII. The cytoplasmic concentration of cyclic AMP, necessary for the induced synthesis of numerous carbohydrate permeases and catabolic enzyme systems, is modulated in response to internal and external stimuli which serve as indicators of carbon and energy sufficiency. These indicators, in turn, control the activities of the cyclic AMP synthetic enzyme and the cyclic nucleotide transport system.

This chapter is not intended to be comprehensive. It selectively deals with subjects which the authors believe provide mechanistic insight into the selective permeability properties of the Gram-negative bacterial cell envelope. Frequent reference to original reearch articles and recent reviews provide the reader with opportunity for more detailed analyses of subjects dealt with only superficially here.

While the authors have restricted their comments to those topics which deal specifically with the membranes of the Gram-negative cell and their permeability properties, some similarities exist between Gram-negative and Gram-positive cells and the processes required to move solute across the cell envelope. For example it has been demonstrated that some Gram-positive organisms possess the phosphoenolpyruvate: sugar phosphotransferase system[136] as well as the ability to utilize proton symport systems.[224] Conversely, the most prominent difference between the Gram-negative and Gram-positive bacteria is the finding that the latter do not possess periplasmic binding proteins and probably do not, therefore, utilize ATP as an energy source for solute transport.[224] For a more complete discussion of transport in Gram-positive organisms the reader is referred to Reference 224.

II. STRUCTURE OF THE GRAM-NEGATIVE BACTERIAL CELL ENVELOPE

The Gram-negative bacterial cell envelope consists of three layers: the inner membrane, the peptidoglycan layer, and the outer membrane.[1-9] Some Gram-negative organisms, e.g., the halophilic bacteria of the genus *Halobacterium,* do not possess a peptidoglycan layer.[10] The inner or cytoplasmic membrane of Gram-negative bacteria is structurally similar to that of Gram-positive organisms and comprises the osmotic barrier of the cell.[5,6] This membrane contains many transport proteins, cytochromes, respiratory chain components, lipopolysaccharide and phospholipid biosynthetic enzymes, and the proton translocating ATPase.[11-15]

The outer membrane contains approximately half of the total protein and phospholipid complement of the cell envelope and provides the cell with a passive diffusion barrier.[16] Treatment of Gram-negative cells with ethylenediaminetetraacetate (EDTA) removes some of the lipopolysaccharide and renders the cell more permeable to certain drugs and enzymes, indicating that some of the permeability properties of the outer membrane are due to this component.[16,17] The outer membrane contains receptors for bacteriophages, colicins, and various transport proteins.[1,18]

The phospholipid compositions of both *Escherichia coli* and *Salmonella typhimurium* are very similar, and the phosphoglycerides are divided roughly equally between the inner and outer membrane.[5,15,19-22] The major phospholipid species in both organisms is phosphatidylethanolamine (66 to 85% of the total phospholipid) followed by phosphatidylglycerol (10 to 22%), and diphosphatidylglycerol (3 to 15%).[15,20-25] The membranes of the halophilic bacteria are composed primarily of phosphatidylglycerol and its sulfate derivative and, to a lesser exent, glycolipids. Present in small amounts are the nonionic lipid, bacterioruberin, and squalenes.[26-28] A more comprehensive discussion of the structure of the Gram-negative cell envelope, including models and electron micrographs of the inner and outer membranes of both Gram-positive and Gram-negative cells, can be found in a number of recent reviews.[1,5,7,8,18,29,30]

III. MEMBRANE PHOSPHOLIPID BIOSYNTHESIS

While the elucidation of the various enzymatic steps involved in the biosynthesis of bacterial membrane phospholipids and their intermediates has been conducted almost exclusively in strains of *E. coli,* these biosynthetic conversions appear to be similar to those found in many other Gram-negative organisms.[31] As there are several current reviews of this subject,[32-36] only an overview of the biosynthetic pathways and the various genetic studies will be presented here.

A. Precursor Requirements

The biosynthesis of membrane phospholipids (Figure 1) is dependent upon the availability of four precursors: *sn*-glycero-3-phosphate, free fatty acids, cytidine 5′-triphosphate (CTP), and L-serine. During growth on glucose, *sn*-glycero-3-phosphate is formed from dihydroxyacetone phosphate by the action of *sn*-glycero-3-phosphate dehydrogenase. This precursor may also be synthesized from *sn*-glycerol by the action of glycerokinase. The typical fatty acid complement of *E. coli* K-12 strains grown under normal conditions includes palmitic acid (25 to 40% of the total cellular fatty acid content), palmitoleic acid (also 25 to 40%), *cis*-vaccenic acid (25 to 35%), and several minor species (all $<5\%$).[35,37,38] While stationary phase cells contain large amounts of cyclopropane fatty acids, they do not appear to be essential to the cells for growth.[39]

Various genetic studies have been conducted with regard to fatty acid biosynthesis and oxidation in *E. coli.*[33-35] The first mutants to be isolated were those unable to synthesize unsaturated fatty acids and can be divided into two classes. While the *fabA* mutants lack the ability to introduce the double bond of the unsaturated fatty acids,[21,40] the *fabB* auxotrophs contain a defective condensing enzyme.[21,41] Further, strains of *E. coli* have been isolated which cannot degrade fatty acids (*fad* mutants).[34,42,43] Mutants containing both the *fab* and *fad* lesions are of great importance since their membrane fatty acid content is determined largely by the fatty acids which are exogenously supplied.[44-49]

As is shown in Figure 1, CTP is required for the formation of cytidine 5′-diphosphate diacylglycerol, an intermediate necessary for the biosynthesis of not only phosphatidylserine and phosphatidylethanolamine but also of phosphatidylglycerol and cardiolipin (diphosphatidylglycerol).[50] The last precursor, L-serine, is required for the formation of the polar head groups of phosphatidylserine and phosphatidylethanolamine.[51]

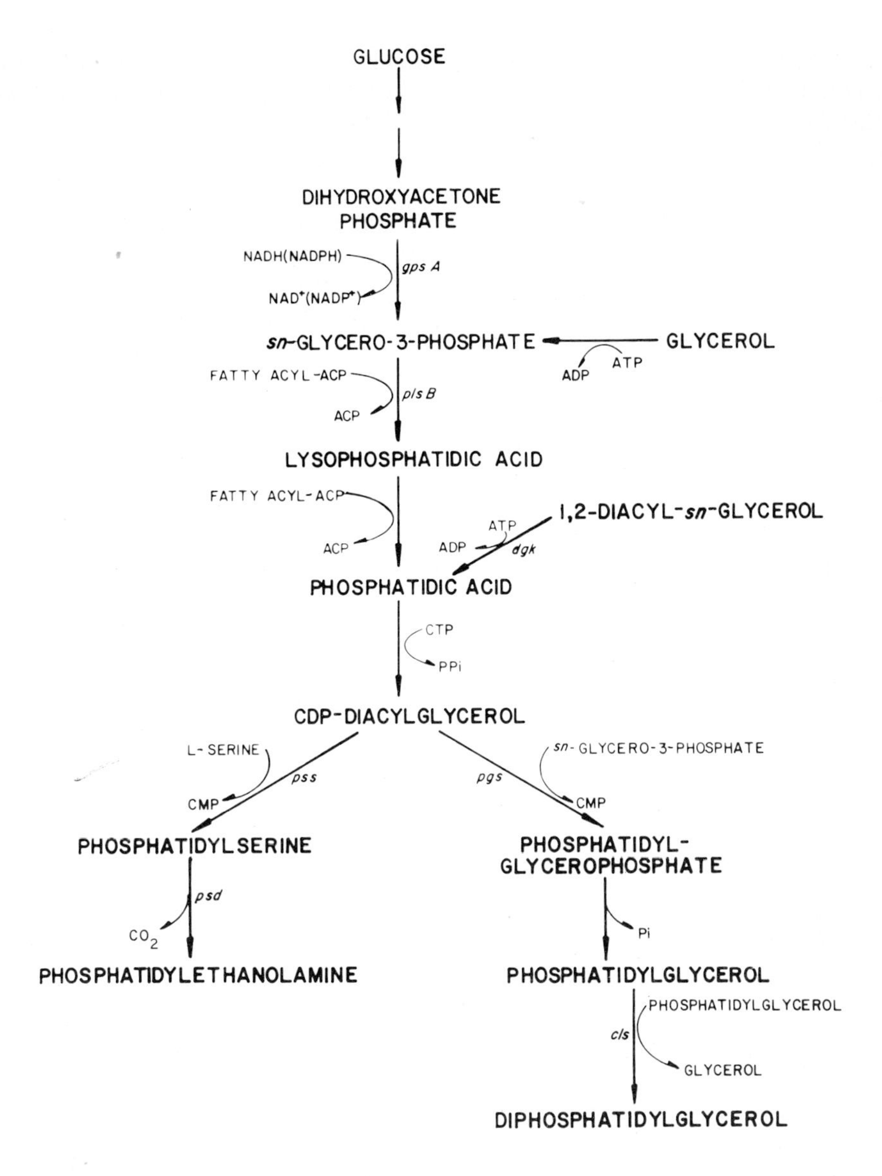

FIGURE 1. Biosynthetic pathways for *E. coli* membrane phospholipids. Various types of mutants are available; these are indicated by the genetic symbol alongside the enzymatic reactions. The abbreviations used are explained in the text. (Redrawn from Raetz, C. R. H., *Microbiol. Rev.,* 42, 614, 1978.)

B. Enzymology and Genetics of Phospholipid Biosynthesis

With the exception of phosphatidylserine synthetase (Figure 1), all of the enzymes of these pathways have been shown to be localized in the cytoplasmic membrane.[14,31,52,53] Phosphatidylserine synthetase, although apparently tightly bound to the ribosomes of most Gram-negative organisms,[31,52] presumably catalyzes the conversion of CDP-diacylglycerol to phosphatidylserine with its substrate firmly attached to the membrane.

1. CDP-Diacylglycerol Biosynthesis

The biosynthesis of this important intermediate not only allows the cell to regulate the fatty acid content of its membrane phospholipids (see below) but also provides the central precursor for the formation of the three major phospholipids of the bacterial membrane. In glucose-grown cells the precursor for all phosphoglycerides, *sn*-glycero-3-phosphate, is formed from dihydroxyacetone phosphate by the action of *sn*-glycero-3-phosphate dehydrogenase.[53,54] (This enzyme has also been referred to as *sn*-glycero-3-phosphate synthetase[55] or biosynthetic *sn*-glycero-3-phosphate dehydrogenase.[56]) It has been recently purified to apparent homogeneity and kinetically characterized.[57-59] The gene which codes for this enzyme has been designated *gpsA* and maps at approximately 81 min on the *E. coli* chromosome.[60] The conversion of *sn*-glycero-3-phosphate to phosphatidic acid is catalyzed by two separate acyltransferases.[61-65] The first step, the formation of lysophosphatidic acid, occurs by the action of *sn*-glycero-3-phosphate acyltransferase, which acylates the substrate at the 1 position. Little lysophosphatidic acid accumulates, however, since it is rapidly acylated at the 2 position by a separate acyltransferase.[61,65] The conversion of *sn*-glycero-3-phosphate to phosphatidic acid by these two reactions represents an obvious site for the cellular regulation of the fatty acid content of the membrane phospholipids.[66]

The initial acyltransferase mutants isolated were those containing a defective *sn*-glycero-3-phosphate acyltransferase and were selected by a tritium suicide procedure.[67] The lesion was mapped at about the 12 min mark on the *E. coli* chromosome,[68] and it was suggested that these *plsA* mutants contained a temperature-sensitive *sn*-glycero-3-phosphate acyltransferase.[67] Further investigation, however, showed that the *plsA* mutants contained a thermolabile adenylate kinase.[33,69] The true structural gene for the *sn*-glycero-3-phosphate acyltransferase, the *plsB* gene, was first identified by Bell.[54] These mutants were among those selected for *sn*-glycero-3-phosphate auxotrophy and, like the *gpsA* mutants, rapidly cease all phospholipid synthesis upon removal of *sn*-glycero-3-phosphate from the medium.[70-72]

Although no mutants have been isolated which lack the 1-acyl-*sn*-glycero-3-phosphate acyltransferase,[32] which catalyzes the second reaction in the synthesis of phosphatidic acid, mutants have been selected which contain a defective diglyceride kinase.[24,73-75] These mutants, designated *dgk*,[32] map at approximately 91 min on the *E. coli* chromosome and are unable to convert *sn*-1,2-diacylglycerol to phosphatidic acid.[24,73-75] This probably represents only a minor pathway for the biosynthesis of phosphatidic acid since these mutants do not exhibit a striking decrease in total phospholipid synthesis.[24,76]

The final reaction in the enzymatic conversion of dihydroxyacetone phosphate to CDP-diacylglycerol is catalyzed by cytidinetriphosphate: phosphatidic acid cytidylyltransferase. This enzyme promotes the condensation of CTP and phosphatidic acid to produce CDP-diacylglycerol and pyrophosphate. No mutants have been isolated which lack this activity.[32]

2. *Phosphatidylethanolamine Biosynthesis*

The conversion of CDP-diacylglycerol to phosphatidylethanolamine is one of the most important of all phospholipid biosynthetic conversions since phosphatidylethanolamine is the major phospholipid of the Gram-negative cell envelope.[15,20-25] The initial reaction in this two step conversion is catalyzed by CDP-diacylglycerol: L-serine *O*-phosphatidyltransferase (phosphatidylserine synthetase). This enzyme was first discovered by Kanfer and Kennedy[51] and has since been purified to homogeneity.[77] While phosphatidylserine synthetase has been reported to be almost totally bound to the 70S ribosomes,[52,77] a significant amount may also be soluble.[78,79]

Raetz[80,81] has isolated several mutants of *E. coli* specifically lacking this enzyme, and the lesion, designated *pss*, maps at approximately 49 min on the *E. coli* chromosome. Two of these isolates, *pss-8* and *pss-21*, are temperature sensitive and cease growth soon after being shifted to the nonpermissive temperature (42°C).[81] After 3 to 6 hr at 42°C the phosphatidylethanolamine content of these mutants is reduced by approximately one-half. Concomitantly, the concentration of cardiolipin rises approximately three fold; the phosphatidylglycerol content of these isolates is also elevated under these conditions.[32,81] Ohta and Shibuya[82] have independently isolated a phosphatidylserine synthetase mutant of *E. coli* which appears to be similar in some respects to those isolated by Raetz.[80]

The final reaction in the conversion of CDP-diacylglycerol is catalyzed by the enzyme phosphatidylserine decarboxylase. The enzyme, which is present in a vast excess compared to the amount of phosphatidylserine synthetase, has been purified to homogeneity[83,84] and contains bound pyruvate which is essential for its activity.[85] Hawrot and Kennedy have isolated conditional lethal mutants of *E. coli* which contain a temperature-sensitive phosphatidylserine decarboxylase,[86] and the lesion maps at approximately 93 min on the *E. coli* chromosome.[60] These mutants replace phosphatidylethanolamine (neutral at physiological pH) with negatively charged phosphatidylserine, and at the nonpermissive temperature the concentration of phosphatidylserine reaches as much as 48%.[83] Hawrot and Kennedy have suggested that this high concentration of negatively charged phospholipid molecules within the membrane may be responsible for the lethal effects observed.[83]

A recent report by Tyhach, et al.[87] described the isolation of a strain of *E. coli* bearing a plasmid and producing 40 to 50 times the normal cellular levels of phosphatidylserine decarboxylase. Further, only about half of the enzyme produced at these high levels was firmly attached to the membrane, the remainder being easily released by sonication. The authors concluded that there may be only a limited number of stable membrane insertion sites available for the enzyme.[87]

3. *Diphosphatidylglycerol Biosynthesis*

This pathway contains the reactions which produce the other major phospholipid species present in the Gram-negative membrane, phosphatidylglycerol and diphosphatidylglycerol (cardiolipin).

The first of three reactions in the synthesis of cardiolipin is catalyzed by phosphatidylglycerol phosphate synthetase (CDP-diglyceride: *sn*-glycero-3-phosphate phosphatidyltransferase) and involves the condensation of CDP-diacylglycerol and *sn*-glycero-3-phosphate with the subsequent release of CMP.[50,51,88,89] Mutants containing diminished levels of the enzyme have been isolated,[80] and the structural gene directing the synthesis of the enzyme, the *pgs* gene, maps at approximately 4 min on the *E. coli* chromosome.[32,81]

Once synthesized, phosphatidylglycerolphosphate is rapidly converted to phosphatidylglycerol by the action of phosphatidylglycerolphosphate phosphatase. This enzyme has not yet been purified, and no mutants lacking it have been reported.[32,88]

Diphosphatidylglycerol (cardiolipin) is synthesized by the condensation of two molecules of phosphatidylglycerol. The reaction is catalyzed by cardiolipin synthetase, and this represents the main route of cardiolipin synthesis.[90-92] Pluschke and co-workers[25] and Isono et al.[93] have selected a mutant which is deficient in cardiolipin synthetase; the lesion, termed *cls,* maps at the 27 min. mark on the *E. coli* chromosome and was reported to confer no specific phenotype to the cell.[25]

IV. FUNCTIONAL INTERACTION OF PHOSPHOLIPIDS WITH TRANSPORT PROTEINS

Because the naturally occurring bacterial membrane is heterogeneous in composition, the relationships between structure, biosynthesis and function are not well understood. In order to study more readily these complex interrelationships, attempts have been made to simplify the composition of the membrane by genetic modification.[1,18,32-35,43] An early approach in this regard was the isolation of mutants unable to synthesize unsaturated fatty acids (*fab* mutants) or unable to degrade fatty acids (*fad* mutants) (see above). Strains containing both of these modifications allow the investigator to control the fatty acid composition of the membrane phospholipids.[44,45] Other mutants, containing defects in the phospholipid biosynthetic pathways, are only now beginning to provide information regarding the effects of specific phospholipid deficiencies on various membrane associated functions.

A. Phospholipid Phase Transitions

Using mutants of defined phospholipid composition, several investigators have concluded that changes in the activities of various permeases and membrane enzymes are due to the occurrence of order-disorder lipid phase transitions.[44,47,48,83,94-111] In particular, mutants containing both *fab* and *fad* mutations that were grown in media containing predominantly elaidic acid (*trans*-18:1$^{\Delta 9}$), oleic acid (*cis*-18:1$^{\Delta 9}$) or *trans*-hexadecenoic acid (*trans*-16:1$^{\Delta 9}$) exhibited distinct and reversible lipid phase transitions as demonstrated by several methods (e.g., dilatometry, X-ray diffraction, electron spin resonance spectroscopy, freeze-etch electron microscopy, fluorimetry and differential scanning calorimetry).[9,112-121] The same types of transitions were also seen in cytoplasmic membranes or phospholipids extracted from mutants grown under the same conditions.[116,117,119,120] These lipid phase transitions are defined by two characteristic temperatures. Whereas the low temperature boundary (t_l) of the transition range (Δt) represents the temperature at which the lipid phase has begun the transition from the ordered, semicrystalline state to a disordered, liquid state, the high temperature boundary (t_h) represents the temperature above which all of the lipid molecules participating in the transition are in the fluid state.[122] (Some membrane phospholipids appear to be closely bound to integral membrane proteins and therefore are not free to participate in the phase transition process.[9]) The width of the transition range (Δt) is defined by $\Delta t = t_h - t_l$, and the midtransition temperature (t_t) is defined as that temperature at which half of the participating lipids are in either phase.[115] Over the temperature range described by t_h and t_l the lipid molecules are believed to be in an equilibrium between the solid and liquid states.[123]

A recent publication by Gent and Ho[124] deals with the fluid state of the membrane and suggests that the integral membrane proteins modify the phase transition. Through the use of ^{19}F-nuclear magnetic resonance spectroscopy, they studied the degree of motion and relative amounts of membrane phospholipid in both the fluidus and solidus phases of the cytoplasmic membrane. Inner membrane vesicles were derived from whole cells of a *fab fad* double mutant grown in the presence of 8,8-difluoromyristate for one generation at 37°C. The results obtained from these studies indicated that the integral membrane proteins significantly modify the phase transition of the bulk of the membrane phospholipids. The temperature range of the phase transition for both whole cells and cytoplasmic membranes was approximately 50% narrower than that observed for either total *E. coli* lipid extracts or purified phospholipids. Gent and Ho suggested that the phospholipid molecules immediately adjacent to the integral membrane proteins would necessarily have less freedom of motion than those lipid molecules which were surrounded by other lipid molecules. However, the former molecules are likely to retain some fluidity, and an approximately constant amount of phospholipid is probably associated with the proteins. These "boundary" lipids may be composed primarily of lower melting lipid molecules which remain in the fluid state even at temperatures below t_l.[9,125] Thus, the remaining, lipid rich regions of the membrane would be more homogeneous and have a higher phase transition temperature than the total membrane lipid.[124]

The conclusion of Gent and Ho that the membrane proteins alter the phase transition profile is in opposition to the results obtained by other investigators.[9,116,121] These workers had suggested that, except for a broadening of the temperature range, the phase transitions for cytoplasmic membranes and phospholipid extracts were essentially the same. However, as Gent and Ho pointed out, their results may not be directly comparable to those obtained by others. The latter investigators[9,116,121] incorporated the unsaturated fatty acid supplement to the extent that it represented between 60 and 90% of the total cellular fatty acid while only about 25% of the fatty acids of the cells and membranes used by Gent and Ho were difluoromyristate.

In a number of early studies, e.g., those on glycoside transport in *E. coli,*[95,96,99,100] only one transition temperature in Arrhenius plots of membrane transport was reported. This was later shown to be due to the collection of insufficient data points, the failure to study the transport rate over a sufficiently broad temperature range, or the insensitivity of the method of detection.[112,115] These problems are now recognized, and the temperature parameters for membrane transport are in better agreement.[115,120] Despite the early criticism of the use of spin labels to determine the boundaries of thermal transitions,[124] this method yields information which is in good agreement with results obtained by fluorescence and X-ray diffraction analyses.[115,120] The available evidence suggests that intrinsic membrane transport proteins and enzymes can act as lipid phase probes, detecting both the beginning and end of the order-disorder transition of membrane phospholipids.

Work from several laboratories has suggested that discrete lipid domains exist within the lipid bilayer of the membrane under physiological conditions.[115,119,126-130] These results were not, however, always confirmed by other investigators.[113] More recently Letellier and co-workers presented evidence which suggests that the lipid and protein components segregate in membranes of *E. coli.*[131] As a consequence of this separation two lipid domains are observed in freeze-fracture electron micrographs. The first domain contains relatively few phospholipids. These are enriched for unsaturated fatty acids and are associated with the integral hydrophobic membrane proteins. The second domain is highly ordered and is enriched in membrane phospholipids containing primarily saturated fatty acids; no proteins are found in this region.[131] The prediction that domains of ordered and disordered membrane lipids should coexist at the midtransition temperature, t_t, of the membrane Arrhenius plots,[118] and that such a membrane state would probably exhibit a high lateral compressibility[118,132] and therefore increased solute permeability,[132] is supported by the finding that liposomes show increased solute transport at the midtransition temperature.[133]

B. Effects of Phospholipid Composition on Sugar Transport

The effects of altering the fatty acid composition of membrane phospholipids on lactose transport in *E. coli* have been thoroughly reviewed by Cronan and Gelman[35] and more recently by Cronan.[33] The transport of lactose and other glycosides has been extensively studied, and the reader interested in examining the manifestations of altered membrane phospholipid content on membrane transport is directed to this literature. Although the early work[99,102,103,100,101] was often conflicting and beset with errors in experimental judgement, these problems and inconsistencies appear to have been corrected in later reports.[107,112,114] The contention that the induction of lactose transport is dependent upon unsaturated fatty acid synthesis.[94,101] was later found to be unsupported.[107,100] Further, the conclusions of Hsu and Fox that the induction of a functional lactose permease is dependent upon *de novo* lipid synthesis[98] and that newly synthesized permease proteins preferentially associate with newly synthesized phospholipids in discrete domains within the membrane[99,103] were disputed in later reports.[108,114] Finally, a recent report by Thilo, Träuble, and Overath[115] clarifies and somewhat supports the early data of Fox and co-workers who concluded that the function of the lactose transport system is dependent upon a fluid membrane state.[102,106,117]

Protein-phospholipid interactions have also been examined employing substrates of the phosphoenolpyruvate: sugar phosphotransferase system (PTS).[134-137] Working with a temperature-sensitive phosphatidylserine synthetase mutant of *E. coli* (see Section III. B. above), provided by C. R. H. Raetz, one of the authors (J. E. L.) has obtained results demonstrating that the loss of phosphatidylethanolamine from the membrane and/or the concomitant accumulation of cardiolipin and phosphatidylglycerol significantly reduces the level of transmembrane transport. This observation holds not only for PTS mediated substrates (e.g., methyl-α-D-glucoside, D-glucose, or mannitol) but also for amino acid transport systems, including both binding protein dependent and independent systems (see below). While similar results have been reported with phosphatidylserine decarboxylase mutants of *E. coli,*[83] comparable results were not obtained with cardiolipin synthetase mutants.[25] At 30°C the transport of these solutes by the mutant strain (RA2021, *pss-21, ts*) approximated that observed for the wild type strain (RA2000). When shifted to 42°C however, the level of uptake in the mutant dropped, and after about 4 hr at the nonpermissive temperature, the levels of transport had decreased from three- to six-fold, depending upon the substrate. That of the parent remained approximately constant throughout the course of the experiment.

Because the genes directing the synthesis of the mannitol Enzyme II and the mannitol-1-phosphate dehydrogenase are present in a single operon,[138] and therefore are under coordinate regulation, the mannitol PTS offered an ideal system in which to study further the observed loss of PTS mediated transport at 42°C. Experiments were designed to answer some initial questions: was the observed decrease in transport due to decreased protein synthesis at 42°C, or was it due to decreased insertion of the newly synthesized Enzyme II protein into the membrane? Alternatively, was the mannitol Enzyme II protein being synthesized and inserted properly at 42°C but simply not as active because of the altered membrane phospholipid composition?

Cultures of the wild type and mutant strains were grown to mid-log phase at 30°C in LB medium[139] containing 0.5% D-mannitol, and a portion of each culture was subsequently harvested by centrifugation and assayed. The remainder of each culture was immediately transferred to a 42°C shaker bath and growth was allowed to continue. At various time intervals during the next 4 hr portions of each culture were harvested and assayed. In some experiments chloramphenicol (50 μg/m*l*) was added at the time of the temperature shift. Both mannitol PTS and mannitol-1-phosphate dehydrogenase activities were assayed in the 30 and 42°C cultures. In the absence of chloramphenicol the wild-type culture, grown at 42°C, showed a two- to three-fold increase in mannitol-1-phosphate dehydrogenase activity compared to the 30°C culture (Figure 2A). The corresponding culture of the phosphatidylserine synthetase mutant also exhibited a two- to three-fold increase in dehydrogenase activity. However, the initial (30°C) and final (4 hr, 42°C) specific activity values were some four-fold lower than the parental values. Moreover, the specific activity of the dehydrogenase did not plateau for approximately 2 hr after temperature shift.

In the absence of protein synthesis (Figure 2B) the specific activity of the dehydrogenase in the parental strain decreased approximately 30% after 4 hr at 42°C (relative to the 30°C culture) while that of the temperature sensitive mutant remained relatively constant. It thus appears that some form of "stringent" control is present in the mutant at 42°C. The data for strain RA2021 support the results of McIntyre et al.,[70] who showed that one or more control mechanisms exist which coordinately regulates macromolecular synthesis and accumulation with net phospholipid biosynthesis.

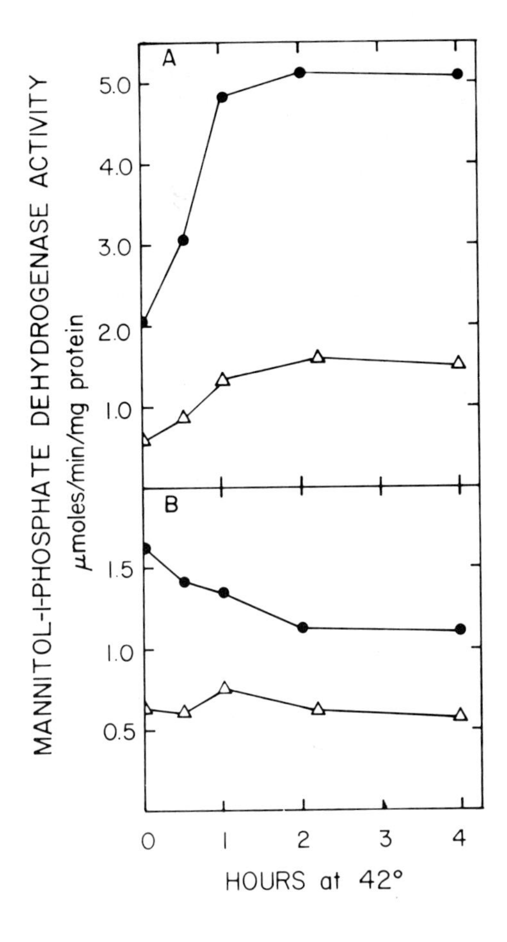

FIGURE 2. Results of mannitol-1-phosphate dehydrogenase assays[140] in strains RA2000 (*pss*$^+$) and RA2021 (*pss, ts*) at 30 and 42°C, without (A) and with (B) 50 µgm/m*l* chloramphenicol. Values on the ordinate are the 30°C control values. Data points represent the average of 3 determinations. Symbols represent: –Δ–Δ–, RA2021; –●–●–, RA2000.

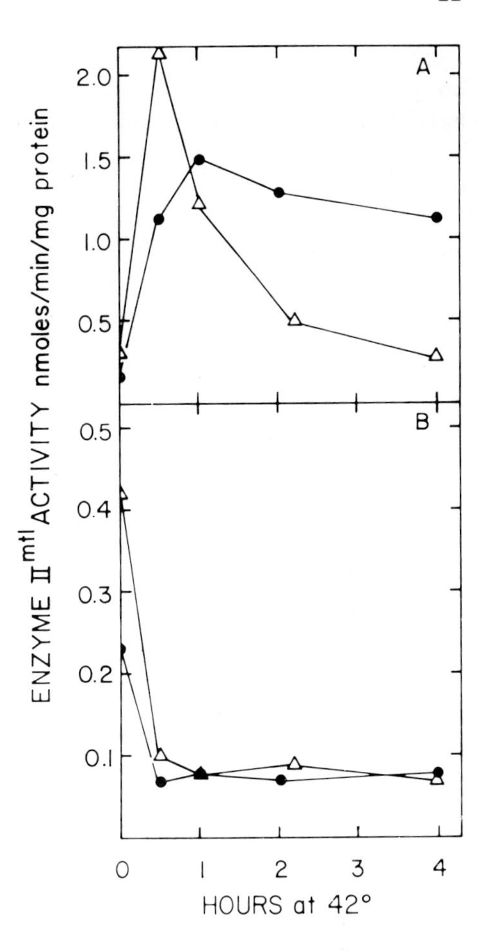

FIGURE 3. Results of mannitol Enzyme II assays[141] in membranes of strains RA2000 (*pss*$^+$) and RA2021 (*pss, ts*) at 30°C and 42°C without (A) and with (B) 50 µg/m*l* chloramphenicol. Values on the ordinate are the 30°C culture values. Data points depict the average of three determinations. Symbols represent: –Δ–Δ–, RA 2021; –●–●–, RA 2000.

Perhaps the regulation of the synthesis of phosphatidylethanolamine represents one such control mechanism.

When the specific activity of the mannitol Enzyme II was studied in the absence of chloramphenicol (Figure 3A), the wild type activity increased rapidly in the 42°C culture (compared to the 30°C control culture), reaching a plateau after approximately 1 hr at the elevated temperature. The specific activity of the mannitol Enzyme II in the wild type culture at 42°C was approximately 6-fold higher than that observed in the 30°C culture. The specific activity of the mannitol Enzyme II in strain RA2021, however, showed a rapid increase after the temperature shift, rising about ten-fold within 30 min. Subsequently, the specific activity of mannitol transport decreased, and after 4 hr at 42°C the level of Enzyme II activity was comparable to that seen in the control culture. This final value was about five-fold below that observed for the parental strain. The initial (30°C) values were very similar for both strains. In the presence of chloramphenicol (Figure 3B) both cultures exhibited a precipitous drop in mannitol

Enzyme II activity from their respective 30°C culture values. Within 30 min. at 42°C both strains had reached the same approximate minimal value and remained at this basal level throughout the remainder of the experiment.

Thus, while there may be some inhibition of protein synthesis in the phosphatidylserine synthetase mutant at 42°C, the mannitol Enzyme II data suggest that the newly synthesized transport protein (Enzyme II^{Mtl}) is being properly inserted into the membrane of the mutant at the nonpermissive temperature, and that once inserted its activity is affected by the altered phospholipid composition. As the ratio of phosphatidylethanolamine to phosphatidylglycerol decreases, so does the level of mannitol Enzyme II activity. Alternatively, the drop in Enzyme II activity observed in the mutant a 42°C may be due to the increased negative charge of the membrane.[25] These results emphasize the importance of phospholipid balance to the normal functioning of transport proteins.

C. Effects of Phospholipid Composition on Amino Acid Transport

Esfahani and co-workers[47] studied the physical state of the phospholipids in the cytoplasmic membrane of an *E. coli* unsaturated fatty acid auxotroph and its relationship to proline transport. Using cytoplasmic membranes from cells grown on elaidic, oleic, or linoleic acid they observed single discontinuities in Arrhenius plots of proline uptake. They concluded that transport was sensitive to the degree of fatty acid unsaturation. These discontinuities in the transport data did not, however, agree with the physical changes in the vesicle phospholipids as detected by X-ray diffraction. The authors attributed this disparity to heterogeneity in membrane composition.

The results obtained by Esfahani et al.,[47] were in direct opposition to those published by Shechter and co-workers.[119] These investigators also correlated the physical state of the cytoplasmic membrane with the discontinuities observed in Arrhenius plots of proline uptake. Using a strain of *E. coli* auxotrophic for unsaturated fatty acids, these workers studied the effects of phospholipids composed primarily of elaidic, oleic, linolenic, and linoleic acids in membrane vesicles. The data reveal a good correlation between the transport of proline and the order-disorder transition detected by X-ray diffraction.[119] Further, freeze-fracture electron micrographs of membranes below the temperature discontinuities show that the integral membrane proteins (like the proline transport carrier protein) were excluded from the ordered lipid domains. The authors concluded that this might account for the high energy of activation calculated for proline transport at temperatures below the thermal discontinuity. For transport to occur normally, the membrane phospholipids must be in a fluid, disordered state. This conclusion is supported by work from other laboratories (see Section IV. B.).[118,132,133]

An early communication by Holden and co-workers[142] showed that proline uptake in whole cells of *E. coli* fatty acid mutants was significantly elevated compared to the parental strains. One isolate studied, mutant E-20, contained an elevated level of *cis*-9, 10-methylene-hexadecanoic acid (a cyclopropane fatty acid). Another strain, mutant E-52L, produced excess phospholipids, including phosphatidylethanolamine. The degree of overproduction was not published for either mutant.[142] While strain E-52L exhibited approximately twice the proline uptake activity of the control strain, mutant E-20 exhibited a three fold enhancement of proline uptake relative to the parent. All cultures were grown at 37°C and assayed at 24 to 27°C. Because the permeases for other amino acids studied (see below) were heterogeneous in their responses to the overproduction of the membrane components, the authors suggested that different membrane proteins might interact differently with various phospholipid molecules, thereby producing the different levels of catalytic activity observed.

Ohta and co-workers[143] inhibited phosphatidylserine decarboxylase[32,86] by growth of

E. coli in the presence of 5 m*M* hydroxylamine. As this enzyme catalyzes the conversion of phosphatidylserine to phosphatidylethanolamine (Figure 1), the levels of phosphatidylserine (usually undetectable in membrane phospholipids of *E. coli*) increased to about 20% of the total cellular phospholipid after growth for 3 hr at 30°C. A concomitant drop in the levels of phosphatidylethanolamine was detected. The viability of the cells was not affected by this treatment. Ohta and co-workers[143] observed that membrane vesicles derived from cells grown for 3 hr at 30° in the presence of 5 m*M* hydroxylamine exhibited an approximately three fold increase in proline uptake when compared to vesicles derived from cells grown in the absence of the inhibitor. While vesicles prepared from nontreated cells contained approximately 78% PE, 11% PG, and 11% diPG, vesicles obtained from hydroxylamine-treated cells contained 74% PE, 5% PG, 9% diPG and 12% PS.

Incubation of hydroxylamine-treated cells in the presence of fresh growth medium containing 50 μg/m*l* liter of chloramphenicol brought about a reversal of phosphatidylserine decarboxylase inhibition. After 4 hr, the level of phosphatidylserine had dropped from approximately 20% of the total phospholipid to about 5%, while that of phosphatidylethanolamine rose from 65% to over 80%. The level of proline uptake observed in vesicles derived from these cells was once again that found in vesicles derived from untreated cells. Thus Ohta et al., concluded that the three fold stimulation of proline transport exhibited by vesicles obtained from hydroxylamine treated cells was due to the increase in phosphatidylserine and/or the decrease in phosphatidylethanolamine.

A recent report by Hawrot and Kennedy[83] described the effect of altering the levels of phosphatidylserine on membrane transport and cell physiology employing a temperature sensitive phosphatidylserine decarboxylase *(psd)* mutant of *E. coli.*[83] At the nonpermissive temperature, as much as 48% of the total membrane phospholipid was phosphatidylserine. Within 5 hr at 42°C, these *psd* mutants typically ceased growth.[83] At 37°C, however, mutant EH470 (*psd*-4) accumulated between 13 and 26% of the total phospholipid as phosphatidylserine. Moreover, at this temperature it showed little alteration of growth rate in liquid culture or plating efficiency on nutrient agar.[83] In order to assess accurately the physiological effects of increased levels of phosphatidylserine (or decreased levels of phosphatidylethanolamine), proline and other transport activities were assayed at this temperature. Proline transport of mutant EH470 grown at 37°C was slightly reduced compared to the control (psd^+) culture. To check the possibility that this reduction in transport activity, which was more pronounced for glutamine uptake, might be due to a reduced rate of insertion of transport proteins into the membrane, the rate of induction at the *lac* permease was checked in mutant EH470 grown at 37°C. The results revealed no differences in the rates of induction in the mutant and control (psd^+) cultures grown and induced under identical conditions.[83] Aside from the experimental differences between the work reported by Ohta et al. and Hawrot and Kennedy, the biochemical basis for the apparently conflicting results is not readily discernible.

As was mentioned above, the temperature-sensitive phosphatidylserine synthetase mutant (RA2021) exhibited a general loss of membrane transport after shift to the nonpermissive temperature. When proline transport was measured in this strain, it was found that after 4 hr at the nonpermissive temperature the level of proline uptake was approximately three fold lower than in the 30°C culture. The wild type strain (RA2000), meanwhile, showed a slight increase in proline transport activity under the same conditions. This response of strain RA2021 was more pronounced than the reduction observed by Hawrot and Kennedy[83] for strain EH470. Whether or not this difference is in part due to the fact that the mutants were grown at different temperatures remains to be examined. It seems likely, however, that the proline carrier is dependent not only upon the fluid state of the membrane but also upon its phospholipid composition.[125]

A later investigation by Holden and co-workers[111] showed that proline transport in an unsaturated fatty acid auxotroph was drastically reduced in cells grown on *trans*-unsaturated fatty acids compared to cells grown on *cis*-vaccenic acid. While the level of proline transport in cells grown on palmitelaidic acid (*trans*-16:1$^{\Delta 9}$) was only 6.2% of that recorded for the control culture, that in elaidate (*trans*-18:1$^{\Delta 9}$) grown cells was approximately 1.5% of that observed for the control. The assay temperature was 21°C. When proline and lysine transport rates were compared over a temperature range of approximately 30°C in cells grown either in vaccenate- or palmitelaidate-supplemented media, the transport systems exhibited opposite responses. While the rate of proline uptake exceeded that observed for lysine over the entire range of assay temperatures for vaccenate grown cells, the exact opposite was true for cells cultured in media containing palmitelaidate as the sole unsaturated fatty acid. An apparently heterogeneous effect of altering the membrane phospholipid composition on amino acid transport rate was interpreted by the authors as differential association of the integral transport proteins with different classes of membrane phospholipids.[111] Alternatively, the authors suggested that the differential effects observed for the uptake of proline and lysine may have been due to the fact that the proline permease was more sensitive to changes in its lipid environment than the lysine carrier protein.

The work thus far reported for proline transport is much like the early work published on the effects of altered membrane composition on lactose uptake. While some reports offer the same interpretation of data obtained from different experimental approaches, most are often incomplete and some reports are contradictory. Nevertheless, using widely differing experimental approaches, several groups have arrived at essentially the same conclusion. All suggest that the activity of the proline permease is sensitive to the surrounding phospholipids and may be dependent upon specific membrane phospholipids.[47,111,142,143]

The accumulation of several amino acids has been studied by Holden et al.[111,142] and by Ohta and co-workers.[143] Amino acids studied by these investigators included alanine, arginine, aspartate, asparagine, glutamate, glutamine, lysine, serine, and threonine. The conclusions expressed for the transport of proline generally applied to the accumulation of these amino acids. Interestingly, however, transport of amino acids through binding protein mediated transport systems may not be as sensitive to the induced alterations of the cytoplasmic membrane phospholipids as is transport through integral membrane transport systems.[111]

As a general comment regarding the reported studies, we think it important to reiterate a criticism first voiced by Cronan and Gelman.[35] In all the amino acid transport studies cited, the investigators have assumed that only the translocation process was perturbed by alterations induced in the membrane phospholipids. While it is known that binding protein dependent and membrane bound transport systems derive their energy from different sources,[144] it must not be taken for granted that the observed alterations in transport are due solely to effects on the membrane associated components of the permease. The energy coupling/generation systems might also be perturbed.

D. Effects of Phospholipid Composition on Bacteriorhodopsin-Mediated Proton Transport

The well-characterized purple membrane of *Halobacterium halobium* has been used to probe the role of the membrane in transmembrane transport. It catalyzes the extrusion of protons from the cytoplasm in response to light. The purple membrane of this extreme halophile occurs in patches within the cell membrane, forming up to 50%

of the cell surface under favorable conditions.[145] In response to light, the proton gradient created by the action of this membrane provides an energy source, a "proton-motive force," which can be used to generate ATP and energize secondary active transport systems.[146,147]

The purple membrane forms a two-dimensional "crystal" composed of lipid (25% by weight) and protein (75%). The sole protein present is bacteriorhodopsin (mol wt = 28,000 daltons), which contains one mole of the chromophore, 11-*cis*-retinal, bound through a Schiff's base linkage to a lysyl residue on the protein. A review of the purple membrane has recently appeared, in which experimental details of its structure and function may be found.[146]

Because of the membrane's natural crystalline state, analysis of its structure by X-ray diffraction and electron microscopy has been possible. From such studies, Henderson and Unwin[148] have developed a model proposing that the membrane consists of trimers of bacteriorhodopsin arranged in a hexagonal lattice and embedded in a lipid bilayer. The bacteriorhodopsin molecules appear to be situated somewhat asymmetrically in the membrane, extending across the planar bilayer in seven alpha-helices.[148,149] They apparently do not rotate across the membrane during proton translocation.[146]

Bacteriorhodopsin undergoes a light-induced cycle of spectroscopic changes during which protons are first released and then taken up again. Using low temperature and flash spectroscopy, workers have identified spectrally distinct intermediates.[146] A current proposal of Lozier and co-workers suggests five intermediates with varying half-lives and absorption maxima.[150] From bacteriorhodopsin, the molecule cycles through K_{590}, L_{550}, M_{412}, (N_{520}), and O_{640}, then back to bacteriorhodopsin with a maximum at 570 nm. The intermediate labeled "N_{520}" is only suspected from temperature-dependent variations in the spectrum of M_{412}.[146] The release of a proton from the bacteriorhodopsin molecule during the photocycle and subsequent uptake of a proton have experimentally been shown to occur from the external and cytoplasmic surfaces, respectively.[150] In reconstituted phospholipid vesicles, however, the orientation of bacteriorhodopsin appeared to be reversed, causing such vesicles to take up protons in response to light.

1. Reconstitution with Purple Membrane Sheets

The photo-induced spectral cycle of bacteriorhodopsin and the physical measurement of the intermediates provide a useful assay of the molecule's function under various conditions. Earlier reconstitution experiments[151-154] involved incorporating purple membrane sheets, complete with halobacterial lipids, into artificial phospholipid vesicles in order to study proton transport in vitro. The inclusion of the purple membrane lipids surrounding the bacteriorhodopsin molecules is of interest because halobacterial lipids are themselves unusual. The polar lipids are diphantyl ether analogs in which the dihydrophytol chains are saturated, branched-chain hydrocarbons. The polar head groups are relatively large and highly anionic at pH 7, being primarily analogs of phosphatidylglycerophosphate and a glycolipid sulfate.[27]

Racker[151] and Racker and Stoeckenius[152] reconstituted bacteriorhodopsin in phospholipid vesicles and measured proton uptake.The reconstitution procedure of adding the purple membrane sheets in 0.15 *M* KCl to a dry sample of phospholipids, followed by sonication, gave active proteoliposomes. The activity of the vesicles, whether reconstituted from soybean lecithin or halobacterial lipids, increased with increasing salt concentration. Although mixtures of soybean phosphatidylcholine (PC) and phosphatidylethanolamine (PE) were as effective as crude soybean phospholipid mixtures in reconstituting active vesicles, pure soybean PC was ineffective, and egg PC showed only about 30% of the activity seen with the mixtures.

Using a similar reconstitution procedure, Happe and co-workers[155] investigated the use of different types of acidic lipids, as well as halobacterial lipids, to form the vesicles. Vesicles reconstituted with cardiolipin from *E. coli* at a pH less than or equal to 2.75 showed a light-dependent acidification of the medium, while those reconstituted at a higher pH alkalinized the medium. These samples contained 31 to 37 mol of lipid per mole of protein. Vesicles reconstituted with phosphatidylglycerol (PG) showed similar behavior, while purified soybean lecithin was relatively ineffective in reconstruction. Halobacterial lipid vesicles gave rise to acidifying vesicles below pH 3.9, and alkalinizing vesicles above this pK. Transition from proton extrusion to uptake occurred in all cases near the pH of the lipid phosphate groups, suggesting a role of the lipids in determining the orientation of bacteriorhodopsin in the membrane.[155]

Using a reconstitution system similar to that of Racker,[151] Hellingwerf and co-workers[156] undertook a series of experiments designed to outline the in vitro conditions under which bacteriorhodopsin could serve as a proton pump. Their results, contrary to earlier investigations,[157] indicated that when dispersed from its two-dimensional lattice by prolonged sonication, bacteriorhodopsin functions as well as when it is in the native hexagonal array. The dissolution of the hexagonal lattice was measured by circular dichroism and by the loss of exiton coupling of the chromatophores of neighboring proteins in the lattice.

Because of the unusual nature of native halobacterial lipids, it was reasoned that bacteriorhodopsin might be inactivated when dispersed from its native lattice in the presence of exogenous phospholipids. The exogenous lipids used in the studies were highly saturated forms of cardiolipin, PC, mixtures of PC and PE, PC and PS, PC and cetyltrimethylammonium bromide (CTAB), and PC and dicetylphosphate.[158] All the lipid preparations used in the reconstitution experiments showed light-driven proton uptake comparable to, or higher than, bacteriorhodopsin vesicles reconstituted with endogenous lipids.[158]

2. Reconstitution with Lipid-Depleted Purple Membrane

Experiments have been reported in which bacteriorhodopsin in the purple membrane was first depleted of its endogenous lipids and then reconstituted with exogenous phospholipids.[158,159] Happe and Overath[26] disrupted the purple membrane with 20 m*M* dodecyl trimethyl ammonium bromide at pH 3.0. The supernatant from this solution contained bacteriorhodopsin from which the endogenous lipids were removed by gel filtration on Sephadex®G-100 or by sucrose gradient centrifugation. Analysis of the resulting bacteriorhodopsin fractions indicated that 90% or more of the endogenous polar lipid was removed, leaving approximately one mole of lipid per mole of bacteriorhodopsin. Thin layer chromatography revealed that the total lipid extracts from the lipid-depleted bacteriorhodopsin and those from native purple membrane showed the same relative amounts of the different species of polar lipids, suggesting that bacteriorhodopsin does not preferentially bind a specific class of polar lipid. The lipid-depleted sample also appears to lose glycolipids and phospholipid analogs equally during the lipid depletion procedure. Lipid depletion did not irreversibly affect the activity of bacteriorhodopsin as a light-driven pump, since proteoliposomes reconstituted from various lipids showed light-activated proton uptake with no significant difference in kinetics or extent of proton uptake per mg bacteriorhodopsin. The phospholipids used included crude soybean lipids, total phospholipid from *E. coli*, and purified PC. Bacteriorhodopsin in these vesicles appeared to have retained the close packing typical of the native hexagonal array, as indicated by the exiton coupling of the retinal chromophores.[26]

Hwang reported another method for lipid depletion of purple membranes.[158] Membranes were treated with 10% deoxycholate and then subjected to sucrose gradient centrifugation. This procedure removed 80% of the endogenous lipid. The lipid-depleted purple membrane appeared indistinguishable from native purple membrane, as determined by physical techniques, such as X-ray diffraction, circular dicroic spectroscopy, and electron microscopy. The photoreaction cycle of bacteriorhodopsin remained the same, although the decay kinetics of the last three spectral intermediates of the cycle, M_{412}, N_{520}, and O_{640}, were slower. This effect could be reversed by increasing the phospholipid to bacteriorhodopsin ratio using exogenous lipids such as egg PC, PE, or total halobacterial lipids. The M_{412} decay constant also appeared to depend on the fluidity of the membrane since Arrhenius plots of this constant revealed a break at the phase transition temperature of the exogenous lipid when dipalmitoylphosphatidylcholine or dimyristoylphosphatidylcholine was added.[157]

Sherman and Caplan[159] also investigated the influence of membrane lipids on the photochemical reaction cycle of bacteriorhodopsin. Their model for the reaction cycle differs somewhat from that of Lozier et al.[150] in suggesting that the cycle moves through an unbranched path from bacteriorhodopsin through K_{590}, L_{550}, M_{410} and back to bacteriorhodopsin with O_{660} occupying a separate pathway. Their reconstitution method involved suspending purple membrane in a KCl solution containing the exogenous lipid and cholate. Sucrose gradient centrifugation followed incubation. After dialysis, the purple band showed, by thin layer chromatography, depletion of more than 90% of the native lipids. Dipalmitoylphosphatidyl choline and egg lecithin were used in separate reconstitution experiments. As with other workers, Sherman and Caplan[159] found that bacteriorhodopsin reconstituted in vesicles of exogenous lipid showed light-induced proton uptake, but decay of the M_{410} intermediate was affected. In native purple membrane, this decay followed first order kinetics, but in the reconstituted systems, an initial rapid decay similar to that of the purple membrane was followed by a slower decay rate. Arrhenius plots of the early decay time revealed a discontinuity with purple membrane lipids at 30°C, as previously reported.[160,161] Jackson and Sturtevant,[162] however, concluded from differential scanning calorimetry studies of the purple membrane that no phase transition occurred between 0 and 75°C. These authors stated that with the parameters used by Sherman and co-workers[160] and Chignell and Chignell,[161] "it is frequently impossible to distinguish between a break in slope at a definite temperature and a broad curvature in the plot throughout the temperature range studied".

Sherman and Caplan[159] found that the Arrhenius plot of reconstituted dipalmitoylphosphatidyl choline vesicles showed two clear breaks at about 25°C and 40°C, while the plot of the egg lecithin vesicles lacked clear discontinuities. They measured the activation energy for synthetic phospholipid membranes and for native purple membranes using the reciprocal of a viscosity proportionality function on the ordinate. The rotational mobility of the chromophore within the purple membrane was significantly slower in the artificial vesicles than in the native purple membrane.[159] When the microviscosity of the various membranes was examined using the fluorophore, 1, 6-diphenyl-1, 3, 5-hexatriene, both the photochemical cycle rate and the chromophore rotational mobility were found to be influenced by lipid microviscosity. The authors[159] concluded that these effects were secondary, and the chromophore was probably not in contact with the membrane lipid molecules.

The evidence reviewed above indicates that bacteriorhodopsin can catalyze light-induced proton uptake when the membrane composition is varied substantially. The range of such variations was studied by Hellingwerf and co-workers.[156] They used sonication to disperse bacteriorhodopsin molecules from the native hexagonal lattice into a monomeric form as a method to expose the protein to exogenous lipids without prior lipid depletion. This method could be questioned on the basis that ten halobacterial lipid molecules per bacteriorhodopsin molecule were still present, and disarray into the monomeric form alone might not guarantee sufficient contact with exogenous lipids. In any case, the proton uptake activity did not appear to be strongly dependent on the lipid environment. A pore mechanism of transport[163] would account for this apparent insensitivity. It should be emphasized, however, that both the lipid to protein ratio and the composition of the lipid environment can influence the decay rate of certain intermediates, particularly M_{412}, and that exogenous phospholipids can decrease the rotational mobility of the chromophore.

The influence of lipid composition on bacteriorhodopsin function could be due to effects on the order-disorder state of the membrane as has been found in other transport systems (see Section IV. B. and C.). Both microviscosity measurements[159] and Arrhenius plots[158,159] indicated that the fluidity of the membrane affects the M_{412} decay constant. An increase in the lipid to protein ratio will also increase the proton uptake activity of bacteriorhodopsin. Since in the native membrane the protein molecules occupy basically fixed positions in an hexagonal lattice, with lipids filling the interstitial spaces at a ratio of one to ten (protein to lipid), a simple decrease in the number of lipid molecules below a certain level may interfere with the cycling of the protein. Hellingwerf and co-workers suggested that an increase in lipid provides increased internal buffering capacity per bacteriorhodopsin molecule without changing the intrinsic properties of the proton pump.[156]

V. SOLUTE PERMEATION THROUGH THE OUTER MEMBRANE

A. Transport of Hydrophilic Compounds through Proteinaceous, Aqueous Channels

The outer membrane of Gram-negative bacteria functions as a diffusion barrier, is involved in conjugation and septum formation during cell division, and contains components of uptake systems for various nutrients. It also contains receptors for bacteriophages and colicins.[1,164] Based on work by Nakae and Nakaido[167], Payne[166] and Payne and Gilvarg[165] first proposed that the outer membrane of Gram-negative bacteria serves as a diffusion barrier. The latter workers established the exclusion limit of reconstituted outer membrane vesicles from *Salmonella typhimurium* at approximately 900 daltons. Shortly thereafter, it was found that the diffusion of small molecules through vesicles composed exclusively of phospholipid and lipopolysaccharide was negligible until a crude preparation of outer membrane proteins from *S. typhimurium* was added.[168] This pioneering work led to the characterization of a group of outer membrane proteins, christened "porins"* by Nakae,[169] that facilitate the movement of small hydrophilic molecules across the outer membrane and into the periplasmic space. Both in vivo and in vitro experimentation has established the identity and physiological function of these proteins.

The observations of Nakumura and Mizushima[168] were later confirmed and extended by Nakae.[170] Nakae found that reconstituted outer membrane vesicles of *S. typhimurium* were impermeable to [^{14}C] sucrose unless a crude mixture of outer membrane proteins was included. Through extraction of the outer membrane with 2% sodium dodecylsulfate (SDS) he obtained an insoluble residue composed of peptidoglycan and protein with no detectable phospholipid or lipopolysaccharide. The SDS-insoluble residue retained the porin activity. Further treatment with lysozyme did not destroy this activity. Following lysozyme treatment the remaining protein residue was fractionated by gel filtration, and four bands were resolved. Only those bands containing 34K, 35K, and 36K molecular weight proteins, as determined by

*The terms porin and matrix protein are used interchangably in the literature and this review.

SDS-polyacrylamide gel electrophoresis, possessed porin activity. Inclusion of these proteins in reconstituted vesicles allowed the penetration of a number of compounds including: galactose, glucosamine, amino acids, uridine, UMP, and GDP. The porins therefore did not appear to show any specificity for either a particular substrate or class of substrates. Inouye[171] proposed that the "Braun lipoprotein" might form aggregates which act as "transmembranous channels." Nakae[170] found that addition of a purified preparation of this protein to impermeable phospholipid-lipopolysaccharide vesicles did not effect permeation. The inactivity of the Braun lipoprotein was later confirmed through the isolation of *lpo* mutants which were deficient in the lipoprotein but showed no defect in in vivo porin activity.[172] However, the bound form of lipoprotein may stabilize the porin protein-peptidoglycan association.[173] At 55°C porin (matrix) protein cannot be dissociated from the peptidoglycan by sodium dodecylsulfate (SDS). Sodium chloride is also necessary, indicating the existence of ionic bonds between the matrix protein and the peptidoglycan. In *lpo* mutants, matrix protein can be dissociated from the peptidoglycan by SDS alone. Similar dissociation occurs in wild type cells when the lipoprotein is first cleaved from the peptidoglycan by trypsin.[173]

Evidence for the physiological function of porins has been obtained through isolation of mutant strains deficient in them. Such strains of *S. typhimurium* had drastically reduced diffusion rates for cephaloridine, a β-lactam antibiotic hydrolyzed by the periplasmic enzyme, β-lactamase.[174,175] Later work demonstrated that porin deficient mutants of *E. coli* B/r had pleiotropic transport deficiencies for a number of substrates, including glucose and other carbohydrates, organic acids, amino acids, sulfate and phosphate, and nucleoside monophosphates. Transport K_ms for these substrates were 30- to 500-fold higher than in the parent strain. Uridine, uracil, and glycerol transport K_ms were not greatly affected by the porin deficiency. Increased K_ms were also reflected by increases in growth K_ms for the corresponding substrates. Growth rates, however, were not reduced when high concentrations of the substrates were present in the culture medium. This led the authors to suggest that porins facilitated the diffusion of small, hydrophilic molecules through the outer membrane.

The loss of the 36.5K protein in *E. coli* B/r resulted in porin deficiency.[176] Through P1 transduction the mutation causing this deficiency (the *kmt* locus) was mapped at 73.7 min between *aroB* and *malT*. It is doubtful that this is the structural gene for porin since revertants of *kmt* strains differed only with respect to the amounts of porin produced. No mutants bearing porins with altered molecular weights or isoelectric points were isolated. Therefore, the *kmt* locus is probably involved in either the regulation, synthesis, or modification of porin. Lutkenhaus[177] in a similar study with porin deficient mutants of *E. coli* B/r, reached similar conclusions regarding the identity and function of the porin proteins. Porin deficient mutants, isolated by tolerance to copper, also exhibited pleiotropic transport deficiencies for a number of nutrients. However, the uptake of uracil, uridine, and maltose was not affected. As will be described later, a specific maltose induced porin has been identified.[178]

For purposes of clarity, the currently known porins found in the outer membranes of *E. coli* and *S. typhimurium* are listed in Table 1 along with some of their identifying characteristics. These characteristics include apparent molecular weights, determined by SDS-polyacrylamide gel electrophoresis, and the genetic loci that either regulate the production of, or are the structural genes for, the porin proteins. Also listed are the various names that have been assigned to them by different investigators. As can be seen, there are differences in porin content between strains within a single species. *E. coli* B appears to have only one porin protein with a molecular weight of 36.5K. It has been labeled protein 1 or 1a, and since it is the major outer membrane protein, it has been called the matrix protein.[176,177,179] Bavoil et al.[176] have shown that the synthesis of this

Table 1
PORINS IN *E. COLI* AND *S. TYPHIMURIUM*

Strain	mol wt (K)	Nomenclature	Genetic loci	Ref.
E. coli	36.5	1 (b, la) (protein I, matrix Protein)	*kmt*	175, 176, 178
E. coli K12	36.5	la, 1b (protein I, matrix protein)	*ompB (kmt)*	178
		la	*tolF(cmlB* or *cry)*	178,179
		1b	*par (meo)*	178,179
E. coli K12 Lysogeny by φPA-2	~34	2	φγPA-2	178
E. coli K12 extragenic pseudorevertants of *ompB* or *tolF*, *par* double mutants	~34		*nmpA*	179
			nmpB	179
			nmpC	179
E. coli K12	44	receptor (maltose porin)	*lamB*	177
S. typhimurium LT-2	34, 35, and 36			181
S. typhimurium LT-2	44	(maltose porin)		180

protein is probably regulated by the product of the *kmt* gene. *E. coli* K12 possesses two outer membrane proteins that have porin activity, proteins la and 1b.[178] Protein la corresponds to protein 1 in *E. coli* B. Pugsley and Schnaitman[179,180] have determined that the *tolF* (*cmlB* or *cry*) gene is the structural gene for protein 1a while the *par* (*meo*) gene codes for protein 1b. Mutations in either of these genes result in partial transport deficiencies for a number of nutrients. Double mutations in the *tolF* and *par* genes or a single mutation in the *ompB* locus results in further loss in transport efficiency. The product of the *ompB* locus regulates both the *tolF* and *par* genes and is probably the same as the *kmt* gene in *E. coli* B.[179] Lutkenhaus[177] has suggested that the appearance of two porins in *E. coli* K12 may be due to gene duplication. This has yet to be substantiated but could be established through determination of the amino acid sequences of the two proteins.

Pugsley and Schnaitman[179] observed that *E. coli* K12 infected with the lambdoid bacteriophage, PA-2, produced an entirely different outer membrane protein, protein 2. This protein had porin activity and was distinct from proteins 1a and 1b. In *E. coli* K12 strains infected with phage PA-2, protein 2 replaced proteins 1a and 1b in the outer membrane. In K12 strains bearing a mutation in the *ompB* gene, protein 2 is able to fully restore permeability. However, in *E. coli* B, protein 2 only partially repairs this defect. Two explanations for partial recovery were offered:[179] (1.) posttranslational modification of porin proteins may be different in B strains as compared to K12 strains and/or (2.) outer membrane proteins of the two strains may be altered to compensate for differences in chemical composition of the lipopolysaccharide of each strain.

Pugsley and Schnaitman[180] have also identified three other genetic loci in *E. coli* K12 that apparently direct the synthesis of porin proteins. These proteins are antigenically distinct from proteins 1a and 1b and are not normally induced in wild type strains. They were discovered in "pseudorevertants" of either *ompB* or *tolF*, *par* mutants. The three loci have been labeled *nmpA, nmpB,* and *nmpC* and map at 82.5, 8.6, and 12 min, respectively, on the *E. coli* chromosome. It has been proposed that *nmp* genes are normally silent and are turned on only under conditions when production of proteins 1a and 1b is turned off. This does not imply that the turning off of the genes directing the production of proteins 1a and 1b turns on the *nmp* genes. However, it seems very plausable that a cell would have backup systems for the production of porins since their activity is essential for normal growth under conditions of limited nutrient concentrations, conditions which would exist often in nature.

It was stated earlier that maltose induces the synthesis of a porin protein. This protein has a molecular weight of 44K and allows the diffusion of maltose and larger maltodextrins through the outer membrane.[178] The maltose porin is also the lambda receptor in *E. coli* K12, a fact that has allowed the isolation of maltose porin deficient mutants by selection for phage resistant colonies. Transport K_m values for maltose and maltotriose in wild type cells are 1 and 2 μM, respectively. In lambda resistant, maltose deficient mutants the maltose transport K_m was 100- to 500-fold higher than in the parent, although the V_{max} value was unchanged. Further, these mutants failed to take up maltotriose. The apparent role of the maltose porin is to aid in the diffusion of the oligosaccharide substrates across the outer membrane, thus assuring ready access to the cytoplasmic membrane transport components. Slower growth rates by maltose porin deficient mutants are only observed at low concentrations of the substrate and can be overcome by the addition of more maltose.[178] At higher concentrations, maltose passes through the matrix protein at rates sufficient to saturate the cytoplasmic transport system. The larger maltodextrins cannot do so because their molecular weights exceed the exclusion limits of these pores. The maltose porin apparently forms larger pores than does the matrix protein (see below).

The maltose porin is encoded by the *lamB* gene which is located in the *malB* regulon on the *E. coli* chromosome.[178] *S. typhimurium* has been found to possess a similar maltose induced porin with approximately the same molecular weight, induction properties, location, and association with peptidoglycan as the *E. coli* protein. It also cross-reacts with antiserum prepared against the lambda receptor, but does not serve as a lambda phage receptor and forms different peptide patterns upon limited proteolysis.[181] Other porin proteins in *S. typhimurium* have been identified and characterized by their ability to allow permeation through reconstituted membrane vesicles and their migration on SDS gels. Porin deficient mutants have also been isolated.[170,181] *S. typhimurium* strain LT-2 possesses 3 matrix proteins with molecular weights of 34, 35, and 36K. As with the *E. coli* strains, qualitative differences in porin content have been noted in various strains of *S. typhimurium*.[182] Location of the genes directing the synthesis of *S. typhimurium* porins has not yet been reported.

Soluble matrix protein has also been found in osmotic shock fluids possibly associated with free lipoprotein.[183] The soluble matrix protein produced the same membrane conductance phenomenon in lipid membranes as the matrix protein isolated from the outer membrane (see below). However, matrix protein is not the only pore-forming protein in osmotic shock fluid since preparations from strains lacking it could also increase membrane conductance.[183]

It is clear that porins are nonspecific with respect to the small hydrophilic molecules that diffuse through them. These molecules include carbohydrates such as galactose and glucosamine, anions such as sulfate and phosphate, and β-lactam antibiotics. Decad and Nikaido[184] have set the exclusion limit of saccharides through the cell wall of Gram-negative organisms at 550 to 650 daltons. Earlier work suggested somewhat higher values.[167] However, both figures are much lower than the exclusion limit (100,000 daltons) for the Gram-positive cell wall of *Bacillus megaterium*.[185] Because dissolution of hydrophilic molecules through hydrophobic membrane bilayers cannot account for the rates at which these molecules pass through the outer membrane, it has been suggested that porin proteins exist as aggregates in the membrane.[182,184] In *E. coli* B there are approximately 10^5 copies of the matrix protein (protein 1) per cell.[186,187] Electron microscopy has revealed that the protein is spread evenly over the surface of the cell in a hexagonal lattice with a repeating distance of 7.7 nm[187] Each lattice appears to contain three porin molecules in negatively strained preparations with a central pit, presumably the pore, filled with stain. These findings are supported by cross-linking experiments carried out with whole cells and isolated cell walls.[188,189] Evidence for this trimeric structure has been reviewed recently.[1]

Experimental support for the contention that the trimeric structure is a water-filled pore has not been easily obtained. However, two recent reports[186,190] have shown that electrical conductance through reconstituted membranes containing matrix proteins is characteristic of conductance through water-filled channels. Benz et al.[190] observed that the addition of isolated matrix protein from *E. coli* K12 to the outer phases of a planar lipid membrane bilayer increased conductance several orders of magnitude. At low porin concentrations membrane conductance increases occurred in discrete steps. This observation suggested that the conductive pathways were localized structures, and by measurement of the incremental increases (approximately 1.9 nS in 1 *M* KCl) the diameter of the pore was estimated at 0.93 μm. The pores showed poor monovalent cation to monovalent anion discrimination, and ion selectivity was low. This lack of selectivity suggested an ohmic vs. voltage current character such as was found by Schindler and Rosenbusch[191] using different methods for inserting the matrix proteins into the planar lipid bilayers. These observations are consistent with the hypothesis that porins form water-filled channels within the outer membrane. It has also been found that addition of the maltose porin from *E. coli* to black lipid membranes causes increases in conductance similar to those caused by the matrix protein. Measurement of the incremental increases has shown that the maltose porin produces larger pores than the matrix proteins.[192] This result is consistent with the observation that only the maltose porin will allow the diffusion of the larger maltodextrins through the outer membrane.

B. Permeability of Hydrophobic Compounds

Hydrophobic compounds should dissolve into the phospholipid bilayer and thus pass through the outer membrane. However, compounds such as actinomycin D,[17] nafcillin,[193] gentian violet,[194] and sodium cholate[195] do not easily permeate the outer membrane. The diffusion of these compounds was increased by genetic alteration of the lipopolysaccharide molecular structure or chemical disruption of the lipopolysaccharide-outer membrane association. Therefore, lipopolysaccharide was thought to be necessary to form the hydrophobic barrier of the outer membrane. Mutants lacking specific outer membrane proteins also exhibited increased permeability to hydrophobic compounds,[18] suggesting that a lipopolysaccharide-protein complex serves as a barrier to hydrophobic molecules.

C. Enterobactin-Mediated Iron Transport

The complex system of iron chelation and transport across the outer membrane has been reviewed.[18] Recently, the outer membrane receptor of the ferric-enterobactin transport system was extracted and characterized.[196] Iron (III) can be chelated by enterobactin, an *E. coli* siderophore, and the complex binds to a receptor on the outer membrane which also recognizes colicins B and D as well as some bacteriophage.[18] Mutants of *E. coli* resistant to colicin B were isolated and found to lack an 80,000- to 90,000-dalton protein seen on sodium dodecyl sulfate polyacrylamide gels of wild type membrane preparations.[196,197] This protein was extracted from the membrane with 2%

Triton X-100 and was shown to bind ferric-enterobactin. Colicin B could compete with ferric-enterobactin for protein binding; the receptor had higher affinity for ferric-enterobactin than for other chelated iron complexes.[196] However, the ferric-enterobactin- and colicin B-binding functions can be separated by mutation[197] and have different temperatures of inactivation.[196]

The ferric-enterobactin transport system is thought to involve both inner and outer membrane components. Transport deficient mutants lacking the outer membrane receptor protein also had altered colicin and bacteriophage recognition.[198] However, the identification of a defect in the inner membrane was not as clear. Recently, a method was found to distinguish between the two possible locations of the defective transport protein.[199] Ferric-enterobactin uptake was measured in mutant cells and in spheroplasts derived from them. Stimulation of uptake by formation of spheroplasts would indicate that the mutation blocked transport at the outer membrane, whereas no stimulation would point to an inner membrane defect. Using this technique the *ton* and *fep* mutations were characterized. *TonA* and *fepA* mutants showed defects in the inner membrane while *tonB* and *fepB* mutants were defective in the outer membrane.[199] The genes for the outer membrane receptor have been mapped at 13 min on the recalibrated *E. coli* linkage map.[197]

D. Vitamins B_{12} Receptor

Vitamin uptake has been reviewed in References 18 and 200. Recently, vitamin B_{12} uptake in *E. coli* was shown to be repressed by growth in the presence of vitamin B_{12}.[201] In contrast, methionine, the end-product of the B_{12}-mediated metabolic pathway, did not affect B_{12} uptake. The difference in rate of uptake of vitamin B_{12} between cells grown with and without B_{12} indicated that the intracellular concentration of B_{12} affects synthesis of the outer membrane B_{12} receptor. Furthermore, the level of the 60,000-dalton receptor protein was lower in the repressed cells.[201]

VI. ACTIVE SOLUTE TRANSPORT ACROSS THE CYTOPLASMIC MEMBRANE

Virtually all actively growing cells maintain a membrane potential, interior negative. Bacteria which have been examined to date maintain this potential by extruding protons from the cytoplasm into the extracellular fluid. Some bacteria drive proton transport employing solar energy in processes dependent on either bacteriorhodopsin[202] (see Section IV. D.) or the bacteriochlorophyll containing photosynthetic apparatus. Most prokaryotic organisms, however, couple proton efflux to oxidative electron transport[203] and/or to ATP hydrolysis.[204,205] Little is known about the mechanisms by which components of the photosynthetic and oxidative electron transfer chains catalyze proton translocation. However, while many fermentative bacteria appear to lack the cytochrome constituents of the electron transport chain, all naturally occurring bacteria so far examined possess a membrane bound ATPase complex. Most of these enzyme complexes are structurally similar to the mitochondrial proton translocating ATPase, consisting of two polypeptide complexes: a peripheral membrane complex, termed F_1, which catalyzes ATP hydrolysis, and an integral membrane complex, termed F_0, which constitutes a transmembrane proton "pore."

A. Proton-Translocating ATPases

Dissociation and reconstitution of the ATPase complex from a thermophilic bacterium, PS3, has been accomplished by Kagawa and his co-workers.[205-207] These and other recent studies have led to a preliminary understanding of the functions of the

Table 2
PROPERTIES OF THE PROTEIN CONSTITUENTS OF THE F_0-F_1 COMPLEX OF THE THERMOPHILIC BACTERIUM PS3 [204-208]

Complex	Protein subunit	Approximate mol wt	Probable stoichio-metries	Demonstrable catalytic function(s) in reconstitution experiments	Presumed functional unit
TF_1	α	56,000	3	Stimulates ATPase activity of β-subunit	ATP-dependent proton "pump" (α, β, γ)
	β	53,000	3	Possesses active site for ADP and ATP binding	
	γ	32,000	1	Stimulates ATPase activity of β-subunit; functions with δ and ε to regulate H^+ transport by F_0	"Gate," regulating proton transport in the channel (γ, δ, ε, TF_1-binding protein)
	δ	16,000	1	Required for binding of F_1 to F_0	
	ε	11,000	1	Required for binding of F_1 to F_0	
TF_0	TF_1-binding protein	13,500	3	Required for binding of F_1 to F_0	
	DCCD-binding proteolipid	5,500	6	Transmembrane proton translocation	"Channel," through which protons traverse the membrane

individual protein constituents of the complex. The F_1 complex contains five dissimilar subunits while the F_0 component appears to consist of two types of proteins. Table 2 summarizes the molecular weights, stoichiometries, and probable functions of the individual protein subunits. The intact F_0–F_1 complex has been purified after extraction from the membrane with detergents and has been inserted into phospholipid vesicle membranes. The purified complex catalyzed proton transport, driven by ATP hydrolysis, as well as ATP synthesis in response to an artificially generated electrochemical proton gradient. Moreover, reconstitution of the energy transducing functions of the complex has been accomplished starting with the separated subunits of the F_1 complex and the two constituent subunits of the F_0 complex.

Studies with various chemical reagents have led to the conclusion that the low molecular weight proteolipid of the F_0 complex (see Table 2) alone functions in transmembrane proton conduction. The larger protein constituent appears to function in the binding of F_1 to the integral membrane complex (Table 2; Reference 207). Surprisingly, treatment of the intact F_0-F_1 complex with dicyclohexylcarbodiimide (DCCD) under conditions which resulted in maximal inhibition of ATPase activity and proton conduction, caused the covalent modification of only one-third of the low molecular weight proteolipid subunits.[207] The same observation had been made previously for the *E. coli* ATPase complex which contains a somewhat larger proteolipid (molecular weight of about 8000). Thus, three (or possibly six) of these proteolipid subunits may function cooperatively in proton translocation.

The presumed functions of the F_1 protein subunits are also summarized in Table 2.[208] While the β subunit possesses the binding site of the complex for ADP and ATP, this subunit must be associated with the α or γ subunit before appreciable ATPase activity is observed. The δ and ε subunits appear to function in the binding of the F_1 complex to F_0 in the membrane. Both of these subunits were shown to interact with F_0 directly with no preferred sequence of binding. The γ, δ, and ε subunits together inhibited F_0 catalyzed proton translocation. A model of the structure of the F_0-F_1 complex has been proposed.[206,207] How proton translocation (the primary function of the F_0 complex) is coupled to ATP synthesis or hydrolysis (the primary function of the F_1 moiety) has yet to be determined.[209]

Recent studies by Clarke and Morris[210] have led to the probability that the proton-translocating ATPase present in vegetative cells of the strict anaerobe, *Clostridium pasteurianum,* has a simpler structure than those in most bacteria studied. The intact F_0-F_1 enzyme complex was purified after detergent solubilization. Sodium dodecyl sulphate/polyacrylamide gel electrophoresis of the enzyme from vegetative cells revealed only four polypeptide species although the enzyme isolated from spores of the same organism was more complex. ATP-dependent transmembrane proton transport was demonstrable when the vegetative enzyme was reconstituted in artificial phospholipid vesicles, and this activity was sensitive to inhibition by dicyclohexylcarbodiimide. The finding that strictly anaerobic bacteria apparently possess a simpler variant of the H^+-transporting ATPase may be of evolutionary significance, since the first (and thus, most primitive) bacteria were undoubtedly strict anaerobes.[211] Conceivably, the four-subunit enzyme is a remnant of the evolutionary precursor of the complex ATPase found in most other bacteria and in eukaryotic organelles. The establishment of an evolutionary relationship between these enzymes will require more detailed structural analyses.

B. Active Transport of Monovalent Cations

The literature on active cation transport has been reviewed through 1976.[212] Since this time, considerable information concerning the mechanisms of Na^+ and K^+ transport has been forthcoming, and some progress has been made in the isolation and characterization of proteins responsible for K^+ transport in *E. coli.* This section will therefore deal with the recent literature concerned with Na^+ and K^+ transport in selected bacterial species.

1. Active Na^+ Transport

Lanyi and MacDonald studied the movements of ions (Na^+, H^+, and K^+) in cell envelope vesicles of *Halobacterium halobium.*[213] In this organism the active extrusion of protons occurs in a light-driven process dependent on bacteriorhodopsin (see Section IV.D.). Employing vesicles which had been loaded with Na^+, it was found that the primary light-induced proton efflux was followed by proton influx, resulting in a decrease in the magnitude of the pH gradient across the vesicular membrane. When the intravesicular Na^+ was depleted, the pH gradient was restored. In the absence of intravesicular Na^+, no such effect was observed. The results led to the conclusion that Na^+ extrusion is directly coupled to H^+ uptake rather than to K^+ countertransport or anion cotransport. In addition, it appeared that Na^+ extrusion occurred in an electrogenic fashion; the ratio of H^+ to Na^+ transported in opposite directions was greater than one. This suggestion was in contrast to conclusions drawn from earlier studies with *E. coli* and *Streptococcus faecalis* where electroneutral H^+/Na^+ exchange had been proposed.[212]

More recent studies of Na^+ transport in *E. coli* vesicles have confirmed the presence of Na^+/H^+ countertransport systems in this organism.[214,215] Employing vesicles of normal orientation, Schuldiner and Fishkes[214] showed that Na^+ efflux from Na^+ preloaded vesicles was induced when a proton electrochemical gradient was generated upon addition of an oxidizable energy source such as D-lactate. Moreover, it was possible to induce alkalinization of the medium by artificially creating a Na^+ concentration gradient across the membrane with the internal Na^+ concentration in excess of that in the external medium. When everted vesicles were employed, addition of Na^+ to the medium of energized vesicles evoked rapid proton efflux. Of several monovalent cations tested, only Li^+ could replace Na^+ as a substrate.[214,215] Whether the system appeared to function as an electroneutral or an electrogenic carrier depended on the experimental conditions employed.[214] It is therefore possible the Na^+ transport system in *E. coli* resembles that in *H. halobium*.

Na^+/H^+ countertransport has also been studied in *Streptococcus faecalis*.[216] Interestingly, $^{22}Na^+/Na^+$ exchange could not be demonstrated in resting cells but was stimulated by energy sources. Possibly a chemical energy source, the membrane potential, or the internal pH plays a role in regulating the activity of the transport system. In this regard, it is interesting that Na^+ efflux from *S. faecalis* cells could be demonstrated after (1) the proton translocating ATPase had been inhibited with dicyclohexylcarbodiimide; (2) an uncoupler had been added to destroy the proton electrochemical gradient; and (3) valinomycin had been added to abolish the K^+ gradient if an energy source such as glucose or arginine was added.[273] It would appear that a primary chemical energy source, such as ATP, can effect the extrusion of Na^+ in this organism.

2. Dual Energy Requirement for K^+ Transport in Streptococcus faecalis

Extensive studies by Rhoads and Epstein have shown that K^+ accumulation in *E. coli* is mediated by four distinct transport systems.[220,221] These have been distinguished by genetic, physiological, and kinetic manipulations.[217] One of these four systems, coded for by the *trkA* gene, is a high efficiency system which exhibits an apparent K_m for K^+ of 1.5 m*M*. The functioning of this transport system is dependent on both chemiosmotic and chemical energy although it is not known how these energy sources function in the transmembrane translocation process.

Bakker and Harold have examined in detail the mechanism by which energy is coupled to K^+ uptake in the Gram-positive fermentative bacterium, *S. faecalis*.[218,219] Metabolizing *S. faecalis* cells accumulate K^+ against a 40,000-fold concentration gradient, apparently as a result of the action of a single transport system. This gradient was considerably in excess of that which would have been expected for a univalent electrogenic K^+ flux driven solely by the membrane potential. In order to account for this anomaly, the K^+ equilibrium distribution ratio ($[K^+]$ in/$[K^+]$ out) was studied in glycolyzing cells as a function of the membrane potential.[212] A plot of the logarithm of this concentration ratio versus the membrane potential gave a slope of $n = 2$, suggesting that K^+ uptake was coupled to the uptake of an additional positive charge. Addition of valinomycin to the cell suspension reduced this slope to a value of $n = 1$ as expected for the equilibration of a univalent cation across the membrane. Indirect evidence, obtained by independently varying the membrane potential and the pH gradient, led to the suggestion that the transport system mediates K^+/H^+ cotransport.

As was noted above for $^{22}Na^+/Na^+$ exchange transport, catalyzed by the Na^+ transport system in *S. faecalis*, $^{42}K^+/K^+$ exchange was not observed in resting cells, but metabolism of glucose or arginine, which enhanced the cellular ATP concentration, promoted this exchange process. When the proton-translocating ATPase was inhibited in glycolyzing cells with dicyclohexylcarbodiimide, K^+ accumulation was not appreciable unless a membrane potential, negative inside, was artificially imposed. Under these same conditions, $^{42}K^+/K^+$ exchange was observed, even in the absence of an appreciable membrane potential. A dependency of K^+ efflux on a chemical energy source could also be demonstrated. From these experiments it was concluded that the proton electrochemical potential provided the driving force for K^+ accumulation while the chemical energy source (presumably ATP) maintained the transport system in a catalytically active state.

3. K^+-Dependent ATPase: A K^+ Pump in E. coli

An entirely different energy coupling mechanism is operative for the high affinity, K^+ repressible, transport system in *E. coli*. This permease system, which can maintain a K^+ concentration gradient in excess of 10^6 to 1, is encoded by three genes *kdpA*, *kdpB*, and *kdpC* within a single operon. The *kdpD* gene is a linked regulatory gene controlling expression of the *kdp* operon. This transport system appears to utilize ATP exclusively to drive K^+ uptake. Laimonis et al.[220] utilized a transducing bacteriophage carrying the *kdp* operon to identify the protein products of the three constituent genes. These genes, carried within the phage genome, were preferentially expressed in UV-irradiated host cells which were defective for DNA repair. The products of the *kdpA, B,* and *C* genes were shown by sodium dodecyl sulfate gel electrophoresis to possess molecular weights of 47,000, 90,000, and 22,000, respectively. All three proteins were associated with the cytoplasmic membrane, and a periplasmic or outer membrane protein constituent appeared to be lacking. Preliminary evidence suggested that the high affinity K^+ binding site was associated with the *kdpA* gene product.

Enzymological studies resulted in the identification of a membrane-bound K^+-stimulated ATPase.[221] The following evidence supported the conclusion that this ATPase was coded for by the *kdp* genes.

1. The ATPase was repressible by K^+, being regulated by the *kdpD* gene.
2. Correlating with the high substrate specificity of the transport system, ATPase activity was stimulated by K^+ but not by Na^+, Rb^+ or Cs^+. In contrast to the Na^+, K^+ ATPase from animal sources, Na^+ appeared to have little or no effect on the activity of the enzyme. Activity was observed only if a divalent cation (Mg^{++}, Mn^{++}, or Co^{++}) was present.
3. Both the transport system and the ATPase exhibited similar K^+ affinity, and specific mutations, probably in the *kdpA* gene, altered the affinities of the transport system and the ATPase for K^+ in parallel.

More recent work on K^+ transport has shown that the 90,000 dalton protein, the product of the *kdpB* gene, is rapidly phosphorylated in the presence of ATP. The intermediate appears to be an acyl phosphate.[274] The identification of a K^+-translocating ATPase in *E. coli* membranes allows one to draw certain parallels with the cation transport enzymes in animal cells and may lead to an appreciation of related mechanistic features.

C. Mechanisms of Nutrient Transport

Requirements for growth of a biological cell include (1.) the generation of a transmembrane electrical potential (interior negative); (2.) the maintenance of appropriate cytoplasmic concentrations of inorganic cations and anions; (3.) the accumulation within the cell of utilizable nutrients; and (4.) the extrusion from the cytoplasm of waste products and other toxic substances. Nutrient accumulation and the egress of end products of metabolism are generally effected by specific transport systems. When the transport process is linked to metabolic energy, cytoplasmic concentrations of the substrate may greatly exceed or be far less than those in the extracellular fluid, depending on the polarity of the "pump". Mechanisms by which nutrient uptake can be coupled to energy have been studied extensively, and excellent reviews on the subject have appeared.[222-224] The active efflux of organic compounds has also been discussed.[225] Consequently, these subjects will not be dealt with here.

Relatively little is known about the mechanism(s) by which solutes traverse the membrane. Does a permease substrate diffuse through the membrane in combination with a proteinaceous substrate binding site (carrier mechanism), or does it pass through the permease from the external surface into the cytoplasm in a free or hydrated state (channel mechanism)? If the latter occurs, do separate solute binding sites exist on the external and internal surfaces of the permease, or does a single binding site suffice for the catalysis of vectorial translocation? Do all nutrients traverse the membrane by mechanistically similar processes, or do several clearly distinguishable transport mechanisms exist? While we are not in a position to answer any of these questions definitively, some recent experiments indicate that at least some permeases possess two distinct substrate binding sites, one localized to the external surface of the permease complex, the other facing the cytoplasm. Three examples of transport systems for which some information about the translocation process is available and for which specific models have been proposed will be considered below. It is worthy of note that for each of these systems (the dicarboxylic acid transport system in *E. coli,* the bacterial phosphoenolpyruvate: sugar phosphotransferase system, and the histidine permease in *Salmonella typhimurium),* a channel type mechanism has been proposed.

1. Dicarboxylic Acid Transport[226-233]

Extensive kinetic and genetic studies indicate that a single permease system transports the three dicarboxylic acids succinate, malate, and fumarate across the cytoplasmic membrane of intact *E. coli* cells. Employing a mutant strain of *E. coli* K12 which could not metabolize succinate, a single K_m value (30 μM) for this substrate was determined. Active uptake of the dicarboxylate anions was apparently driven by the proton electrochemical gradient, and two protons accompanied substrate entry. However, the energy coupling mechanism may be more complex since an involvement of the proton-translocating ATPase was demonstrated when lactate oxidation energized the uptake process in membrane vesicles. Possibly both chemical and chemiosmotic energy are required for the normal functioning of the dicarboxylate transport system as discussed above for K^+ transport in *Streptococcus faecalis* and *E. coli.*

The number of protein components involved in dicarboxylate transport has been determined in genetic and biochemical studies. Three cistrons appear to code for the three protein constituents of the system. The *cbt* gene codes for a periplasmic dicarboxylate binding protein (DBP), while the *dctA* and *dctB* genes code for two integral membrane proteins. All three proteins are required for the uptake of succinate by whole cells, but only the *dctA* and *dctB* gene products mediate uptake into membrane vesicles. Since the outer lipopolysaccharide containing membrane is intact in whole cells but defective in vesicles, it is possible that the periplasmic binding protein functions to transport the dicarboxylate anions across the membrane. The *dctA* and *dctB* gene products may alone translocate the substrates across the cytoplasmic membrane. Relevant to this possibility is the observation that at least some binding proteins may be exposed to the outer surface of the outer membrane of the *E. coli* cell envelope.[229]

In recent studies Lo[230] and Bewick and Lo[231] have shown that about half of the dicarboxylate binding protein which is associated with the *E. coli* cell can be released in a free state during osmotic shock treatment. The shock releaseable DBP fraction, but

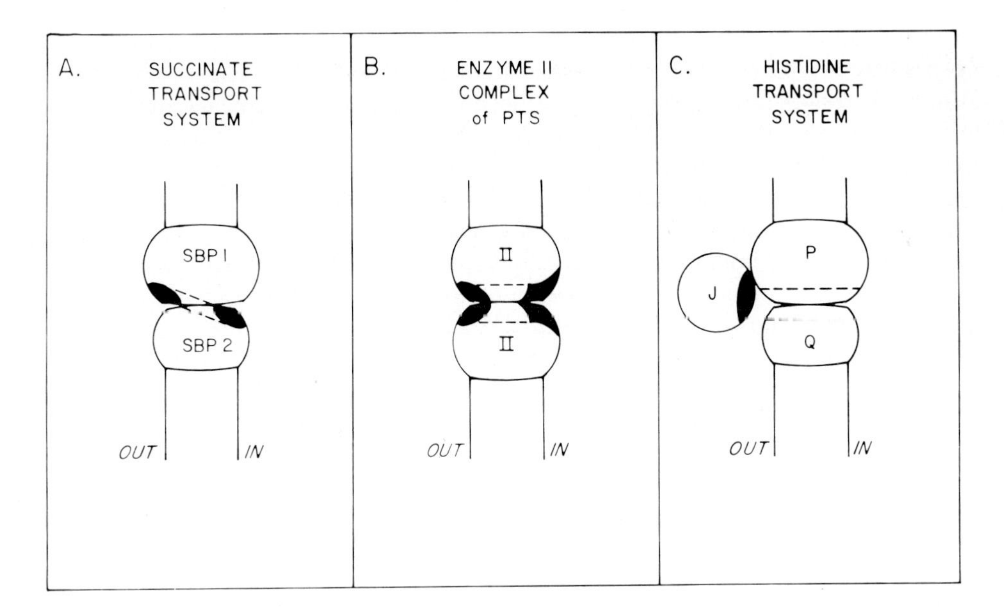

FIGURE 4. Proposed models of bacterial permease systems. (A), the model proposed by Lo[227,230] for the succinate permease in *E. coli*. The proposal suggests that the periplasmic dicarboxylate binding protein (DBP) functions to transport dicarboxylates across the outer membrane and may not interact directly with the two integral membrane constituents of the system, SBP1 and SBP2. The latter two proteins alone may catalyze transport across the cytoplasmic membrane. (B) A model proposed for the structure of an Enzyme II complex of the bacterial phosphotransferase system.[234] (This model should be considered as speculative in view of the fact that chemical cross-linking studies have thus far proved unsuccessful.[136, 279] (C) Model suggested for the high affinity histidine permease in *S. typhimurium*.[239] Dark regions indicate known substrate binding sites; dashed lines indicate presumed channels in the integral membrane proteins through which the solute passes. (Figure 4A redrawn after Lo, *Can. J. Biochem*, 57, 289, 1979.)

not the lysozyme releasable DBP which remained with the cells after cold osmotic shock treatment, appeared to be exposed to the outer surface of the outer membrane. This last conclusion was deduced by treatment of intact cells with surface labeling reagents which could not penetrate the outer membrane of the bacterial cell envelope. These reagents included nonspecific proteases (pronase and papain) and diazo sulfanilic acid.[230,232] Lactoperoxidase radioiodination was also used to surface-label the protein. Since loss of surface DBP, due either to osmotic shock or protease treatment, decreased transport of activity, it was concluded that DBP facilitated transport of the dicarboxylate anions across the outer membrane. It was suggested that DBP might function in conjuction with an outer membrane porin protein.[230,232]

There is, at present, no evidence that DBP plays a direct role in the transport of dicarboxylate anions across the *E. coli* cytoplasmic membrane. As noted above, *cbt* mutants which lacked this protein transported the anions normally, although intact cells were defective in transport function. Substantiation for the conclusion that the two integral constituents of the cytoplasmic membrane, SBP 1 and SBP 2 (see below), alone catalyze succinate transport across this barrier has come from reconstitution experiments.[233] When purified preparations of these proteins were added to rat myoblasts or mouse L-cells, the proteins were incorporated into the animal cell membranes, and the cells gained the ability to transport succinate. Both proteins were required for acquisition of transport function. These results substantiate the model shown in Figure 4A which was redrawn from one presented by Lo.[230]

Active species of the three protein constituents of the dicarboxylate permease have been isolated employing columns of aspartate-coupled Sepharose. The successful use of affinity chromatography for their isolation provided the first evidence that all three proteins possess substrate binding domains. The two membrane components, SBP1 (mol wt = 14,000) and SBP2 (mol wt = 20,000), products of the *dctB* and *dctA* genes, respectively, each bind succinate, fumarate, and malate. While the binding of succinate and malate to SBP1 is characterized by dissociation constants of about 25 μM and 50 μM, respectively (similar to the K_m values of the intact transport system for these substrates), SBP2 binds the anions with tenfold higher affinities. Orientation studies indicated that the substrate-binding site of SBP2 is exposed to the external surface of the membrane while that of SBP1 faces the cytoplasm.[226] Nevertheless, both proteins appear to span the membrane as indicated by studies with nonpenetrating covalent labeling reagents. Furthermore, SBP1 and SBP2 must be in close proximity to one another as indicated by studies with protein cross-linking reagents. Surprisingly, tartaryl diazide, the cross-linking reagent used in these studies, did not inhibit the transport function of the permease system appreciably. Evidently, the reagent did not attack the substrate-binding sites of the transport system. The results suggest that the transport proteins do not undergo major conformational changes during catalysis of solute transport.

Lo has proposed a model for the dicarboxylate transport system which appears to account for most of the data discussed above.[230] A similar model is reproduced in Figure 4A. It is assumed that both transmembrane proteins function in the transport process, and that the free dicarboxylic acid passes from the external substrate-binding site of SBP2 to the binding site of SBP1 on the cytoplasmic surface of the membrane. The two proteins together presumably form a substrate-induced channel through which the dicarboxylic acid can pass.

2. *Enzymes II of the Bacterial Phosphotransferase System*[137,234-236]

Extensive experimental evidence favors the notion that the bacterial phosphotransferase system (PTS) catalyzes the concomitant transport and phosphorylation of its sugar substrates.[137] The sugar recognition component of the system is the sugar-specific integral membrane Enzyme II complex. Until recently, none of these enzymes had been obtained in pure form, and little information was available concerning the mechanism by which they catalyze transmembrane sugar translocation. The discovery of the sugar phosphate: sugar transphosphorylation reactions, catalyzed by the Enzymes II of the PTS, has led to a more detailed concept of the mechanism by which these enzymes act upon their substrates.[234] Recent experiments bearing on this mechanism will be considered in this section.

Extensive evidence supports the contention that the Enzymes II of the PTS catalyze vectorial exchange group translocation as follows:[234]

$$[^{14}C]\ \text{sugar}_{OUT} + \text{sugar-P}_{IN} \rightleftharpoons [^{14}C]\ \text{sugar-P}_{IN} + \text{sugar}_{OUT}$$

A particular Enzyme II exhibits strict specificity toward the sugar as well as the sugar phosphate substrate.

The transphosphorylation reactions catalyzed by the glucose Enzyme II and the mannose Enzyme II have been subjected to detailed kinetic analyses.[235] The reaction mechanism demonstrated for both enzymes was found to be Bi-Bi sequential, indicating that the Enzymes II each possess nonoverlapping binding sites for sugar and sugar phosphate. Although there appears to be no preferred order of substrate binding, association of the two substrates with the enzyme occurred in a positively cooperative fashion. That is, binding of the sugar-P substrate to the Enzyme II complex apparently enhanced the affinity of the enzyme for the sugar substrate and vice versa. A mutant with a defective glucose Enzyme II was isolated which transported methyl α-glucoside and glucose with reduced maximal velocities and higher K_m values. In vitro kinetic studies of the transphosphorylation reaction catalyzed by the mutant enzyme showed decreased maximal velocities and increased K_m values for both the sugar and the sugar phosphate substrates. It was therefore suggested that a single Enzyme II complex catalyzed both transport and transphosphorylation of its sugar substrates.

This conclusion has recently been substantiated with homogeneous mannitol Enzyme II.[236] The purified detergent-solubilized enzyme exhibited a subunit molecular weight of about 60,000. It catalyzed both the phosphoenolpyruvate-dependent phosphorylation of [^{14}C]mannitol in the presence of the soluble enzymes of the PTS (Enzyme I and HPr) as well as mannitol 1-P: [^{14}C]mannitol transphosphorylation in the absence of the soluble enzymes. That the mannitol Enzyme II is the only membrane protein encoded by the mannitol operon in *E. coli* was also demonstrated. In these experiments, a hybrid Col E1 plasmid carrying the mannitol operon was transferred to a minicell-producing strain of *E. coli*. The minicells, which were free of chromosomal DNA, synthesized only two proteins in response to inducer (mannitol) and cyclic AMP. One of these proteins was membrane associated (protomer molecular weight of 60,000; presumed to be the mannitol Enzyme II); while the other was a soluble protein (protomer molecular weight of 40,000; presumed to be mannitol phosphate dehydrogenase). These results lead to the probability that an Enzyme II complex, consisting of a single polypeptide chain (possibly in an oligomeric complex) in association with membrane phospholipids catalyzes phosphoenolpyruvate-dependent sugar uptake and phosphorylation as well as sugar phosphate to sugar exchange group translocation.[234] In some respects the proposed model (Figure 4B) resembles that proposed by Lo for the succinate transport system (Figure 4A).[227] In the case of the Enzyme II complex, however, the nonoverlapping sugar-binding site (on the external surface of the Enzyme II complex) and sugar phosphate-binding site (on the cytoplasmic surface of the Enzyme II) are assumed to be associated with a single polypeptide chain. Although an oligomeric structure is depicted in Figure 4B, no direct experimental evidence is yet available in support of the contention that transport function depends on the association of polypeptide protomers.

As envisaged, sugar phosphorylation is accompanied by transport of the free sugar from the external binding site through a transmembrane "pore" in the Enzyme II complex to the sugar phosphate binding site on the cytoplasmic surface of the Enzyme II complex. Since the affinity of the enzyme for its sugar substrate is about a thousandfold higher than its affinity for the corresponding sugar phosphate,[235] the sugar phosphate should be readily released into the cytoplasm. It should be emphasized that there is, as yet, little direct experimental evidence bearing on the structural aspects of the Enzyme II complex depicted in Figure 4B. Extensive studies will be required to test the validity of this working hypothesis.

3. Histidine Transport System in Salmonella typhimurium[237-239]

A large number of nutrient transport systems in Gram-negative bacteria depend for activity on the presence of a hydrophilic periplasmic solute binding protein as well as hydrophobic integral membrane proteins.[222,224] At least in some cases, the periplasmic binding proteins are thought to associate with the integral membrane constituents of the transport systems. Evidence for this contention resulted from genetic studies with the high affinity histidine permease of *S. typhimurium*.[238] This transport system appears to consist of three distinct proteins, all of which are encoded by cistrons within a single operon. The J protein is a periplasmic histidine binding protein, coded for by the *hisJ* gene, while the P and Q proteins are encoded by the *hisP* and *hisQ* genes, respectively.[239] The P protein (and possibly the Q protein as well) is an integral cytoplasmic membrane protein.

A mutant *S. typhimurium* strain was isolated in which the genetic lesion specifically altered the J protein. The mutant protein still bound histidine with high affinity, but it apparently did not interact normally with the integral membrane constituent(s) of the system. A compensatory mutation in the *hisP* gene altered the P protein such that transport function was restored. This result indicated that an interaction between the J protein and the P protein was required for normal transport activity.[238] These experimental results provided partial verification for an earlier hypothesis advanced by Singer regarding the mechanism of action of bacterial permeases which depend on periplasmic binding proteins.[237]

In a recent communication Ames and Nikaido[239] identified the P protein as a basic membrane protein having an approximate molecular weight of 24,000. The Q protein has not yet been identified. Interestingly, the P protein—Q protein complex probably transports several basic amino acids (L-arginine, L-lysine, L-ornithine, etc., in addition to D- and L-histidine).[239] This observation suggests that the integral membrane constituents of the system (the "pore" proteins) do not exhibit high affinity or strict specificity for the substrate amino acids; rather, the specificity of the system may be dictated by the periplasmic binding proteins. It is possible that several of these proteins can associate with the P protein—Q protein complex and thus mediate the transport of structurally dissimilar amino acids.[239] It is not yet known if the P and Q proteins possess substrate binding sites as do the integral membrane constituents of the succinate transport system and the phosphotransferase system. However, Figure 4C depicts a tentative model of the histidine permease which appears to account for the known properties of this system.

Further studies of the three transport systems illustrated in Figure 4 and of other permeases may reveal a fundamentally unified picture for the transmembrane solute translocation process. If so, the possibility of a common evolutionary ancestry for these membrane proteins would be supported. Alternatively, several distinct transport mechanisms may have independently evolved in parallel. If this was the case, translocation mechanisms may prove to be as diverse as the mechanisms by which energy is coupled to the transport processes.

D. Transport of Antibiotics and Antimicrobial Metabolite Analogs via Nutrient Permease Systems

In recent years, natural populations of pathogenic bacteria have been shown to acquire transmissible extrachromosomal genetic material (plasmids) which confer upon the organisms resistance to commonly employed antibiotics such as penicillin, streptomycin, and tetracycline.[240] This fact has led to widespread recognition of a need for the development of entirely new classes of antimicrobial agents which are both potent against pathogenic bacterial populations and innocuous to the host organism. A rational approach to this problem depends on an understanding of the mechanisms of entry and modes of action of the toxic substances.

In order for most antibiotic agents to exert their toxic effects, they must first cross the outer membrane of the Gram-negative bacterial cell and gain access to the cytoplasmic membrane. Since the vast majority of these agents exert their actions in the cell cytoplasm they must also penetrate the cytoplasmic membrane. In extensive studies in several laboratories it has been shown that the outer membrane acts as a permeability barrier for both hydrophilic and hydrophobic substances (see Section V.). Hydrophilic antibiotics with molecular weights lower than 700 readily diffuse through the outer membrane, even at 0°C, due to the presence of aqueous proteinaceous pores.[1] In contrast, hydrophobic antibiotics of molecular weights in excess of 1200 penetrate the outer membrane in a process which has been shown to be highly temperature dependent.[193] Diffusion of these substances apparently depends on their solubility in the hydrocarbon interior of the outer membrane and is restricted by the hydrophilic "core" region of the lipopolysaccharide present in this structure. As evidence for this conclusion, it was shown that mutants which were defective for the core region of the lipopolysaccharide complex (deep rough mutants) were far more susceptible than the wild type parental organisms to the toxic effects of various hydrophobic antibiotics. In some cases the mutants were as susceptible to the toxic action of these agents as were certain Gram-positive bacteria which lack an outer membrane altogether. It was suggested that the differences in permeability of hydrophobic substances observed in the wild type and deep rough mutants was due to the absence in the former, and presence in the latter, of exposed phospholipid bilayer regions.[193]

1. Hydrophilic Antibiotic Uptake via Known Permeases

Once having gained access to the periplasmic space of the Gram-negative bacterial cell, the antimicrobial agent must intercalate into or penetrate the cytoplasmic membrane. Agents which are sufficiently lipophilic to penetrate the outer membrane by passive diffusion can probably cross the inner membrane by the same mechanism. The more hydrophilic compounds, however, must enter the cytoplasm via solute-specific transport systems. In some cases the toxic agents are accumulated within the cytoplasm against large concentration gradients because the entry process is coupled to metabolic energy. This operation accounts for the fact that the minimal inhibitory concentration is sometimes found to be lower in vivo than in vitro.[200,241] Zähner and Diddens have summarized some of the evidence associating the uptake of various antibiotics with specific permease systems in the cytoplasmic membrane.[242] Table 3 provides a representative sample of the information available. The sideromycins, of which several have been studied, apparently cross the membrane via the sideramine transport system(s). Other antibiotics enter the cytoplasm via permeases which normally transport peptides or amino acids. Showdomycin specifically utilizes the cytidine transport system, while fosfomycin can cross the membrane either via the α-glycerophosphate permease or by the hexose phosphate permease. Finally, nojirimycin (5-aminoglucose) and streptozotocin (an analog of *N*-acetylglucosamine) are transported and phosphorylated by the phosphotransferase system. It is likely that the toxic actions of most hydrophilic antibiotics will prove to be dependent on the activities of nutrient specific transport systems which recognize the toxic agents due to their structural similarities to the natural metabolites.

Virtually all of the permease systems which function in the transport of exogenous sources of carbon and energy are subject to regulation by cyclic AMP.[243,244] Synthesis of the constituent proteins requires that adenylate cyclase (the cyclic AMP synthetic enzyme) and the cyclic AMP receptor protein be functional.[243,244] In the absence of either of these regulatory proteins, transcription of the genes coding for the transport proteins is not activated. Consequently, it should be possible to distinguish antibiotics which cross the cytoplasmic membrane via transport systems which normally function in carbon acquisition from those which function to facilitate the passage of other nutrients. Adenylate cyclase or cyclic AMP receptor protein-deficient mutants should show decreased sensitivity to an antibiotic, relative to the parental strain, if that agent enters the cytoplasm via a permease which functions in carbon acquisition. In a recent series of investigations, Alper and Ames examined the sensitivities of wild type, adenylate

Table 3
UPTAKE OF HYDROPHILIC ANTIBIOTICS VIA NUTRIENT-SPECIFIC TRANSPORT SYSTEMS[242,244]

Antibiotic	Transported by permease system(s) with specificity toward
Sideromycins (albomycin; ferrimycin)	Sideramines
Tripeptides (phosphinothricin-ala-ala; plumbermycin B)	Oligopeptides
Dipeptides (bacilysin)	Dipeptides
D-Cycloserine	D-alanine
5-Methyltryptophan; β-2-thienylalanine; 1-amino-2-phenylethyl-phosphonic acid, azaserine	Aromatic amino acids
L-Azetidine carboxylic acid	Proline
Streptomycin	Polyamines
Showdomycin	Cytidine
Fosfomycin (phosphonomycin)	α-Glycerophosphate; hexose phosphates
Nojirimycin	Glucose
Streptozotocin	*N*-Acetylglucosamine

cyclase-deficient, and cyclic AMP receptor protein-deficient mutants of *S. typhimurium* for sensitivity to growth inhibition by a large number of toxic agents.[244] Of these compounds, 22 antibiotics and 29 inhibitory analogs of normal bacterial carbon sources were more toxic in the parental strain than in the mutants. This observation clearly indicated that the entry of numerous antimicrobial agents is dependent on permeases which normally function in the transport of carbon and energy sources.

2. Toxicity of Substrates of Phosphotransferase System

As stated above, the development of novel classes of antibiotics which are of practical value for the treatment of diseases will require the discovery of nutrient analogs which are transported and/or metabolized by pathogenic bacteria but not by the host organism. An ideal analog would be one which resembles a nutrient which enters the bacterial cell via a permease system and is metabolized via catabolic enzyme systems which differ from those found in the animal or plant host. In this regard, it is worth noting that a large number of bacteria possess the phosphoenolpyruvate-dependent phosphotransferase system (PTS) responsible for the uptake and phosphorylation of many sugars.[136,137] Pathogenic bacteria known to possess this enzyme system include most of the Gram-negative enteric bacteria such as *E. coli* and species of *Salmonella* as well as Gram-positive pathogens such as the *Streptococci, Staphylococci,* and *Lactobacilli.*[136] Species of *Pseudomonas* possess a fructose-specific phosphotransferase system, and at least some spirochetes phosphorylate and transport sugars by a phosphoenolpyruvate-dependent mechanism.[136]

Those organisms which do utilize the phosphotransferase system for sugar uptake frequently metabolize the sugar via a pathway which differs from that employed by mammalian cells. Furthermore, the system phosphorylates sugar analogs which are not substrates of kinases present in mammalian tissues. Table 4 summarizes the initial steps in the metabolic pathways employed for the utilization of sugar substrates of the phosphotransferase system in bacteria, and contrasts these pathways with those

Table 4
DIFFERENCES BETWEEN SUGAR METABOLIC PATHWAYS IN BACTERIA AND MAMMALIAN CELLS[a]

Substrate sugar	Bacteria	Mammalian cells
Fructose	Fructose ↓ PTS Fructose-1-P ↓ Fructose 1-P kinase Fructose 1, 6-di-P	Fructose ↓ Hexokinase Fructose-6-P ↓ P-Fructokinase Fructose 1, 6-di-P
N-Acetylglucosamine	*N*-Acetylglucosamine ↓ PTS *N*-Acetylglucosamine 6-P ↓ NAG-6-P deacetylase Acetate + glucosamine 6-P	*N*-Acetylglucosamine ↓ Deacetylase Acetate + glucosamine ↓ Hexokinase Glucosamine-6-P
Hexitols (glucitol, mannitol galactitol)	Hexitol ↓ PTS Hexitol-P ↓ Hexitol-P dehydrogenase Hexose-P	Hexitol ↓ Hexitol dehydrogenase Hexose ↓ Hexokinase Hexose-6-P
Lactose	Lactose ↓ PTS Lactose-P ↓ P-β Galactosidase Glucose + Galactose 6-P	Lactose ↓ β Galactosidase Glucose + galactose ↓ Galactokinase Galactose-1-P
Sucrose	Sucrose ↓ PTS Sucrose-P	Sucrose ↓ Sucrase Fructose + glucose

[a]Bacteria which do not catabolize the sugar via the phosphotransferase system employ different metabolic pathways.

employed by the animal cell. Some of these differences are noted below. The product of PTS-catalyzed fructose group translocation is fructose 1-P, while mammalian cells phosphorylate fructose in the 6-position. *N*-acetylglucosamine is first phosphorylated and then deacetylated in bacteria while the reverse is true in mammals. The mammalian deacetylase is of low specific activity in serum and most body tissues.[275] Hexitols are directly phosphorylated via the PTS in bacteria, but are initially oxidized by mammalian tissues. Finally, disaccharides such as lactose and sucrose, when transported by the PTS, are directly phosphorylated and then cleaved to the constituent sugar and sugar phosphate in reactions catalyzed by phosphodisaccharidases. By contrast, mammals hydrolyze the disaccharides directly and then phosphorylate the monosaccharides employing ATP-dependent kinases. Lactose (and galactose) metabolism in many Gram-positive bacteria gives rise to galactose 6-P which is converted to tagatose-1, 6-diP, while mammalian tissues utilize galactokinase to phosphorylate the 1-hydroxyl of the sugar, and the phosphate ester is subsequently converted to glucose 1-P and then to glucose 6-P. These metabolic differences suggest that the enzymes responsible for sugar metabolism in bacteria will act upon substrates which are not recognized by the mammalian enzymes.

Limited experimentation has shown that the sugar-specific enzymes of the

Table 5
SOME NONMETABOLIZABLE SUGAR ANALOGS WHICH ARE SUBSTRATES OF THE BACTERIAL PHOSPHOTRANSFERASE SYSTEMS IN *ESCHERICHIA COLI* AND *SALMONELLA TYPHIMURIUM*

Sugar analog	Probably substrate of Enzyme II specific for	Substrate for mammalian hexokinases[a]	Ref.
Methyl α-glucoside	Glucose	–	245,246,247
3-Fluoroglucose	Glucose	?	245
3-*O*-Methylglucose	Glucose	–	246
5-Thioglucose	Glucose	?	246
5-Aminoglucose	Glucose	?	242
2-Deoxyglucose	Mannose	+	245,247
Methyl α-mannoside	Mannose	–	246
Methyl α-2-deoxyglucoside	Mannose	–	246
Iodo *N*-acetylglucosamine	*N*-Acetylglucosamine	–	248,249
Streptozotocin	*N*-Acetylglucosamine	?	242
Methyl α-*N*-acetylglucosamine	*N*-Acetylglucosamine	–	250
L-Sorbose	Fructose	?	251
D-Xylitol	Fructose	–	251
6-Chlorofructose	Fructose	–	246
Glucosaminitol	Sorbitol	–	246
D-Arabinitol	Sorbitol	–	251
2-Deoxyarabinohexitol	Sorbitol	–	246

[a] –, This sugar is not a substrate of hexokinase; +, this sugar is a substrate of hexokinase; ?, the specificity of hexokinase toward this sugar is not known.

phosphotransferase system frequently do act upon sugar analogs which are not substrates of the mammalian metabolic enzymes. A selected list of such sugar analogs is included in Table 5. All of the analogs listed in the table have been shown to be phosphorylated by the phosphotransferase systems in *E. coli* and *S. typhimurium.* All of these compounds also inhibit bacterial growth under appropriate conditions although high concentrations of the analogs are sometimes required. It has not been determined whether the substrates actually kill the bacteria or merely cause growth stasis. Among the most toxic analogs listed to date are the halogenated sugars such as 3-fluoroglucose and halogenated *N*-acetylglucosamine derivatives. These compounds inhibit bacterial growth at micromolar concentrations. It seems likely that some of these or other related compounds will be of medical value if it can be shown that they exhibit highly specific antibacterial activities.

VII. REGULATION OF INTRACELLULAR CYCLIC AMP AND THE SYNTHESIS OF CARBOHYDRATE PERMEASES AND CATABOLIC ENZYMES

The intracellular concentration of cyclic AMP in a bacterial cell can theoretically be regulated by modulating the rates of cyclic AMP synthesis, degradation, and/or transmembrane transport.[243,252,253] The cyclic nucleotide phosphodiesterase is clearly active in intact Gram-negative enteric bacterial cells since genetic loss of this enzyme results in a marked enhancement of the internal levels and net production of cyclic AMP.[243,252] However, the activity of this enzyme does not appear to respond to changes in environmental conditions in a fashion which would account for catabolite repression.

By contrast, the activities of both adenylate cyclase and the cyclic AMP transport system have been shown to depend on exogenous carbon and energy sources available to the cell, and these regulatory interactions appear to be of physiological importance. This section will therefore review recent work dealing with these subjects. The reader is referred to other reviews for comprehensive coverage of the earlier literature.[136,243,252,254]

A. Control of Adenylate Cyclase Activity by the Phosphotransferase System

Several studies have shown that adenylate cyclase is subject to regulation by a mechanism which depends on the activities of the proteins of the phosphotransferase system.[136,247,253-257] The effects of mutations in genes coding for the proteins of the PTS on adenylate cyclase regulation has been studied in some detail, and several conclusions have been drawn from the results reported:

1. Net synthesis of cyclic AMP ceases shortly after cells reach the stationary growth phase. This effect is due to the loss of in vivo adenylate cyclase activity and can be reversed only by growth under conditions which allow active protein synthesis.
2. The net production of cyclic AMP was substantially elevated by genetic loss of the cyclic nucleotide phosphodiesterase but was not altered by a deficiency in Enzyme I due to leaky *ptsI* mutations. Complete loss of Enzyme I function depressed cyclic AMP synthesis. Adenylate cyclase activity was generally dependent on the carbon source and growth conditions employed, being greatest when examined in cells which were grown on a carbohydrate which was rapidly metabolized.
3. Methyl α-glucoside and other sugar substrates of the PTS inhibited cyclic AMP production, and leaky *ptsH* or *ptsI* mutations enhanced sensitivity to this effect. Inhibition of adenylate cyclase was not a transient effect when the sugar was added to freshly washed cell suspensions but lasted for the duration of the experiment (1 hr).
4. Mutation to carbohydrate repression resistance (due to mutations in the *crrA* gene) decreased the quantity of cyclic AMP produced to a level substantially below that found in the wild type strain. *crrA* mutations depressed cyclic AMP synthesis in most strains of *E. coli* and *S. typhimurium* tested[136,256,257] but relatively high adenylate cyclase activity was present in *crrA* mutants isolated in certain *E. coli* genetic backgrounds.
5. A variety of sugars which were not substrates of the PTS also depressed adenylate cyclase activity if the corresponding catabolic enzymes had been induced. However, different mechanisms appeared to be operative since leaky *ptsI* mutations did not alter sensitivity to these effects.[225,247]

Recently it was shown that in a mutant strain of *S. typhimurium* which contained very low Enzyme I activity ($<0.1\%$ of wild type activity) exceptionally low concentrations of methyl α-glucoside ($0.05\ \mu M$) inhibited net cyclic AMP synthesis.[276] Since the apparent K_m for methyl α-glucoside uptake in wild type *S. typhimurium* cells is more than 1000 times this value,[235] this result eliminates a regulatory mechanism involving direct interaction of a methyl α-glucoside receptor (i.e., the Enzyme II) and adenylate cyclase[257] and provides definitive evidence that the PTS proteins function catalytically to regulate adenylate cyclase. A mechanism which appears to account for these results has been proposed.[136,254,256]

B. Effects of *ptsI* Mutations on Transient and Permanent Repression of β-Galactosidase Synthesis in *E. coli*

In a series of unpublished studies,[276] Saier and Feucht examined the induction of β-galactosidase synthesis of *E. coli* employing conditions in which the phenomenon of inducer exclusion was unimportant. Under these conditions a reasonably good

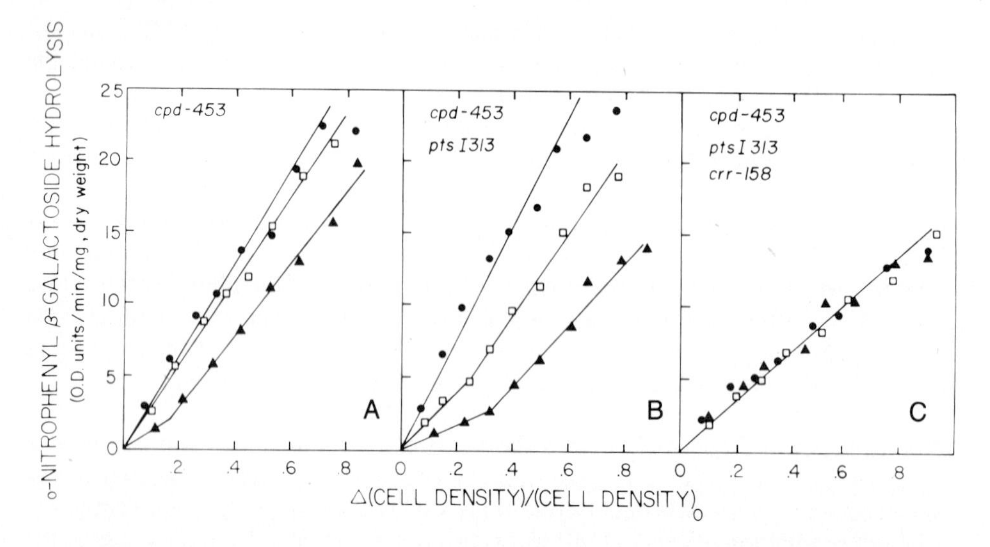

FIGURE 5. Transient and permanent repression of β-galactosidase synthesis in a series of mutant strains of *E. coli.* Cells were grown continuously in medium 63 containing 1% sodium pyruvate. With cells in the exponential growth phase at 34°C, isopropyl β-thiogalactoside (0.5 m*M*) was added alone (O) or together with 5m*M* methyl α-glucoside (□) or 5 m*M* glucose (Δ). Subsequently, aliquots were periodically removed for measurement of cell density and β-galactosidase activity. (A) Strain *cpd-453,* a cyclic AMP phosphodiesterase-negative strain.[257] (B) *cpd-453 ptsI313,* A mutant strain which produces a temperature sensitive Enzyme I with low activity after growth at 34°C.[257] (C) *cpd-'53 ptsI313 crrA158,* A carbohydrate repression-resistant mutant derived from strain *cpd-453 ptsI313.*

correlation was noted between repression of β-galactosidase synthesis and inhibition of cyclic AMP production in vivo by sugar substrates of the PTS in agreement with the results of Epstein et al.[259] However, the results did not distinguish mechanisms in which the *pts* and *crrA* mutations affected transient vs. permanent repression.[260] Consequently repression of β-galactosidase synthesis by glucose and methyl α-glucoside was studied in a series of *E. coli* mutants which lacked cyclic AMP phosphodiesterase (Figure 5). In the parental strain *(cpd-453)* glucose weakly repressed β-galactosidase synthesis, while methyl α-glucoside was almost without effect. Reduction of the Enzyme I content of the cell, due to the *ptsI313* mutation, enhanced both the intensity and duration of transient repression. Sensitivity to permanent repression was also increased. The *crrA158* mutation in the *cpd-453 ptsI313* genetic background depressed the rate of β-galactosidase synthesis, but abolished both transient and permanent repression by glucose. Relative rates of β-galactosidase synthesis correlated approximately with intracellular cyclic AMP both in the presence and absence of sugars.

The results summarized above provide information about the repression of β-galactosidase synthesis by a sugar substrate of the PTS. However, they do not bear on the influence of Enzyme I on transient or permanent repression caused by carbohydrates which are not substrates of the PTS. This problem has been examined by Yang et al. who employed *E. coli* strains which synthesized the glycerol catabolic enzymes and the hexose phosphate permease constitutively. One such strain carried a deletion mutation within the structural gene for Enzyme I.[261] Both transient and permanent repression of β-galactosidase synthesis by glycerol and glucose 6-P were demonstrated in the parental and Enzyme I negative strains. However, the absence of Enzyme I had several effects on β-galactosidase synthesis:

1. The rate of β-galactosidase synthesis in the absence of glycerol or glucose 6-P was reduced, and this rate could be restored by the addition of cyclic AMP.

2. Both the duration and intensity of transient repression were enhanced by the *ptsI* mutation when either glycerol or glucose 6-P served as the repressing carbohydrate.
3. The intensity of permanent repression also appeared to be enhanced by the loss of Enzyme I function although this effect may be strain dependent.[261]

These results appear qualitatively similar to those depicted in Figure 5, in agreement with the suggestion that Enzyme I exerts a primary effect by maintaining adenylate cyclase in an activated state. However, the results do not provide insight into the mechanism of transient repression. Because it was not determined if the deletion mutation encompassed the *crrA* gene[261] it is not clear whether this gene is important to the repression of enzyme synthesis by nonPTS carbohydrates.

As a result of studies on the regulation of hexitol uptake, Lengler and Steinberger have proposed a mechanism to explain the phenomenon of transient repression.[262] They suggest that transient repression of enzyme synthesis (presumably due to lowered internal cyclic AMP, i.e., to inhibition of adenylate cyclase) is a consequence of the transient accumulation of carbohydrate phosphates. The following sequence of events may possibly explain transient repression.

1. Addition of a new carbon source to the growth medium of bacteria which can immediately transport and metabolize that carbohydrate results in increased cellular concentrations of carbohydrate phosphates.
2. Adenylate cyclase is subject to inhibition by intracellular organic phosphate esters so that the activity of this enzyme is depressed, and internal cyclic AMP concentrations fall.
3. Due to decreased internal cyclic AMP as well as to inhibition of inducer uptake, synthesis of the permeases specific for the repressing carbohydrates is partially inhibited.
4. In response to decreased sugar permease activity, cellular sugar-P pools are reduced to a new steady state level. Partial relief of adenylate cyclase and enzyme synthesis from inhibition results.

While the proposed mechanism appears reasonable, it should be pointed out that much experimental evidence will be required to establish or refute its validity. Carbohydrate permeases appear to be sensitive to inhibition by several distinct mechanisms.[135,262,263] A multiplicity of regulatory mechanisms for adenylate cyclase has been postulated[225] but not yet demonstrated. Specifically, physiologically relevant inhibition of adenylate cyclase by sugar phosphates has not been reported. Moreover, relief from transient repression is usually observed to occur suddenly after a specific time interval following addition of the repressing carbohydrate. Gradual relief from transient repression would be expected from the proposed model. Thus, we are still far from a molecular understanding of the phenomenon of transient repression.

C. Physiological Significance of Cyclic AMP Transport to the Regulation of Internal Cyclic AMP and Enzyme Synthesis

In 1965, Makman and Sutherland demonstrated that addition of glucose to a resting suspension of *E. coli* cells in carbohydrate-free medium inhibited net cyclic AMP synthesis and stimulated efflux of the cyclic nucleotide from the cell.[264] The mechanistic basis for the former observation is discussed above, and the explanation for stimulation of cyclic AMP excretion relates to the energy dependency of the process.[265] The physiological significance of cyclic AMP excretion to the regulation of the intracellular cyclic AMP concentration and the synthesis of catabolic enzymes has recently been demonstrated.[266,267] Wright et al. determined internal and extracellular concentrations of cyclic AMP in glucose-limited continuous cultures of *E. coli* as a function of growth rate.[266] At dilution rates between 0.05 and 0.40/hr, the intracellular concentrations of cyclic AMP were constant at about 5 μM. If the rate of cyclic AMP excretion had been constant with time and proportional to the internal concentration of cyclic AMP, the extracellular concentration of the nucleotide at any time during continuous growth should have been proportional to the intracellular concentration divided by the dilution rate. This, however, was not observed. When the dilution rate was large (i.e., when the energy level of the cell was high)[266] extracellular cyclic AMP was present in larger amounts than predicted from the above stated relationship. This result is in accord with expectation for an energy-dependent cyclic AMP transport system.

Epstein et al. had previously measured cyclic AMP efflux as a function of growth rate in batch cultures when different carbon sources served as growth substrates.[259] In their experiment the rate of cyclic AMP efflux was approximately proportional to the internal concentration of the cyclic nucleotide. Although this observation appears to contradict the results of Wright et al.,[266] it should be noted that Wright et al. used energy-depleted cells, while Epstein and co-workers used energy-sufficient cells. If, in the latter study, the cyclic AMP transport system was saturated with energy in all of the cell suspensions examined, proportionality of efflux rate to internal concentration would be expected.

Phillips et al. have recently studied the induction of biodegradative threonine dehydratase in a cyclic AMP phosphodiesterase negative strain of *E. coli*.[267] Induction of the enzyme required anaerobic conditions, and internal cyclic AMP concentrations were measured during the aerobic to anaerobic transition. Intracellular cyclic AMP was approximately constant (5 to 10 μM) during exponential growth under both aerobic and anaerobic conditions. When the aerobically growing culture was depleted of oxygen, the internal cyclic AMP concentration increased to 300 μM within 10 min and subsequently declined to normal. Similarly, when KCN was added to an aerobically growing *E. coli* culture, a transient increase in intracellular cyclic AMP was observed.[267] During the transition period when internal cyclic AMP levels increased, no increase in the rate of cyclic AMP synthesis was observed.[277] Since a cyclic AMP phosphodiesterase negative strain was employed in these studies, the "spike" of internal cyclic AMP could only be attributable to a decrease in the rate of cyclic AMP efflux, which, in turn, was presumably due to a lowering of the cellular energy level.

In response to the elevated internal cyclic AMP concentration, threonine dehydratase was synthesized. It was concluded that

1. Conditions which resulted in a temporary energy deficit caused inhibition of cyclic AMP efflux and a consequent increase in the internal level of cyclic AMP.
2. Increased cyclic AMP in the cell served as a signal for the initiation of threonine dehydratase synthesis.
3. Threonine dehydratase synthesis allowed energy generation by the nonoxidative degradation of threonine.
4. Restoration of the cellular energy level (membrane potential) resulted in increased rates of cyclic AMP efflux and restoration of the intracellular cyclic AMP concentration to normal.

D. Catabolite Repression of β-Galactosidase Synthesis in the Absence of Cyclic AMP

Several studies have indicated that synthesis of carbohydrate catabolic enzymes may be influenced by metabolites other than inducer and cyclic AMP.[225] Recently Dessein and co-workers studied the repression of β-galactosidase synthesis in a *cya* deletion mutant of *E. coli* which was incapable of cyclic AMP synthesis.[268,269] Secondary mutations in this mutant altered the cyclic AMP receptor protein or the promoter region of the lactose operon and allowed more rapid rates of β-galactosidase synthesis. Synthesis of this enzyme in the former double mutants was found to be subject to catabolite repression when the nitrogen source was limiting although carbon source limitation had almost no effect. Surprisingly, the *cya crp** double mutants, with an altered cyclic AMP receptor protein, did not show a corresponding repressive effect in nitrogen-limiting cultures although derepression in media containing poorly utilized carbon sources was observed. The authors suggested that the conformation of the cyclic AMP receptor protein may be directly influenced by a cellular compound termed "catabolite modulator factor" (CMF) as well as by cyclic AMP.[268] Relevant to this suggestion was the observation that while the lactose promoter mutants were sensitive to the repressive effect of CMF, the *cya crp** mutants were relatively insensitive.[269]

The molecular structure of CMF is not known. It is a small heat-stable compound which represses catabolite sensitive operons without exerting effects on catabolite independent systems. Kinetic experiments revealed that the compound was destroyed by intact cells at rates which depended on their physiological state. Although metabolism of CMF was not dependent on cyclic AMP, the degradative rate was slow under conditions of strong repression and high in derepressed cells. Further information concerning the mode of action of CMF must await its chemical identification.

VIII. CONCLUDING REMARKS

This chapter has concentrated on mechanisms of transmembrane solute translocation in bacteria and on functionally important interactions of the permease proteins with the matrix phospholipids. It is clear from the discussion in Section IV. that the activities of a variety of permeases specific for sugars, amino acids, and protons are dependent both on the fatty acid and phospholipid composition of the membrane. While many of the earlier conflicts of interpretation have now been resolved, a number of discrepancies still remain. It seems likely that some of these will prove attributable to differences in experimental procedure. Thus, increasing the membrane concentration of a specific phospholipid may enhance permease activity up to a point but cause inhibition at higher levels (Section IV.; Figure 3). In some cases the various investigators studied a physiological parameter after altering the phospholipid composition of the membrane in vivo to differing extents. Sometimes different mutants or experimental techniques were used to modulate the phospholipid concentration from the normal level. Secondary alterations may not always have been detected or appropriately taken into account. In other cases, where in vitro reconstitution of transport function was examined, different investigators may have removed endogenous phospholipids to differing extents. Few studies have quantified transport activity as a function of the concentration of the membrane phospholipids. Consequently, even qualitative comparisons of results obtained in different laboratories may be misleading. The age-old question of relevance of the in vitro experiment to the in vivo situation also remains with us, and it will persist until phospholipid composition can be rigorously controlled both in the in vivo and the in vitro experiments. Clearly, much work will be required before we can define the specific phospholipid requirements of individual permeases. Even more effort must be expended to provide mechanistic explanations for these dependencies.

Differences in the dependencies of various transport systems on membrane fluidity probably reflect the mechanisms of solute translocation. From the discussions in Sections V. and VI. it seems likely that at least two types of translocation mechanisms will prove to be operative in the Gram-negative bacterial cell: a nonspecific aqueous pore mechanism, operative in the outer membrane, and a highly stereospecific channel mechanism responsible for solute transport across the cytoplasmic membrane. The latter mechanism may involve solute binding to specific sites on both the cytoplasmic and the external surfaces of the transmembrane permease complex. To what extent these two transport processes resemble one another remains to be determined. Nor is sufficient experimental evidence available to allow speculation regarding the number of distinct translocation mechanisms operative in the cytoplasmic membrane. If different modes of transport are involved, these may be distinguishable by measuring sensitivities of the permeases to changes in phospholipid environment. Specifically, one would expect that a mechanism which involves substantial conformational change in the integral membrane permease protein(s) would be more dependent on membrane fluidity than a mechanism involving solute transport through a relatively rigid pore.

The synthesis of a particular permease may be subject to regulation by any of several distinct mechanisms. In Section VII. the complexity of the regulatory interactions controlling the synthesis of carbohydrate permeases by cyclic AMP was considered as a representative example. It should be remembered, however, that permease synthesis may be dependent on cytoplasmic, periplasmic, or extracellular inducer, and evidence has been presented for the direct involvement of permease proteins in the regulation of their own syntheses.[225] Additionally, transcriptional and translational regulatory mechanisms still undreamed of may be revealed by future studies.

It is now clear that the activities as well as the syntheses of permease systems are regulated by several distinct mechanisms. For example, the matrix protein in the *E. coli* outer membrane has recently been shown to form voltage-dependent channels in lipid bilayers.[191] Similarly, the bacterial proton translocating ATPase in the cytoplasmic membrane appears to be gated, admitting protons and synthesizing ATP only when the membrane potential reaches a certain threshold value.[270,271] It also appears that the bacteriocidal protein, colicin K, exerts its toxic effect on *E. coli* cells by forming voltage-controlled channels in the phospholipid bilayer of the cytoplasmic membrane.[272]

Evidence for the operation of other regulatory mechanisms is rapidly accumulating. The phosphotransferase system apparently controls the activities of several other carbohydrate permeases by a mechanism thought to involve allosteric regulation by a protein constituent of the PTS.[135] More recently, evidence has been obtained suggesting that the functioning of one PTS Enzyme II can inhibit the activity of another by a mechanism involving competition for phospho-HPr, the common phosphoryl donor for PTS sugar uptake.[278] Finally, the accumulation of intracellular carbohydrate phosphates appears to inhibit many permeases.[135] Clearly, diverse mechanisms regulate solute uptake. To what extent these are interrelated, mutually exclusive, or cooperative remains to be determined. It is not unreasonable to suppose that any one regulatory mechanism will itself prove to be subject to modulating influences of other regulatory interactions. Regulation must have been imposed on enzymes and transport systems at relatively advanced stages in evolutionary history. The superimposition of regulatory interactions controlling these regulatory mechanisms must have provided the finishing touches which allowed integration of the various constituents of the cellular physiological apparatus into a highly efficient metabolic machine.

ACKNOWLEDGMENTS

We are grateful to Drs. G. Ames, W. Epstein, F. M. Harold, H. Kagawa, P. W. Kent, C. J. Knowles, T. C. Y. Lo, and A. T. Phillips for allowing us to include the results of their experimental work prior to publication.

REFERENCES

1. **DiRienzo, J. M., Nakamura, K., and Inouye, M.,** The outer membrane proteins of Gram-negative bacteria: biosynthesis, assembly, and function, *Ann. Rev. Biochem.*, 47, 481, 1978.
2. **Manning, P. A. and Reeves, P.,** Outer membrane of *Escherichia coli* K-12: differentiation of proteins 3A and 3B on acrylamide gels and further characterization of *con (tolG)* mutants, *J. Bacteriol.*, 127, 1070, 1976.
3. **Datta, D. B., Krämer, C., and Henning, U.,** Diploidy for structural gene specifying a major protein of the outer cell envelope membrane from *Escherichia coli* K-12, *J. Bacteriol.*, 128, 834, 1976.
4. **Smit, J., Kamino, Y., and Nikaido, H.,** Outer membrane of *Salmonella typhimurium:* chemical analysis and freeze-fracture studies with lipopolysaccharide mutants, *J. Bacteriol.*, 124, 942, 1975.
5. **Osborn, M. J., Rick, P. D., Lehman, V., Rupprecht, E., and Singh, M.,** Structure and biogenesis of the cell envelope of Gram-negative bacteria, *Ann. N.Y. Acad. Sci.*, 235, 52, 1974.
6. **Bayer, M. E.,** Ultrastructure and organization of the bacterial envelope, *Ann. N.Y. Acad. Sci.*, 235, 6, 1974.
7. **Sleytr, U. B.,** Regular arrays of macromolecules on bacterial cells: structure, chemistry, assembly, and function, in *International Review of Cytology,* Vol. 53, Bourne, G. H. and Danielli, J. F., Eds., Academic Press, New York, 1978, 1.
8. **Salton, M. R. J. and Owen, P.,** Bacterial membrane structure, *Ann. Rev. Microbiol.*, 30, 451, 1976.
9. **Träuble, H., and Overath, P.,** The structure of *Escherichia coli* membranes studied by fluorescence measurements of lipid phase transitions, *Biochim. Biophys. Acta,* 307, 491, 1973.
10. **Mescher, M. F., Strominger, J. L., and Watson, S. W.,** Protein and carbohydrate composition of the cell envelope of *Halobacterium salinarium, J. Bacteriol.*, 120, 945, 1974.
11. **Osborn, M. J.,** Structure and biosynthesis of the bacterial cell wall, *Ann. Rev. Biochem.*, 38, 501, 1969.
12. **Wilson, D. B.,** Cellular transport mechanisms, *Annu. Rev. Biochem.*, 47, 933, 1978.
13. **Schnaitman, C. A.,** Protein composition of the cell wall and cytoplasmic membrane of *Escherichia coli, J. Bacteriol.*, 104, 890, 1970.
14. **White, D. A., Albright, F. A., Lennarz, W. J., and Schnaitman, C. A.,** Distribution of the phospholipid-synthesizing enzymes in the wall and membrane subfractions of the envelope of *Escherichia coli, Biochim. Biophys. Acta,* 249, 636, 1971.
15. **Osborn, M. J., Gander, J. E., Parisi, E., and Carson, J.,** Mechanism of assembly of the outer membrane of *Salmonella typhimurium:* isolation and characterization of cytoplasmic and outer membrane, *J. Biol. Chem.*, 247, 3962, 1972.
16. **Nakae, T. and Nikaido, H.,** Outer membrane as a diffusion barrier in *Salmonella typhimurium:* penetration of oligo- and polysaccharides into isolated membrane vesicles and cells with degraded peptidoglycan layer, *J. Biol. Chem.*, 250, 7359, 1975.
17. **Leive, L.,** Release of lipopolysaccharide by EDTA treatment of *E. coli, Biochem. Biophys. Res. Commun.*, 21, 290, 1965.
18. **Kadner, R. J. and Bassford, P. J., Jr.,** The role of the outer membrane in active transport, in *Bacterial Transport,* **Rosen, B. P., Ed., Marcel Dekker,** New York, 1978, 413.
19. **Osborn, M. J., Gander, J. E., and Parisi, E.,** Mechanism of assembly of the outer membrane of *Salmonella typhimurium, J. Biol. Chem.*, 247, 3973, 1972.
20. **Lugtenberg, E. J. J. and Peters, J.,** Distribution of lipids in cytoplasmic and outer membranes of *Escherichia coli, Biochim. Biophys. Acta,* 441, 38, 1976.
21. **Finnerty, W. R. and Makula, R. A.,** Microbial lipid metabolism, *CRC Crit. Rev. Microbiol.*, 4, 1, 1975.
22. **White, D. A., Lennarz, W. J., and Schnaitman, C. A.,** Distribution of lipids in the wall and cytoplasmic membrane subfractions of the cell envelope of *Escherichia coli, J. Bacteriol.*, 109, 686, 1972.
23. **Zuchowski, C. and Pierucci, O.,** Phospholipid turnover during the division cycle of *Escherichia coli, J. Bacteriol.*, 133, 1533, 1978.
24. **Raetz, C. R. H. and Newman, K. F.,** Neutral lipid accumulation in the membranes of *Escherichia coli* mutants lacking diglyceride kinase, *J. Biol. Chem.*, 253, 3882, 1978.
25. **Pluschke, G., Hirota, Y., and Overath, P.,** Function of phospholipid in *Eschericha coli.* Characterization of a mutant deficient in cardiolipin synthesis, *J. Biol. Chem.*, 253, 5048, 1978.
26. **Happe, M. and Overath, P.,** Bacteriorhodopsin depleted of purple membrane lipids, *Biochem. Biophys. Res. Commun.*, 72, 1504, 1976.
27. **Placky, W. Z., Lanyi, J. K., and Kates, M.,** Lipid interactions in membranes of extremely halophilic bacteria. I. Electron spin resonance and dilatometric studies of bilayer structure, *Biochemistry,* 13, 4906, 1974.
28. **Lanyi, J. K., Placky, W. Z., and Kates, M.,** Lipid interactions in membranes of extremely halophilic bacteria. II. Modification of the bilayer structure by squalene, *Biochemistry,* 13, 4912, 1974.
29. **Kotyk, A. and Janáček, K.,** Membranes, in *Biomembranes,* Vol. 9, Manson, L. A., Ed., Plenum Press, New York, 1977, 11.

30. **Anderson, H. C.,** Probes of membrane structure, *Annu. Rev. Biochem.,* 47, 359, 1978.
31. **Dutt, A. and Dowhan, W.,** Intracellular distribution of enzymes of phospholipid metabolism in several Gram-negative bacteria, *J. Bacteriol.* 132, 159, 1977.
32. **Raetz, C. R. H.,** Enzymology, genetics, and regulation of membrane phospholipid synthesis in *Escherichia coli, Microbiol. Rev.,* 42, 614, 1978.
33. **Cronan, J. E., Jr.,** Molecular biology of bacterial membrane lipids, *Annu. Rev. Biochem.,* 47, 163, 1978.
34. **Silbert, D. F.,** Genetic modification of membrane lipid, *Annu. Rev. Biochem.,* 44, 315, 1975.
35. **Cronan, J. E., Jr. and Gelmann, E. P.,** Physical properties of membrane lipids: biological relevance and regulation, *Bacteriol. Rev.,* 39, 232, 1975.
36. **Bloch, K. and Vance, D.,** Control mechanisms in the synthesis of saturated fatty acids, *Annu. Rev. Biochem.,* 46, 263, 1977.
37. **Cronan, J. E., Jr. and Vagelos, P. R.,** Metabolism and function of the membrane phospholipids of *Escherichia coli, Biochim. Biophys. Acta,* 265, 25, 1972.
38. **Raetz, C. R. H. and Foulds, J.,** Envelope composition and antibiotic hypersensitivity of *Escherichia coli* mutants defective in phosphatidylserine synthetase, *J. Biol. Chem.,* 252, 5911, 1977.
39. **Taylor, F. and Cronan, J. E., Jr.,** Selection and properties of *Escherichia coli* mutants defective in the synthesis of cyclopropane fatty acids, *J. Bacteriol.,* 125, 518, 1976.
40. **Silbert, D. F. and Vagelos, P. R.,** Fatty acid mutant of *E. coli* lacking a β-hydroxydecanoyl thioester dehydrase, *Proc. Natl. Acad. Sci. U.S.A.,* 58, 1579, 1967.
41. **Rosenfield, J. S., D'Agnolo, G., and Vagelos, P. R.,** Synthesis of unsaturated fatty acids and the lesion in *fabB* mutants, *J. Biol. Chem.,* 248, 2452, 1973.
42. **Overath, P., Pauli, G., and Schairer, H. U.,** Fatty acid degradation in *Escherichia coli:* an inducible acyl-CoA synthetase, the mapping of *old*-mutations, and the isolation of regulatory mutants, *Eur. J. Biochem.,* 7, 559, 1969.
43. **Silbert, D. F., Cronan, J. E., Jr., Beacham, I. R., and Harder M. E.,** Genetic engineering of membrane lipid, *Fed. Proc. Fed. Am. Soc. Exp. Biol.,* 33, 1725, 1974.
44. **Overath, P., Schairer, H. U., and Stoffel, W.,** Correlation of *in vitro* and *in vivo* phase transitions of membrane lipids in *Escherichia coli, Proc. Natl. Acad. Sci. U.S.A.,* 67, 606, 1970.
45. **Silbert, D. F. and Vagelos, P. R.,** Systems for membrane alterations: unsaturated fatty acid auxotrophs of *Escherichia coli, Methods Enzymol.,* 32B, 856, 1974.
46. **Davis, M-T. B. and Silbert, D. F.,** Changes in cell permeability following a marked reduction of saturated fatty acid content in *Escherichia coli* K-12, *Biochim. Biophys. Acta,* 363, 1, 1974.
47. **Esfahani, M., Limbrick, A. R., Knutton, S., Oka, S., and Wakil, S. J.,** The molecular organization of lipids in the membrane of *Escherichia coli:* phase transitions, *Proc. Natl. Acad. Sci. U.S.A.,* 68, 3180, 1971.
48. **Esfahani, M., Crowfoot, P. D., and Wakil, S. J.,** Molecular organization of lipids in *Escherichia coli.* II. Effect of phospholipids on succinic-ubiquinone reductase activity, *J. Biol. Chem.,* 247, 7251, 1972.
49. **Silbert, D. F., Ladenson, R. C., and Honeggar, J. L.,** The unsaturated fatty acid requirement in *Escherichia coli:* temperature dependence and total replacement by branched chain acids, *Biochim. Biophys. Acta,* 311, 349, 1973.
50. **Raetz, C. R. H. and Kennedy, E. P.,** Function of cytidine diphosphate-diglyceride and deoxycytidine diphosphate-diglyceride in the biogenesis of membrane lipids in *Escherichia coli, J. Biol. Chem.,* 248, 1098, 173.
51. **Kanfer, J. N. and Kennedy, E. P.,** Metabolism and function of bacterial lipids. II. Biosynthesis of phospholipids in *Escherichia coli, J. Biol. Chem.,* 239, 1720, 1964.
52. **Raetz, C. R. H. and Kennedy, E. P.,** The association of phosphatidylserine synthetase with ribosomes in extracts of *Escherichia coli, J. Biol. Chem.,* 247, 2008, 1972.
53. **Wiener, J. H. and Heppel, L. A.,** Purification of the membrane-bound and pyridine nucleotide-independent L-glycerol-3-phosphate dehydrogenase from *Escherichia coli, Biochem. Biophys. Res. Commun.,* 47, 1360, 1972.
54. **Bell, R. M.,** Mutants of *Escherichia coli* defective in membrane phospholipid synthesis: macromolecular synthesis in an *sn*-glycerol-3-phosphate acyltransferase Km mutant, *J. Bacteriol.,* 117, 1065, 1974.
55. **Lin, E. C. C.,** Glycerol dissimilation and its regulation in bacteria, *Ann. Rev. Microbiol.,* 30, 535, 1976.
56. **Cronan, J. E., Jr., and Bell, R. M.,** Mutants of *Escherichia coli* defective in membrane phospholipid synthesis: mapping of the structural gene for L-glycerol-3-phosphate dehydrogenase, *J. Bacteriol.,* 118, 598, 1974.
57. **Edgar, J. R., and Bell, R. M.,** Biosynthesis in *Escherichia coli* of *sn*-glycerol-3-phosphate, a precursor of phospholipid: purification and physical characterization of wild type and feedback-resistant forms of the biosynthetic *sn*-glycerol-3-phosphate dehydrogenase, *J. Biol. Chem.,* 253, 6348, 1978.

58. **Edgar, J. R. and Bell, R. M.,** Biosynthesis in *Escherichia coli* of *sn*-glycerol-3-phosphate, a precursor of phospholipid: kinetic characterization of wild type and feedback-resistant forms of the biosynthetic *sn*-glycerol-3-phosphate dehydrogenase, *J. Biol. Chem.,* 253, 6354, 1978.
59. **Edgar, J. R. and Bell, R. M.,** Biosynthesis in *Escherichia coli* of *sn*-glycerol-3-phosphate, a precursor of phospholipid: palmitoyl-CoA inhibition of the biosynthetic *sn*-glycerol-3-phosphate dehydrogenase, *J. Biol. Chem.,* 254, 1016, 1979.
60. **Bachmann, B. J., and Low, K. B.,** Linkage map of *Escherichia coli* K-12, Edition 6, *Microbiol. Rev.,* 44, 1, 1980.
61. **Okuyama, H. and Wakil, S. J.,** Positional specificities of acyl coenzyme A: glycerophosphate and acyl coenzyme A: monoacylglycerophosphate acyltransferases in *Escherichia coli, J. Biol. Chem.,* 248, 5197, 1973.
62. **Ray, T. K., Cronan, J. E., Jr., Mavis, R. D., and Vagelos, P. R.,** The specific acylation of glycerol-3-phosphate to monoacyl-glycerol-3-phosphate in *Escherichia coli:* evidence for a single enzyme conferring this specificity, *J. Biol. Chem.,* 245, 6442, 1970.
63. **van den Bosch, H. and Vagelos, P. R.,** Fatty acyl-CoA and fatty acyl-acyl carrier protein as donors in the synthesis of lysophosphatidate and phosphatidate in *Escherichia coli, Biochim. Biophys. Acta,* 218, 233, 1970.
64. **Snider, M. D. and Kennedy, E. P.,** Partial purification of glycerophosphate acyltransferase from *Escherichia coli, J. Bacteriol.,* 130, 1072, 1977.
65. **Bell, R. M.,** Mutants of *Escherichia coli* defective in membrane phospholipid synthesis: Properties of wild type and Km defective *sn*-glycerol-3-phosphate acyltransferase activities, *J. Biol. Chem.,* 250, 7147, 1975.
66. **Sinensky, M.,** Temperature control of phospholipid synthesis in *Escherichia coli, J. Bacteriol.,* 106, 449, 1971.
67. **Cronan, J. E., Jr., Ray, T. K., and Vagelos, P. R.,** Selection and characterization of an *E. coli* mutant defective in membrane lipid biosynthesis, *Proc. Natl. Acad. Sci. U.S.A.,* 65, 737, 1970.
68. **Nakamura, H., Tojo, T., and Greenberg, J.,** Interaction of the expression of two membrane genes, *acrA* and *plsA,* in *Escherichia coli, J. Bacteriol.,* 122, 874, 1975.
69. **Glaser, M., Nulty, W., and Vagelos, P. R.,** Role of adenylate kinase in the regulation of macromolecular biosynthesis in a putative mutant of *Escherichia coli* defective in membrane phospholipid biosynthesis, *J. Bacteriol.,* 123, 128, 1975.
70. **McIntyre, T. M., Chamberlain, B. K., Webster, R. E., and Bell, R. M.,** Mutants of *Escherichia coli* defective in membrane phospholipid synthesis. Effects of cessation and reinitiation of phospholipid synthesis on macromolecular synthesis and phospholipid turnover, *J. Biol. Chem.,* 252, 4487, 1977.
71. **McIntyre, T. M. and Bell, R. M.,** Mutants of *Escherichia coli* defective in membrane phospholipid synthesis: effect of cessation of net phospholipid synthesis on cytoplasmic and outer membranes, *J. Biol. Chem.,* 250, 9053, 1975.
72. **Bell, R. M. and Cronan, J. E., Jr.,** Mutants of *Escherichia coli* defective in membrane phospholipid synthesis: phenotypic suppression of *sn*-glycerol-3-phosphate acytransferase Km mutants by loss of feedback inhibition of the biosynthetic *sn*-glycerol-3-phosphate dehydrogenase, *J. Biol. Chem.,* 250, 7153, 1975.
73. **Pieringer, R. A. and Kunnes, R. S.,** The biosynthesis of phosphatidic acid and lysophosphatidic acid by glyceride phosphokinase pathways in *Escherichia coli, J. Biol. Chem.,* 240, 2833, 1965.
74. **Schneider, E. G. and Kennedy, E. P.,** Phosphorylation of ceramide by diglyceride kinase preparations from *Escherichia coli, J. Biol. Chem.,* 248, 3739, 1973.
75. **Thomas, E. L., Weissbach, H., and Kaback, H. R.,** Further studies on metabolism of phosphatidic acid of isolated *E. coli* membrane vesicles, *Arch. Biochem. Biophys.,* 150, 797, 1972.
76. **Weissbach, H., Thomas, E., and Kaback, H. R.,** Studies on the metabolism of ATP by isolated bacterial membranes: formation and metabolism of membrane-bound phosphatidic acid, *Arch. Biochem. Biophys.,* 147, 249, 1971.
77. **Larson, T. J. and Dowhan, W.,** Ribosomal-associated phosphatidyl-serine synthetase from *Escherichia coli:* purification by substrate-specific elution from phosphocellulose using cytidine 5′-diphospho-1,2-diacyl-*sn*-glycerol, *Biochemistry,* 15, 5212, 1976.
78. **Ishinaga, M. and Kito, M.,** Participation of soluble phosphatidylserine synthetase in phosphatidylethanolamine biosynthesis in *Escherichia coli* membrane, *Eur. J. Biochem.,* 42, 483, 1974.
79. **Bell, R. M., Mavis, R. D., Osborn, M. J., and Vagelos, P. R.,** Enzymes of phospholipid metabolism: localization in the cytoplasmic and outer membrane of the cell envelope of *Escherichia coli* and *Salmonella typhimurium, Biochim. Biophys. Acta,* 249, 628, 1971.
80. **Raetz, C. R. H.,** Isolation of *Escherichia coli* mutants defective in enzymes of membrane lipid synthesis, *Proc. Natl. Acad. Sci. U.S.A.,* 72, 2274, 1975.
81. **Raetz, C. R. H.,** Phosphatidylserine synthetase mutants of *Escherichia coli:* genetic mapping and membrane phospholipid composition, *J. Biol. Chem.,* 251, 3242, 1976.

82. **Ohta, A. and Shibuya, I.,** Membrane phospholipid synthesis and phenotypic correlation of an *Escherichia coli pss* mutant, *J. Bacteriol.,* 132, 434, 1977.
83. **Hawrot, E. and Kennedy, E. P.,** Phospholipid composition and membrane function in phosphatidylserine decarboxylase mutants of *Escherichia coli, J. Biol. Chem.,* 253, 8213, 1978.
84. **Dowhan, W., Wickner, W. T., and Kennedy, E. P.,** Purification and properties of phosphatidylserine decarboxylase from *Escherichia coli, J. Biol. Chem.,* 249, 3079, 1974.
85. **Satre, M. and Kennedy, E. P.,** Identification of bound pyruvate essential for the activity of phosphatidylserine decarboxylase of *Escherichia coli, J. Biol. Chem.,* 253, 479, 1978.
86. **Hawrot, E. and Kennedy, E. P.,** Biogenesis of membrane lipids: mutants of *Escherichia coli* with temperature-sensitive phosphatidylserine decarboxylase, *Proc. Natl. Acad. Sci. U.S.A.,* 72, 1112, 1975.
87. **Tyhach, R. J., Hawrot, E., Satre, M., and Kennedy, E. P.,** Increased synthesis of phosphatidylserine decarboxylase in a strain of *Escherichia coli* bearing a hybrid plasmid, *J. Biol. Chem.,* 254, 627, 1979.
88. **Chang, Y. -Y. and Kennedy, E. P.,** Biosynthesis of phosphatidyl-glycerol phosphate in *Escherichia coli, J. Lipid Res.,* 8, 447, 1967.
89. **Hirabayashi, T., Larson, T. J., and Dowhan, W.,** Membrane-associated phosphatidylglycerophosphate synthetase from *Escherichia coli:* purification by substrate affinity chromatography on cytidine 5′-diphospho-1,2-diacyl-*sn*-glycerol Sepharose, *Biochemistry,* 15, 5205, 1976.
90. **Stanacev, N. Z., Chang, Y. -Y., and Kennedy, E. P.,** Biosynthesis of cardiolipin in *Escherichia coli, J. Biol. Chem.,* 242, 3018, 1967.
91. **Tunaitis, E. and Cronan, J. E., Jr.,** Characterization of the cardiolipin synthetase activity of *Escherichia coli* envelopes, *Arch. Biochem. Biophys.,* 155, 420, 1973.
92. **Hirschberg, C. B. and Kennedy, E. P.,** Mechanism of the enzymatic synthesis of cardiolipin in *Escherichia coli, Proc. Natl. Acad. Sci. U.S.A.,* 69, 648, 1972.
93. **Isono, K., Krauss, I., and Hirota, Y.,** Isolation and characterization of temperature-sensitive mutants of *Escherichia coli* with altered ribosomal proteins, *Molec. Gen. Gent.,* 149, 297, 1976.
94. **Fox, C. F.,** A lipid requirement for induction of lactose transport in *Escherichi coli, Proc. Natl. Acad. Sci. U.S.A.,* 63, 850, 1969.
95. **Schairer, H. U. and Overath, P.,** Lipids containing *trans*-unsaturated fatty acids change the temperature characteristic of thiomethylgalactoside accumulation in *Escherichia coli, J. Mol. Biol.,* 44, 209, 1969.
96. **Wilson, G., Rose, S. P., and Fox, C. F.,** The effect of membrane lipid unsaturation on glycoside transport, *Biochem. Biophys. Res. Commun.,* 38, 617, 1979.
97. **Milner, L. S. and Kaback, H. R.,** The role of phosphatidylglycerol in the vectorial phosphorylation of sugar by isolated bacterial membrane preparations, *Proc. Natl. Acad. Sci. U.S.A.,* 65, 683, 1970.
98. **Hsu, C. C. and Fox, C. F.,** Introduction of the lactose transport system in a lipid-synthesis-defective mutant of *Escherichia coli, J. Bacteriol.,* 103, 410, 1970.
99. **Wilson, G. and Fox, C. F.,** Biogenesis of microbial transport systems: evidence for coupled induction of newly synthesized lipids and proteins into membrane, *J. Mol. Biol.,* 55, 49, 1971.
100. **Overath, P., Hill, F. F., and Lamnek-Hirsch, I.,** Biogenesis of *E. coli* membrane: evidence for randomization of lipid phase, *Nat. (London) New Biol.,* 234, 264, 1971.
101. **Robbins, A. R. and Rotman, B.,** Inhibition of methylgalactoside transport in *Escherichia coli* upon the cessation of unsaturated fatty acid biosynthesis, *Proc. Natl. Acad. Sci. U.S.A.,* 69, 2125, 1972.
102. **Tsukagoshi, N. and Fox, C. F.,** Abortive assembly of the lactose transport system in *Escherichia coli, Biochemistry,* 12, 2816, 1973.
103. **Tsukagoshi, N. and Fox, C. F.,** Transport system assembly and the mobility of membrane lipids in *Escherichia coli, Biochemistry,* 12, 2822, 1973.
104. **Beacham, I. F. and Silbert, D. F.,** Studies on the uridine diphosphategalactose: lipopolysaccharide galactosyltransferase reaction using a fatty acid mutant of *Escherichia coli, J. Biol. Chem.,* 248, 5310, 1973.
105. **Rose, S. P. and Fox, C. F.,** The β-glucoside system of *Escherichia coli.* III. Properties of a P-HPr: β-glucoside phosphotransferase extracted from membranes with detergent, *J. Supramol. Struct.,* 1, 565, 1973.
106. **Linden, C. D. and Fox, C. F.,** A comparison of characteristic temperatures for transport in two unsaturated fatty acid auxotrophs of *Escherichia coli, J. Supramol. Struct.,* 1, 535, 1973.
107. **Nunn, W. D. and Cronan, J. E., Jr.,** Unsaturated fatty acid synthesis is not required for induction of lactose transport in *Escherichia coli, J. Biol. Chem.,* 249, 724, 1974.
108. **Weisberg, L. J., Cronan, J. E., Jr., and Nunn, W. D.,** Induction of lactose transport in *Escherichia coli* during the absence of phospholipid synthesis, *J. Bacteriol.,* 123, 492, 1975.
109. **Therisod, H., Letellier, L., Weil, R., and Schechter, E.,** Functional *lac* carrier proteins in cytoplasmic membrane vesicles isolated from *Escherichia coli.* 1: Temperature dependence of dansyl galactoside binding and β-galactoside transport, *Biochemistry,* 16, 3772, 1977.

110. **Letellier, L., Weil, R., and Schechter, E.,** Functional *lac* carrier proteins in cytoplasmic membrane vesicles isolated from *Escherichia coli.* 2: Experimental evidence for a segregation of the *lac* carrier proteins induced by a conformational transition of the membrane lipids, *Biochemistry,* 16, 3777, 1977.
111. **Holden, J. T., Bolen, J., Easton, J. A., and de Groot, J.,** Heterogeneous amino acid transport rate changes in *E. coli, Biochem. Biophys. Res. Commun.,* 81, 588, 1978.
112. **Linden, C. D., Keith, A. D., and Fox, C. F.,** Correlations between fatty acid distribution in phospholipids and the temperature dependence of membrane physical state, *J. Supramol. Struct.,* 1, 523, 1973.
113. **Baldassare, J. J., Rhinehart, K. B., and Silbert, D. F.,** Modification of membrane lipid: physical properties in relation to fatty acid structure, *Biochemistry,* 15, 2986, 1976.
114. **Thilo, L. and Overath, P.,** Randomization of membrane lipids in relation to transport system assembly in *Escherichia coli, Biochemistry*, 15, 328, 1976.
115. **Thilo, L., Träuble, H., and Overath, P.,** Mechanistic interpretation of the influence of lipid phase transitions on transport functions, *Biochemistry,* 16, 1283, 1977.
116. **Overath, P. and Träuble, H.,** Phase transitions in cells, membranes, and lipids of *Escherichia coli:* detection by fluorescent probes, light scattering, and dilatometry, *Biochemistry,* 12, 2625, 1973.
117. **Linden, C. D., Wright, K. L., McConnell, H. M., and Fox, C. F.,** Lateral phase separations in membrane lipids and the mechanism of sugar transport in *Eschericha coli, Proc. Natl. Acad. Sci. U.S.A.,* 70, 2271, 1973.
118. **Shimshick, E. J. and McConnell, H. M.,** Lateral phase separation in phospholipid membranes, *Biochemistry,* 12, 2351, 1973.
119. **Schechter, E., Letellier, L., and Gulik-Krzywicki, T.,** Relations between structure and function in cytoplasmic membrane vesicles isolated from an *Escherichia coli* fatty acid auxotroph, *Eur. J. Biochem.,* 49, 61, 1974.
120. **Linden, C. D., Blasie, J. K., and Fox, C. F.,** A confirmation of the phase behavior of *Escherichia coli* cytoplasmic membrane lipids by X-ray diffraction, *Biochemistry,* 16, 1621, 1977.
121. **Jackson, M. B. and Sturtevant, J. M.,** Studies of the lipid phase transitions of *Escherichia coli* by high sensitivity differential scanning calorimetry, *J. Biol. Chem.,* 252, 4749, 1977.
122. **Sandermann, H., Jr.,** Regulation of membrane enzymes by lipid, *Biochim. Biophys. Acta,* 515, 209, 1978.
123. **Shimshick, E. J., Kleemann, W., Hubbell, W. L., and McConnell, H. M.,** Lateral phase separations in membranes, *J. Supramol. Struct.,* 1, 285, 1973.
124. **Gent, M. P. N. and Ho, C.,** Fluorine-19 nuclear magnetic resonance studies of lipid phase transitions in model and biological membranes, *Biochemistry,* 17, 3023, 1978.
125. **Overath, P., Brenner, M., Gulik-Drzywicki, T., Shechter, E., and Letellier, L.,** Lipid phase transitions in cytoplasmic and outer membranes of *Escherichia coli, Biochim. Biophys. Acta,* 389, 358, 1975.
126. **Lee, A. G., Birdsall, N. J. M., Metcalfe, J. C., Toon, P. A., and Warren, G. P.,** Clusters in lipid bilayers and the interpretation of thermal effects in biological membranes, *Biochemistry,* 13, 3699, 1974.
127. **Morrisett, J. D., Pownall, H. J., Plumlee, R. T., Smith, L. C., Zehner, Z. E., Esfahani, M., and Wakil, S. J.,** Multiple thermotropic phase transitions in *Escherichia coli* membranes and membrane lipids: a comparison of results obtained by nitroxyl stearate paramagnetic resonance, pyrene excimer fluorescence, and enzyme activity measurements, *J. Biol. Chem.,* 250, 6969, 1975.
128. **van Heerikhuizen, H., Kwak, E., van Bruggen, E. F. J., and Witholt, B.,** Characterization of a low density cytoplasmic membrane subfraction isolated from *Escherichia coli, Biochim. Biophys. Acta,* 413, 177, 1975.
129. **Haest, C. W. M., Verkleij, A. J., De Gier, J., Scheek, R., Ververgaert, P. H. J., and van Deenen, L. L. M.,** The effect of lipid phase transitions on the architecture of bacterial membranes, *Biochim. Biophys. Acta,* 356, 17, 1974.
130. **Kleemann, W. and McConnell, H. M.,** Lateral phase separations in *Escherichia coli* membranes, *Biochim. Biophys. Acta,* 345, 220, 1974.
131. **Letellier, L., Moudden, H., and Shechter, E.,** Lipid and protein segregation in *Escherichia coli* membrane: morphological and structural study of different cytoplasmic membrane fractions, *Proc. Natl. Acad. Sci. U.S.A.,* 74, 452, 1977.
132. **Nagle, J. F. and Scott, H. L., Jr.,** Lateral compressibility of lipid mono- and bilayers: theory of membrane permeability, *Biochim. Biophys. Acta,* 513, 236, 1978.
133. **van Dijck, P. W. M., Ververgaert, P. H. S., Verkleij, A. J., van Deenen, L. L., and de Gier, J.,** Influence of Ca^{+2} and Mg^{+2} on the thermotropic behavior and permeability properties of liposomes prepared from dimyristoyl phosphatidylglycerol and mixtures of dimyristoyl phosphatidylglycerol and dimyristoyl phosphatidylcholine, *Biochim. Biophys. Acta,* 406, 465, 1975.
134. **Hays, J. B.,** Group translocation transport systems, in *Bacterial Transport,* Rosen, B. P., Ed., Marcel Dekker, New York, 1978, 43.

135. **Saier, M. H., Jr. and Moczydlowski, E. G.,** The regulation of carbohydrate transport in *Escherichia coli* and *Salmonella typhimurium,* in *Bacterial Transport,* Rosen, B. P., Ed., Marcel Dekker, New York, 1978, 103.
136. **Saier, M. H., Jr.,** Bacterial phosphoenolpyruvate: sugar phosphotransferase systems: structural, functional, and evolutionary interrelationships, *Bacteriol. Rev.,* 41, 856, 1977.
137. **Postma, P. W. and Roseman, S.,** The bacterial phosphoenolpyruvate: sugar phosphotransferase system, *Biochim. Biophys. Acta,* 457, 213, 1976.
138. **Cordaro, C.,** Genetics of the bacterial phosphoenolpyruvate: glucose phosphotransferase system, *Annu. Rev. Genet.* 10, 341, 1976.
139. **Miller, J. H.,** Appendix I, in *Experiments in Molecular Genetics,* Cold Spring Harbor Laboratory, Cold Spring Harbor, New York, 1977, 431.
140. **Brown, A. T. and Wittenberger, C. L.,** Mannitol and sorbitol catabolism in *Streptococcus mutans, Arch. Oral Biol.,* 18, 117, 1973.
141. **Saier, M. H., Jr., Feucht, B. U., and Mora, W. K.,** Sugar phosphate: sugar transphosphorylation and exchange group translocation catalyzed by Enzyme II complexes of the bacterial phosphoenolpyruvate: sugar phosphotransferase system, *J. Biol. Chem.,* 252, 8899, 1977.
142. **Holden, J. T., Utech, N. M., Hegeman, G. D., and Kenyon, C. N.,** Heterogeneous changes in amino acid transport activity in *E. coli* lipid overproducing mutants, *Biochem. Biophys, Res. Commun.,* 50, 266, 1973.
143. **Ohta, T., Okuda, S., and Takahashi, H.,** Relationship between phospholipid compositions and transport activities of amino acids in *Escherichia coli* membrane vesicles, *Biochim. Biophys. Acta,* 466, 44, 1977.
144. **Rosen, B. P. and Kashket, E. R.,** Energetics of active transport, in *Bacteriol. Transport,* Rosen, B. P., Ed., Marcel Dekker, New York, 1978, 559.
145. **Oesterhelt, D., and Stoeckenius, W.,** Functions of a new photoreceptor membrane, *Proc. Natl. Acad. Sci. U.S.A.,* 70, 2853, 1973.
146. **Henderson, R.,** The purple membrane from *Halobacterium halobium, Ann. Rev. Biophys. Bioeng.,* 6, 87, 1977.
147. **Harold, F. M.,** Vectorial metabolism, in *The Bacteria,* Vol. VI, Sokatch, J. R. and Ornston, L. N., Eds., Academic Press, New York, 1977, 463.
148. **Henderson, R. and Unwin, P. N. T.,** Three-dimensional model of purple membrane obtained by electron microscopy, *Nature (London)* 257, 28, 1975.
149. **Blaurock, A. E. and King, G. I.,** Asymmetric structure of the purple membrane, *Science,* 196, 1101, 1977.
150. **Lozier, R. H., Niederberger, W., Bogomolni, R. A., Hwang, S-B, and Stoeckenius, W.,** Kinetics and stoichiometry of light-induced proton release and uptake from purple membrane fragments, *Halobacterium halobium* cell envelopes, and phospholipid vesicles containing oriented purple membranes, *Biochim. Biophys. Acta,* 440, 545, 1976.
151. **Racker, E.,** A new procedure for the reconstitution of biologically active phospholipid vesicles, *Biochem. Biophys. Res. Commun.,* 55, 224, 1973.
152. **Racker, E. and Stoeckenius, W.,** Reconstitution of purple membrane vesicles catalyzing light driven proton uptake and adenosine triphosphate formation, *J. Biol. Chem.,* 249, 662, 1974.
153. **Racker, E. and Hinkle, P. C.,** Effect of temperature on the function of a proton pump, *J. Membrane Biol.,* 17, 181, 1974.
154. **Kayushin, L. P. and Skulachev, V. P.,** Bacteriorhodopsin as an electrogenic proton pump: reconstitution of bacteriorhodopsin proteoliposomes generating $\Delta\Psi$ and ΔpH, *FEBS Letters,* 39, 39, 1974.
155. **Happe, M., Teather, R. M. Overath, P., Knobling, A., and Oesterhelt, R. D.,** Direction of proton translocation in proteoliposomes formed from purple membrane and acid lipids depends on the pH during reconstitution, *Biochim. Biophys. Acta,* 465, 415, 1977.
156. **Hellingwerf, K. J., Scholte, B. J., and van Dam, K.,** Bacteriorhodopsin vesicles: an outline of the requirements for light-dependent H^- pumping, *Biochim. Biophys. Acta,* 513, 66, 1978.
157. **Cherry, R. J., Heyn, M. P., and Oesterhelt, D.,** Rotational diffusion and exiton coupling of bacteriorhodopsin in the cell membrane of *Halobacterium halobium, FEBS Letters,* 78, 25, 1977.
158. **Hwang, S-B,** Phospholipid requirement of the bacteriorhodopsin photoreaction cycle, *Biophysical J.,* 21, 182a, 1978.
159. **Sherman, W. V. and Caplan, S. R.,** Influence of membrane lipids on the photochemistry of bacteriorhodopsin in the purple membrane of *Halobacterium halobium, Biochim. Biophys. Acta,* 1502, 222, 1978.
160. **Sherman, W. V., Korenstein, R., and Caplan, S. R.,** Energetics and chronology of phototransients in the light response of the purple membrane of *Halobacterium halobium, Biochim. Biophys. Acta,* 430, 454, 1976.

161. **Chignell, C. F. and Chignell, D. A.,** A spin-label study of purple membranes from *Halobacterium halobium, Biochem. Biophys. Res. Commun.,* 62, 136, 1975.
162. **Jackson, M. B. and Sturtevant, J. M.,** Phase transitions of the purple membrane of *Halobacterium halobium, Biochemistry,* 17, 911, 1978.
163. **Konishi, T. and Packer, L.,** Light-dark conformational states in bacteriorhodopsin, *Biochem. Biophys. Res. Comm.,* 72, 1437, 1976.
164. **Costeron, J. W., Ingram, J. M., and Cheng, K. J.,** Structure and function of the cell envelope of Gram-negative bacteria, *Bacteriol. Rev.,* 38, 87, 1974.
165. **Payne, J. W. and Gilvarg, C.,** Size restriction on peptide utilization in *Escherichia coli, J. Biol. Chem.,* 243, 6291, 1968.
166. **Payne, J. W.,** Oligopeptide transport in *Escherichia coli, J. Biol. Chem.,* 243, 3395, 1968.
167. **Nakae, T. L. and Nikaido, H.,** Outer membrane as a diffusion barrier in *Salmonella typhimurium, J. Biol. Chem.,* 250, 7359, 1975.
168. **Nakumura, K., and Mizushima, A.,** *In vitro* reassembly of the membranous vesicle from *Escherichia coli* outer membrane components: role of individual components and magnesium ions in reassembly, *Biochim. Biophys. Acta,* 413, 371, 1975.
169. **Nakae, T. L.,** Outer membrane of *Salmonella typhimurium:* reconstitution of sucrose-permeable membrane vesicles, *Biochem. Biophys. Res. Commun.,* 64, 1224, 1975.
170. **Nakae, T. L.,** Outer membrane of *Salmonella,* isolation of protein complex that produces transmembrane channels, *J. Biol. Chem.,* 251, 2176, 1976.
171. **Inouye, M.,** The three-dimensional molecular assembly model of a lipoprotein from the *Escherichia coli* outer membrane, *Proc. Natl. Acad. Sci. U.S.A.,* 71, 2396, 1974.
172. **Nikaido, H., Bavoil, P., and Hirota, Y.,** Outer membrane of Gram-negative bacteria. XV. Transmembrane diffusion rates in lipoprotein-deficient mutants of *Escherichia coli, J. Bacteriol.,* 133, 329, 1978.
173. **DeMartini, M. and Inouye, M.,** Interaction between two major outer membrane proteins of *Escherichia coli:* the matrix protein and the lipoprotein, *J. Bacteriol.,* 133, 329, 1978.
174. **Nurminen, M., Lounatmaa, K., Sarvas, M., Mäkelä, P. H., and Nakae, T. L.,** Bacteriophage-resistant mutants of *Salmonella typhimurium* deficient in two major outer membrane proteins, *J. Bacteriol.,* 127, 941, 1976.
175. **Nikaido, H., Song, S. A., Shaltiel, L., and Nurminen, M.,** Outer membrane of *Salmonella.* XIV. Reduced transmembrane diffusion rates in porin-deficient mutants. *Biochem. Biophys. Res. Commun.,* 76, 324, 1977.
176. **Bavoil, P., Nikaido, N., and von Meyenberg, K.,** Pleiotropic transport mutants of*Escherichia coli* lack porin, a major outer membrane protein, *Molec. Gen. Genet.,* 158, 23, 1977.
177. **Lutkenhaus, J. F.,** Role of a major outer membrane protein in *Escherichia coli, J. Bacteriol.,* 131, 631, 1977.
178. **Szmelcan, S., Schwartz, M., Silhavy, T. J., and Boos, W.,** Maltose transport in *E. coli* K-12, *Eur. J. Biochem.,* 65, 13, 1976.
179. **Pugsley, A. P. and Schnaitman, C. A.,** Outer membrane proteins of *Escherichia coli.* VII. Evidence that bacteriophage-directed protein 2 functions as a pore, *J. Bacteriol.,* 133, 1181, 1978.
180. **Pugsley, A. P. and Schnaitman, C. A.,** Identification of three genes controlling production of new outer membrane pore proteins in *Escherichia coli* K-12, *J. Bacteriol.,* 135, 1118, 1978.
181. **Palva, E. T.,** Major outer membrane protein in *Salmonella typhimurium* induced by maltose, *J. Bacteriol.,* 136, 286, 1978.
182. **Nakae, T. L. and Ishii, J.,** Transmembrane permeability channels in vesicles reconstituted from single species of porins from *Salmonella typhimurium, J. Bacteriol.,* 133, 1412, 1978.
183. **Benz, R., Boehler-Kohler, B. A., Dieterle, R., and Boos, W.,** Porin activity in the osmotic shock fluid of *Escherichia coli, J. Bacteriol.,* 135, 1080, 1978.
184. **Decad, G. M. and Nikaido, H.,** Outer membrane of Gram-negative bacteria. XII. Molecular sieving function of cell wall, *J. Bacteriol.,* 128, 325, 1976.
185. **Scherrer, R. and Gerhardt, P.,** Molecular sieving by the *Bacillus megaterium* cell wall and protoplast, *J. Bacteriol.,* 107, 718, 1971.
186. **Rosenbusch, J. P.,** Characterization of the major envelope protein from *Escherichia coli:* regular arrangement on the peptidoglycan and unusual dodecyl sulfate binding, *J. Biol. Chem.,* 249, 8019, 1974.
187. **Steven, A. C., ten Heggler, B., Muller, R., Kistler, J., and Rosenbusch, J. P.,** Ultrastructure of a periodic protein layer in the outer membrane of *Escherichia coli, J. Cell Biol.,* 72, 292, 1977.
188. **Palva, E. T., and Randall, L. L.,** Nearest-neighbor analysis of *Escherichia coli* outer membrane proteins using cleavable cross-links, *J. Bacteriol.,* 127, 1558, 1976.
189. **Reitheimer, R. A. F. and Bragg, P. D.,** Cross-linking of the proteins in the outer membrane of *Escherichia coli, Biochim. Biophys. Acta,* 466, 245, 1977.

190. **Benz, R., Janko, K., Boos, W., and Läuger, P.,** Formation of large, ion-impermeable membrane channels by the matrix protein (porin) of *Escherichia coli, Biochim. Biophys. Acta,* 511, 305, 1978.
191. **Schindler, H. and Rosenbusch, J. P.,** Matrix protein from *Escherichia coli* outer membranes forms voltage-controlled channels in lipid bilayers, *Proc. Natl. Acad. Sci. U.S.A.,* 75, 3751, 1978.
192. **Boehler-Kohler, B. A., Boos, W., Dieterle, R., and Benz, R.,** The receptor for λ phage of *Escherichia coli* forms larger pores in black lipid membranes than the matrix protein (porin), *J. Bacteriol.,* 138, 33, 1979.
193. **Nikaido, H.,** Outer membrane of *Salmonella typhimurium* transmembrane diffusion of some hydrophobic substances, *Biochim. Biophys. Acta,* 433, 118, 1976.
194. **Gustafsson, P., Nordstrom, K., and Normark, S.,** Outer penetration barrier of *Escherichia coli* K-12: kinetics of the uptake of gentian violet by wild type and envelope mutants, *J. Bacteriol.,* 116, 893, 1973.
195. **Eriksson-Grennberg, K. G., Nordstrom, K., and Englund, P.,** Resistance of *Escherichia coli* to penicillins. IX. Genetics and physiology of class II ampicillin-resistant mutants that are galactose negative or sensitive to bacteriophage C21, or both, *J. Bacteriol.,* 108, 1210, 1971.
196. **Hollifield, W. C., Jr. and Neilands, J. B.,** Ferric enterobactin transport system in *Escherichia coli* K-12. Extraction, assay, and specificity of the outer membrane receptor, *Biochemistry,* 17, 1922, 1978.
197. **McIntosh, M. A., Chenault, S. S., and Earhart, C. F.,** Genetic and physiological studies on the relationship between colicin B and ferric enterobactin uptake in *Escherichia coli* K-12, *J. Bacteriol.,* 137, 653, 1979.
198. **Wookey, P. and Rosenberg, H.,** Involvement of inner and outer membrane components in the transport of iron and in colicin B action in *Escherichia coli, J. Bacteriol.,* 133, 661, 1978.
199. **Braun, V., Hancock, R. E. W., Hantke, K., and Hartmann, A.,** Functional organization of the outer membrane of *Escherichia coli:* phage and colicin receptors as components of iron uptake systems, *J. Supramol. Struct.,* 5, 37, 1976.
200. **Kadner, R. J.,** Transport of vitamins and antibiotics, in *Bacterial Transport,* Rosen, B. P., Ed., Marcel Dekker, New York, 1978, Chap. 10.
201. **Kadner, R. J.,** Repression of synthesis of the vitamin B_{12} receptor in *Escherichia coli, J. Bacteriol.,* 136, 1050, 1978.
202. **Stoeckenius, W., Lozier, R. H., and Bogomolni, R. A.,** Bacteriorhodopsin and the purple membrane of *Halobacteria, Biochim. Biophys. Acta,* 505, 3, 1978.
203. **Haddock, B. A. and Jones, C. W.,** Bacterial Respiration, *Bacteriol Rev.,* 41, 47, 1977.
204. **Downie, J. A., Gibson, F., and Cox, G. B.,** Membrane adenosine triphosphatases of prokaryotic cells, *Annu. Rev. Biochem.,* 48, 103, 1979.
205. **Kagawa, Y.,** Reconstitution of the energy transformer, gate, and channel subunit reassembly, crystalline ATPase and ATP synthesis, *Biochim. Biophys. Acta,* 505, 45, 1978.
206. **Yoshida, M., Okamoto, H., Sone, N., Hirata, H., and Kagawa, Y.,** Reconstitution of thermostable ATPase capable of energy coupling from its purified subunits, *Proc. Natl. Acad. Sci. U.S.A.,* 74, 936, 1977.
207. **Sone, N., Yoshida, M., Hirata, H., and Kagawa, Y.,** Resolution of the membrane moiety of the H^+-ATPase complex into two kinds of subunits, *Proc. Natl. Acad. Sci. U.S.A.,* 75, 4219, 1978.
208. **Yoshida, M., Sone, N., Hirata, H., and Kagawa, Y.,** Reconstitution of adenosine triphosphatase of thermophilic bacterium from purified individual subunits, *J. Biol. Chem.,* 252, 3480, 1977.
209. **Boyer, P. D., Chance, B., Ernster, L., Mitchell, P., Racker, E., and Slater, E. C.,** Oxidative phosphorylation and photophosphorylation, *Annu. Rev. Biochem.,* 46, 955, 1977.
210. **Clarke, D. J., and Morris, J. G.,** Reconstitution of a functional proton translocating adenosine triphosphatase from the obligately anaerobic bacterium, *Clostridium pasteurianum, Biochem. Soc. Trans.* 5, 140, 1977.
211. **Riley, M. and Anilionis, A.,** Evolution of the bacterial genome, *Annu. Rev. Microbiol.,* 32, 519, 1978.
212. **Silver, S.,** Transport of cations and anions, in *Bacterial Transport,* Rosen, B. P., Ed., Marcel Dekker, New York, 1978, 221.
213. **Lanyi, J. K. and MacDonald, R. E.,** Existence of electrogenic hydrogen ion/sodium ion antiport in *Halobacterium halobium* cell envelope vesicles, *Biochemistry,* 15, 4608, 1976.
214. **Schuldiner, S. and Fishkes, H.,** Sodium-proton antiport in isolated vesicles of *Escherichia coli, Biochemistry,* 17, 706, 1978.
215. **Brey, R. N. Beck, J. C., and Rosen, B. P.,** Cation/proton antiport systems in *Escherichia coli, Biochem. Biophys. Res. Commun., 83, 1588, 1978.*
216. **Harold, F. M. and Altendorf, K.,** Cation transport in bacteria: K^+, Na^+, and H^+, *Curr. Top. Membr. Transp.,* 5, 1, 1974.
217. **Rhoads, D. B. and Epstein, W.,** Energy coupling to net K^+ transport in *Escherichia coli* K-12, *J. Biol. Chem.,* 252, 1394, 1977.
218. **Bakker, E. P. and Harold, F. M.,** Energy coupling to potassium transport in *Streptococcus faecalis.* I. Interplay of ATP and the protonmotive force, *J. Biol. Chem.,* 255, 433, 1980.

219. **Bakker, E. P. and Harold, F. M.,** Energy coupling to potassium transport in *Streptococcus faecalis.* II. Evidence for potassium/proton symport, *J. Biol. Chem.,* 255, 433, 1980.
220. **Laimonis, L. A., Rhoads, D. B., Altendorf, K., and Epstein, W.,** Identification of the structural proteins of an ATP-driven potassium transport system in *Escherichia coli, Proc. Natl. Acad. Sci. U.S.A.,* 75, 3216, 1978.
221. **Epstein, W., Whitelaw, V. and Hesse, J.,** A K^+ transport ATPase in *Escherichia coli, J. Biol Chem.,* 253, 6666, 1978.
222. **Rosen, B. P., Ed.,** *Bacterial Transport,* Marcel-Dekker, New York, 1978.
223. **Harold, F. M.,** Conservation and transformation of energy by bacterial membranes, *Bacteriol. Rev.,* 36, 172, 1972.
224. **Dills, S. S., Schmidt, M., and Saier, M. H., Jr.,** Carbohydrate transport in bacteria, *Microbiol. Rev.,* 44, 385, 1980.
225. **Saier, M. H., Jr.,** The role of the cell surface in regulating the internal environment, in *The Bacteria,* Vol. 7, Sokatch, J. R. and Ornston, L. N., Eds., Academic Press, New York, 1979, 167.
226. **Kay, W. W.,** Transport of carboxylic acids, in *Bacterial Transport,* Rosen, B. P., Ed., Marcel-Dekker, New York, 1978, 385.
227. **Lo, T. C. Y.,** The molecular mechanism of dicarboxylic acid transport in *Escherichia coli* K-12, *J. Supramol. Struct.,* 7, 463, 1977.
228. **Lo, T. C. Y. and Bewick, M. A.,** The molecular mechanism of dicarboxylic acid transport in *Escherichia coli* K-12. The role and orientation of the two membrane-bound dicarboxylate binding proteins, *J. Biol. Chem.,* 253, 7826, 1978.
229. **Argast, M., Schumacher, G., and Boos, W.,** Characterization of a periplasmic protein related to *sn*-glycerol-3-phosphate transport in *Escherichia coli, J. Supramol. Struct.,* 6, 135, 1977.
230. **Lo, T. C. Y.,** The molecular mechanisms of substrate transport in Gram-negative bacteria, *Can. J. Biochem.,* 57, 289, 1979.
231. **Bewick, M. A. and Lo, T. C. Y.,** Localization of dicarboxylate binding protein in the cell envelope of *Escherichia coli* K-12, *Can. J. Biochem.,* 58, 885, 1980.
232. **Bewick, M. A. and Lo, T. C. Y.,** Dicarboxylic acid transport in *Escherichia coli* K-12: involvement of a binding protein in the translocation of dicarboxylic acids across the outer membrane of the cell envelope, *Can. J. Biochem.,* 57, 653, 1979.
233. **Lo, T. C. Y.,** The transfer of a bacterial transmembrane function to eukaryotic cells, *J. Biol. Chem.,* 254, 591, 1979.
234. **Saier, M. H., Jr.,** Sugar transport mediated by the bacterial phosphoenolpyruvate-dependent phosphotransferase system, in *Microbiology,* 72, 1979.
235. **Rephaeli, A. W. and Saier, M. H., Jr.,** Kinetic analyses of the sugar phosphate: sugar transphosphorylation reaction catalyzed by the glucose Enzyme II complex of the bacterial phosphotransferase system, *J. Biol. Chem.,* 253, 7595, 1978.
236. **Jacobson, G. R., Lee, C. A., and Saier, M. H., Jr.,** Purification of the mannitol specific Enzyme II of the *Escherichia coli* phosphoenolpyruvate: sugar phosphotransferase system, *J. Biol. Chem.,* 254, 249, 1979.
237. **Singer, S. J.,** The molecular organization of membranes, *Ann. Rev. Biochem.,* 43, 805, 1974.
238. **Ames, G. F.-L. and Spudich, E. N.,** Protein-protein interaction in transport: periplasmic histidine binding protein J interacts with P protein, *Proc. Natl. Acad. Sci. U.S.A.,* 73, 1877, 1976.
239. **Ames, G. F.-L. and Nikaido, K.,** Identification of a membrane protein as a histidine transport component in *Salmonella typhimurium, Proc. Natl. Acad. Sci. U.S.A.,* 75, 5447, 1978.
240. **Unowsky, J. and Rachmeler, M.,** Mechanisms of antibiotic resistance determined by resistance-transfer factors, *J. Bacteriol.,* 92, 358, 1966.
241. **Franklin, T. J.,** Antibiotic transport in bacteria, *CRC Crit. Rev. Microbiol.,* 30, 253, 1973.
242. **Zähner, H. and Diddens, H.,** Some experiments with semisynthetic sideromycins, *Jpn. J. Antibiotics,* 30 (Suppl. S201), 1977.
243. **Pastan, I. and Adahya, S.,** Cyclic adenosine 3′, 5′-monophosphate in *Escherichia coli, Bacteriol. Rev.,* 40, 527, 1976.
244. **Alper, M. D. and Ames, B. N.,** Transport of antibiotics and metabolite analogues by systems under cyclic AMP control: positive selection of *Salmonella typhimurium cya* and *crr* mutants, *J. Bacteriol.,* 133, 149, 1978.
245. **Kornberg, H. L.,** Nature and regulation of hexose uptake by *Escherichia coli,* in *The Molecular Basis of Biological Transport,* Woessner, J. F. and Huijing, Eds., Academic Press, New York, 1972, 157.
246. **Saier, M. H., Jr., Jones-Mortimer, M. C., and Kornberg, H. L.,** unpublished results.
247. **Saier, M. H., Jr., Feucht, B. U., and Hofstadter, L. J.,** Regulation of carbohydrate uptake and adenylate cyclase activity mediated by the Enzymes II of the phosphoenolpyruvate: sugar phosphotransferase system in *Escherichia coli, J. Biol. Chem.,* 251, 883, 1976.

248. **Kent, P. W., Ackers, J. P., and White, R. J.,** *N*-iodoacetyl-D-glucosamine, an inhibitor of growth and glycoside uptake in *Escherichia coli, Biochem. J.*, 118, 73, 1970.
249. **White, R. J. and Kent, P. W.,** An examination of the inhibitory effects of *N*-iodoacetylglucosamine on *Escherichia coli* and isolation of resistant mutants, *Biochem. J.*, 118, 81, 1970.
250. **White, R. J.,** The role of the phosphoenolpyruvate phosphotransferase by *Escherichia coli, Biochem. J.*, 118, 89, 1970.
251. **Reiner, A. M.,** Xylitol and D-arabitol toxicities due to derepressed fructose, galactitol and sorbitol phosphotransferases of *Escherichia coli, J. Bacteriol.*, 132, 166, 1977.
252. **Rickenberg, H. V.,** Cyclic AMP in prokaryotes, in *Annu. Rev. Microbiol.*, 28, 353, 1974.
253. **Peterkofsky, A. and Gazdar, C.,** Glucose inhibition of adenylate cyclase in intact cells of *Escherichia coli, Proc. Natl. Acad. Sci. U.S.A.*, 71, 2324, 1974.
254. **Gonzalez, J. E. and Peterkofsky, A.,** The mechanism of sugar-dependent repression of synthesis of catabolic enzymes in *Escherichia coli, J. Supramol. Struct.*, 6, 495, 1977.
255. **Harwood, J. P. and Peterkofsky, A.,** Glucose-sensitive adenylate cyclase in toluene-treated cells of *Escherichia coli* B, *J. Biol. Chem.*, 250, 4656, 1975.
256. **Saier, M. H., Jr. and Feucht, B. U.,** Coordinate regulation of adenylate cyclase and carbohydrate permeases by the phosphoenolpyruvate: sugar phosphotransferase system in *Salmonella typhimurium, J. Biol. Chem.*, 250, 7078, 1975.
257. **Castro, L., Feucht, B. U., Morse, M. L., and Saier, M. H., Jr.,** Regulation of carbohydrate permeases and adenylate cyclase in *Escherichia coli.* Studies with mutant strains in which Enzyme I of the phosphoenolypyruvate: sugar phosphotransferase system is thermolabile, *J. Biol. Chem.*, 251, 5522, 1976.
258. **Peterkofsky, A. and Gazdar, C.,** Interaction of Enzyme I of the phosphoenolpyruvate: sugar phosphotransferase system with adenylate cyclase of *E. coli, Proc. Natl. Acad. Sci. U.S.A.*, 72, 2920, 1975.
259. **Epstein, W., Rothman-Denes, L. B., and Hesse, J.,** Adenosine 3′,5′-cyclic monophosphate as mediator of catabolite repression in *Escherichia coli, Proc. Natl. Acad. Sci. U.S.A.*, 72, 2300, 1975.
260. **Magasanik, B.,** Glucose effects: inducer exclusion and repression, in *The Lactose Operon,* Beckwith, J. R. and Zipser, D., Eds., Cold Spring Harbor Laboratory, Cold Spring Harbor New York, 1970, 189.
261. **Yang, J. K., and Bloom, R. W., and Epstein, W.,** Catabolite and transient repression in *Escherichia coli* do not require Enzyme I of the phosphotransferase system, *J. Bacteriol.*, 138, 275, 1979.
262. **Lengeler, J. and Steinberger, H.,** Analysis of regulatory mechanisms controlling the activity of the hexitol transport systems in *Escherichia coli* K-12, *Molec. Gen. Genet.*, 167, 75, 1978.
263. **Kornberg, H. L.,** Fine control of sugar uptake by *Escherichia coli, Symp. Soc. Expl. Biol.*, 27, 175, 1973.
264. **Makman, R. S. and Sutherland, E. W.,** Adenosine 3′, 5′-phosphate in *Escherichia coli, J. Biol. Chem.*, 240, 1309, 1965.
265. **Saier, M. H., Jr., Feucht, B. U., and McCaman, M. T.,** Regulation of intracellular adenosine cyclic 3′, 5′-monophosphate levels in *Escherichia coli* and *Salmonella typhimurium.* Evidence for energy dependent excretion of the cyclic nucleotide, *J. Biol. Chem.*, 250, 7593, 1975.
266. **Wright, L. F., Milne, B. P., and Knowles, C. J.,** The regulatory effects of growth rate and cyclic AMP levels on carbon catabolism and respiration in *Escherichia coli* K-12, *Biochim. Biophys. Acta,* 1979.
267. **Phillips, A. T., Egan, R. M., and Lewis, B.,** Control of biodegradative threonine dehydratase inducibility by cyclic AMP in energy-restricted *Escherichia coli, J. Bacteriol.*, 135, 828, 1978.
268. **Dessein, A., Schwartz, M., and Ullmann, A.,** Catabolite repression in *Escherichia coli* mutants lacking cyclic AMP, *Molec. Gen. Genet.*, 162, 83, 1978.
269. **Dessein, A., Tillier, F., and Ullmann, A.,** Catabolite modulator factor: physiological properties and *in vivo* effects, *Molec. Gen. Genet.*, 162, 89, 1978.
270. **Maloney, P. C.,** Obligatory coupling between proton entry and the synthesis of adenosine 5′-triphosphate in *Streptococcus lactis, J. Bacteriol.*, 132, 564, 1977.
271. **Schönfeld, M. and Neumann, J.,** Proton conductance of the thylakoid membrane: modulation by light, *FEBS Letters,* 73, 51, 1977.
272. **Schein, S. J., Kagen, B. L., and Finkelstein, A.,** Colicin K acts by forming voltage-dependent channels in phospholipid bilayer membranes, *Nature (London),* 276, 159, 1978.
273. **Harold, F. M.,** personal communication.
274. **Epstein, W.,** personal communication.
275. **Kent, P. W.,** personal communication.
276. **Saier, M. H. and Feucht, B. U.,** unpublished results.
277. **Phillips, A. T.,** personal communication.
278. **Saier, M. H., Jr. and Kornberg, H. L.,** unpublished observations.
279. **Jacobson, G. R.,** unpublished results.

Chapter 2

THE ROLE OF THE MEMBRANE IN THE BIOENERGETICS OF BACTERIAL CELLS

Terry A. Krulwich, Arthur A. Guffanti, and Kenneth G. Mandel

TABLE OF CONTENTS

I. THE PROTONMOTIVE FORCE

A. Introduction

The relative impermeability of the cytoplasmic membrane to protons, and at least some other cations, may critically relate to the energization of a variety of processes, including ATP synthesis, many solute transport systems, and bacterial motility. According to the chemiosmotic hypothesis of Mitchell,[1-3] a protonmotive force (PMF) is generated by the extrusion of protons during electron transport or during ATP hydrolysis. This proton extrusion results in the establishment of a transmembrane pH gradient (ΔpH, outside acid) and a transmembrane electrical potential ($\Delta\Psi$, outside positive). These two gradients comprise the PMF and are postulated to represent the form of energy whereby ATP synthesis, etc., are driven. The magnitude of the PMF has accordingly been studied in a variety of bacterial cells, and membrane vesicles have been obtained therefrom. Establishment of the PMF and its possible role in various processes have also been examined. In this section reports on the magnitude of the PMF and factors involved in its establishment will be reviewed. Some of the controversies surrounding these determinations will be discussed. Excellent reviews by others have been very helpful in providing a framework.[4-8]

B. Measurements in Whole Cells and Membrane Vesicles

The PMF values summarized in Tables 1 and 2 for whole cells and vesicles, respectively, have been obtained using many different methods that are cited in the individual references. The theoretical basis of and some of the practical considerations involved in those determinations have been reviewed by Maloney, Kashket, and Wilson.[50] Values for the PMF in whole cells fall in a wide range, with large variations in the relative proportions of ΔpH and ΔΨ. Clearly, the growth substrate, conditions of aerobiosis, pH, and ionic conditions are important factors (see Table 1).[51] In general, however, acidophiles and alkalophiles represent opposite extremes, both with respect to the magnitude and nature of the PMF. Acidophilic organisms have the highest PMFs reported in bacteria, and these PMFs are generated entirely as a large ΔpH. In fact, the ΔΨ in such organisms is the reverse of the usual direction, i.e., is positive inside. By contrast, alkalophiles (even on nonfermentable carbon sources) exhibit rather low PMFs which result from substantial ΔΨs and "reversed" ΔpHs whereby the cytoplasmic pH is maintained at pH 9.5 or below. In at least several species, the ΔΨ rises considerably with increasing external pH in whole cells, while the ΔΨ in vesicles exhibits much less or even no change as the external pH is raised. In other respects, PMF values observed in vesicles generally relate to the patterns and magnitudes found with whole cells. The ionic constituents are extremely important in vesicle studies, at least in part because of antiport activities reviewed below. In general, as Table 2 illustrates, Na^+-loaded vesicles exhibit higher ΔΨs than K^+-loaded vesicles. The large spectrum of membrane-active agents that affect the PMF will not be discussed here but are used as controls as well as probes in the individual studies cited.

C. Primary Events in the Establishment of the PMF

Transmembrane, electrogenic, proton movements are the primary event in establishment of the PMF. In bacteria, they occur: (1) during respiration; (2) upon hydrolysis of ATP by the Ca^{++}, Mg^{++}-ATPase; (3) upon illumination of bacteriorhodopsin; and (4) upon illumination of photosynthetic membranes. The latter two processes are reviewed in this and/or other volumes of this series. The properties and H^+-pumping roles of the bacterial respiratory chain have also been summarized by others.[52-56] The concept of loops of redox carriers which mediate proton extrusion has found congenial ground in bacteria. A variety of recent studies have extended the number of bacterial species whose respiratory chains are at least partially described and have revealed a wide range of organisms possessing alternate terminal oxidases, especially cyanide-resistant cytochrome *d*.[57-62] Formation of respiratory chains in facultative and obligate anaerobes has also received attention,[63-66] and proton pumping has been demonstrated during the flow of electrons to fumarate and nitrate in *Escherichia coli*.[67,68] In view of Mitchell's proposal of a ubiquinone cycle,[69] studies have been conducted on quinone-deficient *E. coli* strains; a model based on Mitchell's hypothesis has been presented,[70] and reconstitution experiments demonstrating the mobility of ubiquinone have been conducted.[71] Jones et al.[72] have reported the presence of at least two proton-conducting redox loops in nine unrelated species of bacteria. The current controversies centered upon the number of protons extruded per site of phosphorylation in the mitochondrial respiratory chain have raised the same issues in bacteria. H^+/O ratios of 2, 3, and 4 have been reported in various systems in which different species, mutations, electron acceptors, and methods have been employed.[73-79]

Recent studies of the bacterial ATPase, as indicated below, have focused on purification and reconstitution of ATP synthetic activity, and on mutations that can yield specific information on function. The role of the ATPase in establishing a PMF by hydrolysis of ATP has been documented in bacteria for some time[21,80] and is confirmed in more recent work.[81]

Table 1
DETERMINATIONS OF THE PROTONMOTIVE FORCE USING WHOLE CELLS

Optimal pH range for growth	Organism	Growth substrate	External pH	Protonmotive force measurements: ΔpH (pH in—pH out)	Δψ (mvolts)	$\Delta\bar{\mu}H+$ (mvolts)	Ref.
Neutral	*Arthrobacter pyridinolis*	D-Gluconate	5.5	1.3	−87	−164	9
			7.5	0	−87	−87	
	Bacillus megaterium	D-Fructose/ L-malate	7.4	0.64	−80	−118	10
	Bacillus subtilis	Undefined		—	—	−90	11
	Clostridium pasteurianum	Glucose	7.1	0.4	—	—	12
	Escherichia coli	Glycerol or succinate	6.0	2		−242	13,14
			7.6	0	−122—−140	~−122	
			9.0	−0.5		~−92	
	Halobacterium halobium (illuminated)	Undefined	6.0	1.1—1.2	−120	−186	15
		Undefined	6.6	0.6—0.8	−100	−140	16
	Paracoccus denitrificans	Glucose	6.3	0.9	−17	−70	17
			7.1	0.3	−40	−58	
	Staphylococcus aureus	Undefined	6.5	—	—	−230	18
	Staphylococcus epidermis	Undefined	5.0	1—1.5	−40—−50	−100—−140	19
			6.0	0	−80—−90	−80—−90	
	Staphylococcus lactis	D-Glucose	7.0	0.75	−35.5	−79.9	20
		L-Arginine	7.0	0	−39	−39	
	Streptococcus faecalis	D-Glucose	7.0	0.7	−150—−200	−191—−241	21,22

Table 1 (continued)
DETERMINATIONS OF THE PROTONMOTIVE FORCE USING WHOLE CELLS

Optimal pH range for growth	Organism	Growth substrate	External pH	Protonmotive force measurements: ΔpH (pH in—pH out)	$\Delta\psi$ (mvolts)	$\Delta\bar{\mu}H^+$ (mvolts)	Ref.
Acid	*Bacillus acidocaldarius*	Lactose	2	4.15	+34	−249	23,24
			4.5	1.81	+34	−109	
	Thermoplasma acidophila	Undefined	1.7	3.8	—	—	25
		Glucose/yeast ext	2.0	4.4—4.9	+109—+125	~−170	26,27
	Thiobacillus ferrooxidans	CO_2, Fe^{2+}	2.0	4.5	+10	−256	28
Alkaline	*Bacillus alcalophilus*	L-Malate	9.0	0	−83	−83	29
			11.0	−1.4	−145	−63	
		Lactose	8.5	−0.7	−135	−93	30
			10	−1.4	−182	−100	
	Bacillus circulans	Lactose	9	0	−63	−63	31
	Bacillus firmus RAB	L-Malate	8.5	0	−105	−105	32
			10.5	−1.2	−140	−66	

Table 2
DETERMINATIONS OF THE PROTONMOTIVE FORCE USING MEMBRANE VESICLES

Organism	Membrane	External pH	Monovalent cation	Protonmotive force measurements: ΔpH (pH in—pH out)	Δψ (mvolt)	$\Delta\bar{\mu}H^+$ (mvolt)	Ref.
Azotobacter vinelandii	Cytoplasmic	7.0	K^+	—	−75 to −80	—	33
			Na^+	—	−104	—	
Bacillus alcalophilus	Cytoplasmic	8.0	K^+	0	−125	−125	34
		10.5		0	−98	−98	
		8.0	Na^+	−1.3	−124	−46	
		10.5		−0.4	−126	−102	
		8.0	2-Amino-2-methyl propandiol	0.73	−64	−108	
		10.0		—	−96	≥−96	
Escherichia coli	Cytoplasmic	5.5	K^+	1.83—2.0	−70—−80	−180—−200	35,36,37
		7.0	K^+	0.5	−75—−100	−105—−130	
		7.5	K^+	0	−75	−75	
	Anaerobic Cytoplasmic	6.6	K^+	1.2—1.5	−90	−160	38,39
	Everted Cytoplasmic	7.5	K^+	−3.3—−3.5	—	—	40,41
Halobacterium halobium	Cytoplasmic	5.0	Na^+	0.77	−99	−145	42,43
		6.8	Na^+	1.83	−120	−229	
		6.8	K^+	2.02	−34	−153	
		6.8	$Na^+ + K^+$	1.86—2.0	−62—−72	−172—−190	
Paracoccus denitrificans	Everted Cytoplasmic	7.3	Trisacetate	−0.5—−1.5	+90—+145	—	44
Rhodopseudomonas sphaeroides	Cytoplasmic	7.0	K^+	0.6	−70	−110	45,46
	Chromatophore	6.0	K^+	−0.98	+50	+110	
		7.3	Choline	−1.4	+140	+225	
Rhodospirillum rubrum	Chromatophore	8.0	Choline/Na^+	—	+90—+110	—	47
			Na^+ or K^+	−2.47	+109	+257	48
			Trisacetate	—	+80—+100	—	49

D. Role of Cation/Proton Antiporters

The existence of cation/proton antiporters was predicted by Mitchell and a Na^+/H^+ antiporter was first described in *E. coli* by West and Mitchell.[82] An electroneutral cation/proton antiporter can function in the interconversion of ΔpH and ΔΨ,[83] (e.g., could act in conjunction with respiration to allow maintenance of a ΔΨ at an alkaline pH at which the ΔpH might be zero). An electrogenic cation/proton antiporter could also function in this way, although some dissipation of the ΔΨ would be involved in its activity; it should be noted that electrogenicity of an antiporter might relate to the ratio of the entities transported and/or to properties of the carrier itself. Cation/proton antiporters have also been proposed to play a role in the maintenance of cytoplasmic pH,[13] and at least in some organisms, that is an obviously important role. In *Bacillus alcalophilus,* in which the cytoplasm must be maintained at a considerably more acid pH than the milieu,[29,34] an electrogenic Na^+/H^+ antiporter appears to be required for growth at alkaline pH. A third role for the cation/proton antiporters is in extrusion of the cations from the cytoplasm; this may be to compensate for overshoot of cation uptake systems (a possible role for the K^+/H^+ antiporter?), to eliminate cytotoxic ions, or to establish functional gradients of cations (e.g., $\Delta\bar{\mu}_{Na+}$ to energize transport reactions).[84,85] The properties of bacterial cation/proton antiporters are summarized in Table 3.

E. Problems

Controversies with respect to proton/site ratios have been mentioned. These controversies are being most forcefully expressed by the mitochondriologists, but are important for all who are interested in energy-transducing membranes because of the related issues of respiratory mechanisms, proton pumps, and mechanisms of energy-requiring processes. A major problem raised by the data summarized in Table 1 and 2 is whether the magnitude of the PMF has been accurately assessed. The variability of the values obtained in similar organisms is worrisome alone, but another cogent problem is the difficulty in formulating mechanisms whereby very small PMFs can often energize processes such as ATP synthesis. This question has also been raised *vis a vis* mitochondria.[105] Intramembrane gradients have been suggested as alternatives to the PMF as the primary form of energy derived from respiration.[106-108] A neochemical intermediate mechanism for ATP synthesis in *B. megaterium* has also been proposed.[10] On the other hand, proposals have been made for stoichiometric considerations which would facilitate the conceptualization of energization of relevant processes by low PMFs.[109] A true resolution of the problem awaits a greater understanding of the mechanisms with continued attention to the development and corroboration of methods and detailed studies of the systems which exhibit particularly small PMFs.

II. PROCESSES THAT ARE DRIVEN BY THE ENERGIZED MEMBRANE STATE

A. Introduction

Among the processes in bacterial cells that are believed to be energized by the PMF are ATP synthesis, energy-linked transhydrogenase, many active transport systems, photosynthesis, and motility; the latter three topics are reviewed, at least in part, in other chapters of this volume and will not be discussed here. Indeed, the proposed roles of the PMF will be illustrated by examples in which the precise mechanism whereby energization occurs is still quite unresolved, i.e., ATP synthesis and transhydrogenase. However, even in some of the ostensibly clearer examples, such as certain active transport systems involving proton symport, the details of the mechanism are still to be elucidated.

Table 3
CATION/PROTON ANTIPORTERS IN BACTERIA

Antiporter	Organism	Specificity	Stoichiometry	Electrogenic/ electroneutral	Km	Inhibitors	Ref.
Na^+/H^+	*Alteromonas haloplanktis* cells	Na^+, Li^+	1:1	Electroneutral	—		86
	Anacystis nidulans cells			—	—	100—200 μMDCCD 3 m*M* NEM	87
	Azotobacter vinelandi membrane vesicles			Electroneutral	44 m*M*		88
	Bacillus alcalophilus membrane vesicles	Na^+, Li^+	$H^+/Na^+ > 1$	Electrogenic	0.7 m*M*	1 m*M* NEM	34
	Bacillus firmus (ATCC 14575) cells			Electroneutral	—		32
	Bacillus firmus (RAB)			Electrogenic	—		32
	Escherichia coli						
	Cells			Electroneutral (pH 7.1)	—		82
	Membrane vesicles	Li^+, Na^+		Electroneutral (pH 6.6) Electrogenic (pH 7.5)	3 m*M*	5 m*M* LiCl	89,90,91,92
	Halobacterium halobium membrane vesicles and –subbacterial particles		$H^+/Na^+ > 1$	Electrogenic	—		93,94,95,96
	Salmonella typhimurium membrane vesicles	Na^+/Li^+		Electrogenic at pH 7.5	—		97

Table 3 (continued)
CATION/PROTON ANTIPORTERS IN BACTERIA

Antiporter	Organism	Specificity	Stoichiometry	Electrogenic/ electroneutral	Km	Inhibitors	Ref.
Na^+/H^+	*Streptococcus faecalis* cells			Electroneutral	—		80
Ca^{2+}/H^+	*Azotobacter vinelandii* everted membrane vesicles			Electroneutral	48 μM (influx) 14 mM (efflux)		98,99
	Escherichia coli everted membrane vesicles	$Ca^{2+} \simeq Mn^{2+} > Sr^{2+} > Ba^{2+}$	$H^+/Ca^{2+} > 2$	Electrogenic		50 mM Mg^{2+} or 50 μM La^{3+} →90% inhib.; 1 mM NEM 2 mM Sr^{2+}	90,93,100 101,102
Ca^{2+}/Na^+	*Halobacterium halobium* membrane vesicles		mNa^+/nCa^{2+} where $m/n \geq 2$	—	71 μM		103
K^+/H^+	*Escherichia coli* everted vesicles	K^+, Rb^+, Na^+		—	—		90
	Streptococcus lactis cells			—	—		104
	Bacillus alcalophilus vesicles	K^+, Rb^+		Electroneutral	—		34

B. Studies of ATP Synthesis and ATPases

Many excellent reviews[110-113] have summarized various aspects of the enormous body of literature on the membrane-bound Ca^{2+}, Mg^{2+}-ATPase which catalyzes ATP synthesis in the presence of a PMF or establishes a PMF during ATP hydrolysis. Synthesis of ATP upon establishment of a PMF has been demonstrated using starved whole cells,[114-116] natural membrane vesicles,[117,118] and artificial vesicles into which purified ATPase has been reconstituted.[119] The BF_1, or soluble part, of the bacterial ATPase has been purified from several species and has been studied with respect to subunit structure and function. There are five different subunits in the BF_1: the α and β subunits are both involved in catalytic activity;[120-123] the γ subunit is needed to restore activity after cold inactivation;[122-124] the δ subunit functions in membrane attachment, for which Mg^{2+} is also required;[122,125-129] and the ϵ subunit inhibits the hydrolytic activity of ATPase.[122] During the past few years, enormous progress has been made in characterization of the intrinsic membrane protein part of the ATPase, the BF_0. The DCCD-binding protein (lipoprotein) has been isolated from *E. coli* [130-132] and *Mycobacterium phlei*.[133] A molecular weight of 9,000 was reported for the former [132] and three subunits with molecular weights of 24,000, 18,000, and 8,000 were reported for the latter.[133] The most extensive, and extraordinarily elegant, work on the BF_0 has been that of Kagawa and his associates on thermophilic bacterium PS3. The BF_0 from that species is a lipoprotein consisting of two subunits of 13,500 (band 6) and 5,400 (band 8) molecular weights; proton conduction by the BF_0 has been achieved in reconstitution experiments and is inhibited by specific antibody and DCCD.[134-136] Band 8 is a proteolipid which binds DCCD[136,137] and band 6 binds TF_1, the F_1 portion of the ATPase in that organism.[137] The proton channel is proposed to be a trimer of band 8.[138]

In *E. coli* several types of mutation in the proton-conducting ATPase have been mapped and characterized; these strains are unable to grow on nonfermentable carbon sources. Unc A mutants have been isolated largely by selection for neomycin resistance.[139] They have been reported to have an altered active site [140] with defects in the α subunit of the BF_1.[141-143] The defect renders them leaky to protons, a leak that is corrected by DCCD;[144-146] secondary mutations in the BF_0 can also reduce the proton leak.[147,148] Unc B strains have mutations in a separate gene locus which result in proton impermeable BF_0s [149,150] and DCCD insensitivity;[151-153] these mutants, unlike unc A mutants, retain the hydrolytic activity of the ATPase. A different mutation, unc C, is a BF_0 mutation in which both ATP hydrolytic activity and DCCD sensitivity are retained.[154,155] Finally, mutations in the β subunit, designated as unc D, have been described. These mutations cause a loss of ATP hydrolytic activity.[156,157] Uncoupler-resistant,[10,158] as well as neomycin-resistant [159] ATPase mutants of *B. Megaterium* have been described, and DCCD-resistant mutants of *Streptococcus faecalis* have been isolated.[160] A series of interesting mutations in *E. coli* that are related to energy-coupling for active transport, although probably not the ATPase *per se*, have been characterized by Hong and his coworkers.[161-163]

A bacterial ATPase which does not translocate protons but functions in K^+ transport has been studied in *E. coli*.[164-166] Three structural genes coding for three proteins in the inner *E. coli* membrane are required.[165,166]

C. Other Processes

An energy-dependent pyridine nucleotide transhydrogenase has been described in *E. coli*,[167,168] *R. rubrum*,[169] and a variety of other bacterial species.[170] Evidence for energization by a PMF, established via respiration or ATP hydrolysis, has been obtained using mutants impaired in respiration or in the ATPase.[171,172] Studies of an *E. coli* mutant which lacks transhydrogenase activity indicate that the enzyme is not an essential source of NADPH in vivo.[173]

The PMF may interact with aspects of chemotaxis in bacteria, although that process, unlike motility, depends upon ATP *per se*.[174] Hyperpolarizing effects of attractants [175] and other changes in $\Delta\Psi$ [176] have been related to chemotactic behavior in a preliminary way.

III. SPECIAL MEMBRANOUS STRUCTURES

In view of the many bioenergetic functions in bacteria, a plethora of specialized membranous structures, adapted in particular ways, might have been expected to exist. However, evidence for large numbers of functionally specialized membranes in bacteria has not been obtained. The purple membrane, photosynthetic membranes in bacteria, and special structures in N_2-fixing organisms are among the few examples that have been described. The first two structures will not be reviewed here. The aerobic N_2-fixing organisms, e.g., species of *Azotobacter* and blue-green algae, are faced with the particular problem of conducting anaerobic biochemistry in an aerobic organism. The heterocysts of blue-green algae are proposed to create an anaerobic environment in which N_2-fixation can occur.[177] In *Azotobacter* species, complex intracytoplasmic membranes are formed in N_2-fixing cells [178,179] and are virtually absent from cells grown on combined nitrogen.[180] These membranes as well as enhanced respiration [181] may protect the nitrogenase system.

IV. A FEW RELATED NOTES ABOUT EUKARYOTES

With the growth of interest in ion gradients—their generation, maintenance, and roles—in bacteria, the whole field of bacterial bioenergetics has moved much closer to areas of eukaryotic bioenergetics. The problems and approaches to the PMF, ATP synthesis, and transporters of the mitochondria are, of course, a direct eukaryotic counterpart to much of what has been reviewed here. Prokaryotic and eukaryotic photosynthetic mechanisms are similarly related, and the recent work on the K^+ transporting ATPase evokes analogies with the eukaryotic Na^+,K^+-ATPase. Moreover, the finding that bacteria use a "sodium motive force" as well as a PMF to energize at least some transport reactions, might represent a precursor to the wide use of a Na^+ currency for eukaryotic transport.[182] There follows a brief sampling, which is by no means comprehensive, of some other eukaryotic systems in which considerations of transmembrane electrochemical potentials are currently being pursued.

In the yeast *Rhodotorula gracilis,* both a transmembrane ΔpH, resulting from active proton extrusion,[183] and a $\Delta\Psi$ [184] have been reported. Energization of sugar transport by the $\Delta\bar{\mu}_{H^+}$ has been suggested.[185] Proton symport solute uptake systems utilizing actively extruded protons have also been described in *Saccharomyces*, with a compensatory K^+ efflux.[186,187] Other protists have been investigated during the past few years with similar findings.[188,189] Extensive studies of electrogenic ion pumping and resulting gradients across the plasma membrane of *Neurospora crassa* have been conducted.[190-193]

In isolated chromaffin granules, a proton-translocating ATPase has been reported to establish a transmembrane ΔpH;[194-197] a $\Delta\Psi$ has also been measured.[198] Evidence for a role of the ΔpH in catecholamine uptake has been presented,[199-200] although there is also a suggestion that the $\Delta\Psi$ rather than the ΔpH is the relevant energy source.[201] Electrical potentials across plasma membranes have been found in various mammalian cells [202-205] (in addition to nerve cells) and have raised intriguing possibilities with respect to roles.

V. CONCLUSIONS

Reviews of the magnitude, mechanisms of maintenance, and roles of the PMF tend to be out-of-date as fast as they are published. The selective permeability of the bacterial membrane to protons and other ions is now viewed as a focal point from which a better understanding of bioenergetic and other biochemical processes will emerge. Elucidation of the mechanistic details of such processes will continue to depend upon methodological advances in measurements of the relevant gradients and stoichiometries, and in the studies of purified and reconstituted intrinsic membrane proteins. The many possible regulatory effects that may be exerted by transmembrane gradients on membrane proteins and, conversely, possible effectors of membrane permeability to selected ions are, as yet unclear. The current investigative trends in various areas of bacterial membrane bioenergetics have created important fields of overlapping interests with those working on similar processes in eukaryotes. Work in prokaryotes may make as substantial a contribution to the general field of bioenergetics as has been made in the field of molecular biology.

REFERENCES

1. **Mitchell, P.,** Coupling of phosphorylation to electron and hydrogen transfer by a chemiosmotic type of mechanism, *Nature (London)*, 191, 144, 1961.
2. **Mitchell, P.,** Chemiosmotic coupling in oxidation and photosynthetic phosphorylation, *Biol. Rev.*, 41, 445, 1966.
3. **Mitchell, P.,** Chemiosmotic coupling in energy transduction: a logical development of biochemical knowledge, *J. Bioenerg.*, 4, 63, 1973.
4. **Harold, F. M.,** Membranes and energy transduction in bacteria, *Curr. Topics Bioenerg.*, 6, 83, 1977.
5. **Harold, F. M.,** Conservation and transformation of energy by bacterial membranes, *Bacteriol. Rev.*, 36, 172, 1972.
6. **Garland, P. B.,** Energy transduction and transmission in microbial systems, in *Microbial Energetics*, Haddock, B. A. and Hamilton, W. A., Eds., Cambridge University Press, London, 1977, 1.
7. **Skulachev, V. P.,** Transmembrane electrochemical H^+-potential as a convertible energy source for the living cell, *FEBS Lett.*, 74, 1. 1977.
8. **Rosen, B. P. and Kashket, E. R.,** Energetics of active transport, in *Microbial Transport,* Rosen, B. P., Ed., Marcel Dekker, New York, 1978, 559.
9. **Mandel, K. G. and Krulwich, T. A.,** D-Gluconate transport in *Arthrobacter pyridinolis:* metabolic trapping of a protonated solute, *Biochim. Biophys. Acta,* 552, 478, 1979.
10. **Decker, S. J. and Lang, D. R.,** Membrane bioenergetic parameters in uncoupler-resistant mutants of *Bacillus megaterium, J. Biol. Chem.*, 253, 6738, 1978.
11. **Shioi, J.-1., Imae, Y., and Oosawa, F.,** Protonmotive force and motility of *Bacillus subtilis, J. Bacteriol.*, 133, 1083, 1978.
12. **Riebeling, V., Thauer, R. K., and Jungerman, K.,** The internal-alkaline pH gradient sensitive to uncoupler and ATPase inhibitor in growing *Clostridium pasteurianum, Eur. J. Biochem.*, 55, 445, 1975.
13. **Padan, E., Zilberstein, D., and Rottenberg, H.,** The proton electrochemical gradient in *Escherichia coli* cells, *Eur. J. Biochem.*, 63, 533, 1976.
14. **Griniuviene, B., Chmieliauskaite, V., Melvydas, V., Dzheja, P., and Grinius, L.,** Conversion of *Escherichia coli* cell-produced metabolic energy into electric form, *Bioenergetics,* 7, 17, 1975.
15. **Michel, H. and Oesterhelt, D.,** Light-induced changes of the pH-gradient and the membrane potential in *H. halobium, FEBS Lett.*, 65, 175, 1976.
16. **Bakker, E. P., Rottenberg, H., and Caplan, S. R.,** An estimation of the light-induced electrochemical potential difference of protons across the membrane of *Halobacterium halobium, Biochim. Biophys. Acta,* 440, 557, 1976.

17. **Deutsch, C. J. and Kula, T.,** Transmembrane electrical and pH gradients of *Paracoccus denitrificans* and their relationship to oxidative phosphorylation, *FEBS Lett.,* 87, 145, 1978.
18. **Collins, S. H. and Hamilton, W. A.,** Magnitude of the protonmotive force in respiring *Staphylococcus aureus* and *Escherichia coli, J. Bacteriol.,* 126, 1224, 1976.
19. **Horan, N. J., Midgley, M., and Dawes, E. A.,** Anaerobic transport of serine and 2-aminoisobutyric acid by *Staphylococcus epidermidis. J. Gen. Micro.,* 109, 119, 1978.
20. **Kashket, E. R. and Wilson, T. H.,** Protonmotive force in fermenting *Streptococcus lactis* 7962, *Biochem. Biophys. Res. Commun.,* 59, 879, 1974.
21. **Harold, F. M., Pavlasova, E., and Baarda, J. R.,** The transmembrane pH gradient in *Streptococcus faecalis:* origin and dissipation by proton conductors and N,N′-dicyclohexylcarbodiimide, *Biochim. Biophys. Acta,* 196, 235, 1970.
22. **Harold, F. M. and Papineau, D.,** Cation transport and electrogenesis by *Streptococcus faecalis.* I. The membrane potential, *J. Membrane Biol.,* 8, 27, 1979.
23. **Krulwich, T. A., Davidson, L. F., Filip, S. F., Zuckerman, R. F., and Guffanti, A. A.,** The protonmotive force and β-galactoside transport in *Bacillus acidocaldarius, J. Biol. Chem.,* 253, 4599, 1978.
24. **Oshima, T., Arakawa, H., and Baba, M.,** Biochemical studies on an acidophilic, thermophilic bacterium, *Bacillus acidocaldarius:* isolation of bacteria, intracellular pH, and stabilities of biopolymers, *J. Biochem.,* 81, 1107, 1977.
25. **Searcy, D. G.,** *Thermoplasma acidophilus:* intracellular pH and potassium concentration, *Biochim. Biophys. Acta,* 451, 278, 1976.
26. **Hsung, J. C. and Haug, A.,** Membrane potential of *Thermoplasma acidophila, FEBS Lett.,* 73, 47, 1977.
27. **Hsung, J. C. and Haug, A.,** Intracellular pH of *Thermoplasma acidophila, Biochim. Biophys. Acta,* 389, 477, 1975.
28. **Cox, J. C., Nicholls, D. G., and Ingledew, J. W.,** Transmembrane electrical potential and transmembrane pH gradient in the acidophile *Thiobacillus ferrooxidans, Biochem. J.,* 178, 195, 1979.
29. **Guffanti, A. A., Susman, P., Blanco, R., and Krulwich, T. A.,** The protonmotive force and α-aminoisobutyric acid transport in an obligately alkalophilic bacterium, *J. Biol. Chem.,* 253, 708, 1978.
30. **Guffanti, A. A., Blanco, R. B., and Krulwich, T. A.,** A requirement for ATP for β-galactoside transport by *Bacillus alcalophilus, J. Biol. Chem.,* 254, 1033, 1979.
31. **Guffanti, A. A., Monti, L. G., Blanco, R., Ozick, D., and Krulwich, T. A.,** β-Galactoside transport in an alkaline-tolerant strain of *Bacillus circulans, J. Gen. Micro.,* 112, 161, 1979.
32. **Guffanti, A. A., Blanco, R., Benenson, R. A., and Krulwich, T. A.,** Bioenergetic properties of alkaline-tolerant and alkalophilic strains of *Bacillus firmus,* J. Gen. Microbiol., 119, 79, 1980.
33. **Bhattacharyya, P., Shapiro, S. A., and Barnes, E. M., Jr.,** Generation of a transmembrane electric potential during respiration by *Azotobacter vinelandii* membrane vesicles, *J. Bacteriol.,* 129, 756, 1977.
34. **Mandel, K. G., Guffanti, A. A., and Krulwich, T. A.,** Monovalent cation/proton antiporters in membrane vesicles of *Bacillus alcalophilus,* J. Biol. Chem., 255, 7391, 1980.
35. **Hirata, H., Altendorf, K., and Harold, F. M.,** Role of an electrical potential in the coupling of metabolic energy to active transport by membrane vesicles of *Escherichia coli, Proc. Natl. Acad. Sci., USA,* 70, 1804, 1973.
36. **Ramos, S., Schuldiner, S., and Kaback, H. R.,** The electrochemical gradient of protons and its relationship to active transport in *Escherichia coli* membrane vesicles, *Proc. Nat. Acad. Sci., USA,* 73, 1892, 1976.
37. **Ramos, S., and Kaback, H. R.,** The electrochemical proton gradient in *Escherichia coli* membrane vesicles, *Biochemistry,* 16, 848, 1977.
38. **Boonstra, J. and Konings, W. N.,** Generation of an electrochemical proton gradient by nitrate respiration in membrane vesicles from anaerobically grown *Escherichia coli, Eur. J. Biochem.,* 78, 861, 1977.
39. **Konings, W. N. and Boonstra, J.,** Anaerobic electron transfer and active transport in bacteria, in *Current Topics in Membranes and Transport,* Vol. 9, Bronner, F. and Kleinzeller, A., Eds., Academic Press, New York, 1977, 177.
40. **Singh, A. P. and Bragg, P. D.,** Effects of inhibitors on the substrate-dependent quenching of 9-aminoacridine fluorescence in inside-out membrane vesicles of *Escherichia coli, Eur. J. Biochem.,* 67, 177, 1976.
41. **Singh, A. P. and Bragg, P. D.,** ATP synthesis driven by a pH gradient imposed across the cell membranes of lipoic acid and unsaturated fatty acid auxotrophs of *Escherichia coli, FEBS, Lett.,* 98, 21, 1979.

42. **Renthal, R. and Lanyi, J. K.,** Light-induced membrane potential and pH gradient in *Halobacterium halobium* envelope vesicles, *Biochemistry,* 15, 2136, 1976.
43. **Lanyi, J. K.,** Coupling of aspartate and serine transport to the transmembrane electrochemical gradient for sodium ions in *Halobacterium halobium.* Translocation stoichiometries and apparent cooperativity, *Biochemistry,* 17, 3011, 1978.
44. **Kell, D. B., John, P., and Ferguson, S. J.,** The protonmotive force in phosphorylating membrane vesicles from *Paracoccus denitrificans.* Magnitude, sites of generation and comparison with the phosphorylation potential, *Biochem. J.,* 174, 257, 1978.
45. **Michels, A. M. and Konings, W. N.,** The electrochemical proton gradient generated by light in membrane vesicles of chromatophores from *Rhodopseudomonas sphaeroides, Eur. J. Biochem.,* 85, 147, 1978.
46. **Ferguson, S. J., Jones, O. T. G., Kell, D. B., and Sorgato, M. C.,** Comparison of permeant ion uptake and carotenoid band shift as methods for determining the membrane potential in chromatophores from *Rhodopseudomonas sphaeroides, Biochem. J.,* 180, 75, 1979.
47. **Pick, U. and Avron, M.,** Measurement of transmembrane potentials in *Rhodospirillum rubrum* chromatophores with an oxacarbyocyanine dye, *Biochim. Biophys. Acta,* 440, 189, 1976.
48. **Leiser, M. and Gromet-Elhanan, Z.,** Comparison of the electrochemical proton gradient and phosphate potential maintained by *Rhodospirillum rubrum* chromatophores in the steady state, *Arch. Biochem. Biophys.,* 178, 79, 1977.
49. **Kell, D. B., Ferguson, S. J., and John, P.,** Measurement by a flow dialysis technique of the steady-state proton-motive force in chromatophores from *Rhodospirillum rubrum, Biochim. Biophys. Acta,* 502, 111, 1978.
50. **Maloney, P. C., Kashket, E. R., and Wilson, T. H.,** Methods for studying transport in bacteria, in *Methods in Membrane Biology,* Korn, E. D., Ed., Plenum Press, New York, 1975, 1.
51. **Barker, S. L. and Kashket, E. R.,** Effects of sodium ions on the electrical and pH gradients across the membrane of *Streptococcus lactis* cells, *J. Supramol. Struc.,* 6, 383, 1977.
52. **Haddock, B. A. and Jones, C. W.,** Bacterial respiration, *Bacteriol. Rev.,* 41, 47, 1977.
53. **Harold, F. M.,** Ion currents and physiological functions in microorganisms, *Ann. Rev. Microbiol.,* 31, 181, 1978.
54. **Jones, C. W.,** Aerobic respiratory systems in bacteria, in *Microbial Energetics,* Haddock, B. A. and Hamilton, W. A., Eds., Cambridge University Press, London, 1977, 23.
55. **Kruger, A.,** Phosphorylative electron transport with fumarate and nitrate as terminal hydrogen acceptors, in *Microbial Energetics,* Haddock, B. A. and Hamilton, W. A., Eds., Cambridge University Press, London, 1977, 61.
56. **Papa, S.,** Proton translocation reactions in the respiratory chains, *Biochim. Biophys. Acta,* 456, 39, 1976.
57. **Henry, M. F., and Vignais, P. M.,** Induction by cyanide of cytochrome *d* in the plasma membrane of *Paracoccus denitrificans, FEBS Lett.,* 100, 41, 1979.
58. **Hogarth, C., Wilkinson, B. J., and Ellar, D. J.,** Cyanide-resistant electron transport in sporulating *Bacillus megaterium* KM, *Biochim. Biophys. Acta,* 461, 109, 1977.
59. **Hollander, R.,** The cytochromes of *Thermoplasma acidophilum, J. Gen. Microbiol.,* 108, 165, 1978.
60. **Kenimer, E. A., and Lapp, D. F.,** Effects of selected inhibitors on electron transport in *Neisseriae gonorrhoeae, J. Bacteriol.,* 134, 537, 1978.
61. **Pelliccione, N. J., Jaffin, B. J., Sobel, M. E., and Krulwich, T. A.,** Induction of the phosphoenol-pyruvate: hexose phosphotransferase system associated with relative anaerobiosis in an obligate aerobe, *Eur. J. Biochem.,* 95, 69, 1979.
62. **Sweet, W. J. and Peterson, J. A.,** Changes in cytochrome content and electron transport patterns in *Pseudomonas putida* as a function of growth phase. *J. Bacteriol.,* 133, 217, 1978.
63. **Macy, J., Kulla, H., and Gottschalk, G.,** H_2-dependent anaerobic growth of *Escherichia coli* on L-malate: succinate formation, *J. Bacteriol.,* 125, 423, 1976.
64. **Ritchey, T. W. and Seeley, H. W.,** Distribution of cytochrome-like respiration in streptococci, *J. Gen. Microbiol.,* 93, 195, 1976.
65. **Sperry, J. F. and Wilkins, T. D.,** Cytochrome spectrum of an obligate anaerobe, *Eubacterium lentum, J. Bacteriol.* 125, 905, 1976.
66. **Reddy, C. A. and Peck, H. D., Jr.,** Electron transport phosphorylation coupled to fumarate reduction by H_2- and Mg^{2+}-dependent adenosine triphosphatase activity in extracts of the rumen anaerobe *Vibrio succinogenes, J. Bacteriol.,* 134, 982, 1978.
67. **Haddock, B. A. and Kendall-Tobias, M. W.,** Functional anaerobic electron transport linked to the reduction of nitrate and fumarate in membranes from *Escherichia coli* as demonstrated by quenching of atebrin fluorescence, *Biochem. J.,* 152, 655, 1975.
68. **Gutowski, S. J. and Rosenberg, H.,** Proton translocation coupled to electron flow from endogenous substrates to fumarate in anaerobically grown *Escherichia coli* K12, *Biochem. J.,* 164, 265, 1977.

69. **Mitchell, P.,** Possible molecular mechanisms of the protonmotive function of cytochrome systems, *J. Theoret. Biol.,* 62, 327. 1976.
70. **Downie, J. A. and Cox, G. B.,** Sequence of *b* cytochromes relative to ubiquinone in the electron transport chain of *Escherichia coli, J. Bacteriol.,* 133, 477, 1978.
71. **Stroobant, P. and Kaback, H. R.,** Reconstitution of ubiquinone-linked functions in membrane vesicles from a double quinone mutant of *Escherichia coli, Biochemistry,* 18, 226, 1979.
72. **Jones, C. W., Brice, J. M., Downs, A. J., and Drozd, J. W.,** Bacterial respiration-linked proton translocation and its relationship to respiratory-chain composition, *Eur. J. Biochem.,* 52, 265, 1975.
73. **Cox, J. C. and Haddock, B. A.,** Phosphate transport and the stoichiometry of respiratory driven proton translocation in *Escherichia coli, Biochem. Biophys. Res. Commun.,* 82, 46, 1978.
74. **Garland, P. B., Downie, J. A., and Haddock, B. A.,** Proton translocation and the respiratory nitrate reductase of *Escherichia coli, Biochem. J.,* 152, 547, 1975.
75. **Lawford, H. G. and Haddock, B. A.,** Respiration-driven proton translocation in *Escherichia coli, Biochem. J.,* 136, 217, 1973.
76. **Lawford, H. G.,** Proton translocation coupled to ubiquinol oxidation in *Paracoccus denitrificans, Can. J. Biochem.,* 57, 172, 1979.
77. **Meijer, E. M., van Verseveld, H. W., van Der Beek, E. G., and Southamer, A. H.,** Energy conservation during aerobic growth in *Paracoccus denitrificans, Arch. Microbiol.,* 112, 25, 1977.
78. **O'Keefe, D. T. and Anthony, C.,** The stoichiometry of respiration-driven proton translocation in *Pseudomonas AM 1* and in a mutant lacking cytochrome *c, Biochem. J.,* 170, 561, 1978.
79. **Pritchard, G. G. and Wimpenny, J. W. T.,** Cytochrome formation, oxygen-induced proton extrusion and respiratory activity in *Streptococcus faecalis* var. *zymogenes* grown in the presence of haematin, *J. Gen. Microbiol.,* 104, 15, 1978.
80. **Harold, F. M. and Papineau, D.,** Cation transport and electrogenesis by *Streptococcus faecalis.* II. Proton and sodium extrusion, *J. Membrane Biol.,* 8, 45, 1972.
81. **Brookman, J. J., Downie, J. A., Gibson, F., Cox, G. B., and Rosenberg, H.,** Proton translocation in cytochrome-deficient mutants of *E. coli, J. Bacteriol.,* 137, 705, 1979.
82. **West, I. C. and Mitchell, P.,** Proton/sodium antiport in *Escherichia coli, Biochem. J.,* 144, 87, 1974.
83. **Skulachev, V. P.,** Membrane linked energy buffering as the biological function of Na^+/K^+ gradient, *FEBS Lett.,* 87, 171, 1978.
84. **Lanyi, J. K.,** Light energy conversion in *Halobacterium halobium, Microbiol. Rev.,* 42, 682, 1978.
85. **Lanyi, J. K.,** The role of sodium ions in bacterial membrane transport, *Biochim. Biophys. Acta,* 559, 377, 1979.
86. **Niven, D. F. and MacLeod, R. A.,** Sodium ion-proton antiport in a marine bacterium, *J. Bacteriol.,* 134, 737, 1978.
87. **Paschinger, H.,** DCCD induced sodium uptake by *Anacystis nidulans, Arch. Microbiol,* 113, 285, 1977.
88. **Bhattacharyya, P. and Barnes, E. M., Jr.,** Proton-coupled sodium uptake by membrane vesicles from *Azotobacter vinelandii, J. Biol. Chem.,* 253, 3848, 1978.
89. **Schuldiner, S. and Fishkes, H.,** Sodium-proton antiport in isolated membrane vesicles of *Escherichia coli, Biochemistry,* 17, 706, 1978.
90. **Brey, R. N., Beck, J. C., and Rosen, B. P.,** Cation/proton antiport systems in *Escherichia coli, Biochem. Biophys. Res. Commun.,* 83, 1588, 1978.
91. **Beck, J. C. and Rosen, B. P.,** Cation/proton antiport systems in *Escherichia coli:* properties of the sodium/proton antiporter, *Arch. Biochem. Biophys.,* 194, 208, 1979.
92. **Tsuchiya, T. and Takeda, K.,** Calcium/proton and sodium/proton antiport systems in *Escherichia coli, J. Biochem.,* 85, 943, 1979.
93. **Lanyi, J. K., Renthal, R., and MacDonald, R. E.,** Light-induced glutamate transport in *Halobacterium halobium* envelope vesicles. II. Evidence that the driving force is a light-dependent sodium gradient, *Biochemistry,* 15, 1603, 1976.
94. **Lanyi, J. K. and MacDonald, R. E.,** Existence of electrogenic hydrogen ion/sodium ion antiport in *Halobacterium halobium* cell envelope vesicles, *Biochemistry,* 15, 4608, 1976.
95. **Lanyi, J. K.,** Transport in *Halobacterium halobium:* light-induced cation-gradients, amino acid transport kinetics, and properties of transport carriers, *J. Supramol. Struc.,* 6, 169, 1977.
96. **Eisenbach, M., Cooper, S., Garty, H., Johnstone, R. M., Rottenberg, H., and Caplan, S. R.,** Light-driven sodium transport in sub-bacterial particles of *Halobacterium halobium, Biochim. Biophys. Acta,* 465, 599, 1977.
97. **Tokuda, H. and Kaback, H. R.,** Sodium-dependent methyl-1-thio β-D-galactopyranoside transport in membrane vesicles isolated from *Salmonella typhimurium, Biochemistry,* 16, 2130, 1977.
98. **Bhattacharyya, P. and Barnes, E. M., Jr.,** ATP-Dependent calcium transport in isolated membrane vesicles from *Azotobacter vinelandii, J. Biol. Chem.,* 251, 5614, 1976.

99. **Barnes, E. M., Jr., Roberts, R. R., and Bhattacharyya, P.,** Respiration-coupled calcium transport by membrane vesicles from *Azotobacter vinelandii, Memb. Biochem.,* 1, 73, 1978.
100. **Tsuchiya, T. and Rosen, B. P.,** Characterization of an active transport system for calcium in inverted membrane vesicles of *Escherichia coli, J. Biol. Chem.,* 250, 7687, 1975.
101. **Tsuchiya, T. and Rosen, B. P.,** Calcium transport driven by a proton gradient in inverted membrane vesicles of *Escherichia coli, J. Biol. Chem.,* 251, 962, 1976.
102. **Brey, R. N. and Rosen, B. P.,** Cation/proton antiport systems in *Escherichia coli:* properties of the calcium/proton antiporter, *J. Biol. Chem.,* 254, 1957, 1979.
103. **Belliveau, J. W. and Lanyi, J. K.,** Calcium transport in *Halobacterium halobium* envelope vesicles, *Arch. Biochem. Biophys.,* 186, 98, 1978.
104. **Kashket, E. R. and Berker, S. L.,** Effects of potassium ions on the electrical and pH gradients across the membrane of *Streptococcus lactis* cells, *J. Bacteriol.,* 130, 1017, 1977.
105. **Boyer, P. D., Chance, B., Ernster, L., Mitchell, P., Racker, E., and Slater, E. C.,** Oxidative phosphorylation and photophosphorylation, *Ann. Rev. Biochem.,* 46, 955, 1977.
106. **Williams, R. J. P.,** The multivarious couplings of energy transduction, *Biochim. Biophys. Acta,* 505, 1, 1978.
107. **Williams, R. J. P.,** The history and the hypotheses concerning ATP-formation by energized protons, *FEBS Lett.,* 85, 9, 1978.
108. **Gould, J. M.,** Respiration-linked proton transport changes in pH, and membrane energization in cells of *Escherichia coli, J. Bacteriol.,* 138, 176, 1979.
109. **Rottenberg, H.,** The driving force for proton(s) metabolites cotransport in bacterial cells, *FEBS Lett.,* 66, 159, 1976.
110. **Kagawa, Y.,** Reconstitution of the energy transformer, gate and channel subunit reassembly, crystalline ATPase and ATP synthesis, *Biochim. Biophys. Acta,* 505, 45, 1978.
111. **Monteil, H. and Serrahima-Zieger, M.,** Les ATPases, bactériennes: propríetés moléculaires et fonctions, *Bull. Inst. Pasteur Paris,* 76, 207, 1978.
112. **Simoni, R. D. and Postma, P. W.,** The energetics of bacterial active transport, *Ann. Rev. Biochem.,* 44, 523, 1975.
113. **Smith, J. B., Sternweiss, P. C., Larson, R. J., and Heppel, L. A.,** Subunits of the bacterial proton-pump ATPase: a synopsis, *J. Cell. Physiol.,* 89, 567, 1976.
114. **Wilson, D. M., Alderte, J. F., Maloney, P. C., and Wilson, T. H.,** Protonmotive force as the source of energy for adenosine 5′-triphosphate synthesis in *Escherichia coli, J. Bacteriol.,* 126, 327, 1976.
115. **Maloney, P. C.,** Coupling between H^+ entry and ATP formation in *Escherichia coli, Biochem. Biophys. Res. Commun.,* 83, 1496, 1978.
116. **Doddema, H. J., Hutten, T. J., van der Drift, C., and Vogels, G. D.,** ATP hydrolysis and synthesis by the membrane-bound ATP synthetase complex of *Methanobacterium thermoautotrophicum, J. Bacteriol.,* 136, 19, 1978.
117. **Tsuchiya, T. and Rosen, B. P.,** Adenosine 5′-triphosphate synthesis energized by an artificially imposed membrane potential in membrane vesicles of *Escherichia coli, J. Bacteriol.,* 127, 154, 1976.
118. **Tsuchiya, T.,** Adenosine 5′-triphosphate synthesis driven by a protonmotive force in membrane vesicles of *Escherichia coli, J. Bacteriol.,* 129, 763, 1977.
119. **Sone, N., Yoshida, M., Hirata, H., and Kagawa, Y.,** Adenosine triphosphate synthesis by electrochemical proton gradient in vesicles reconstituted from purified adenosine triphosphatase and phospholipids of thermophilic bacterium, *J. Biol. Chem.,* 252, 295, 1977.
120. **Vogel, G. and Steinhart, R.,** ATPase of *Escherichia coli:* Purification, dissociation, and reconstitution of the active complex from the isolated subunits, *Biochemistry,* 15, 208, 1976.
121. **Leimgruber, R. M., Jensen, C., and Abrams, A.,** Accessibility of the α chains in membrane-bound and solubilized bacterial ATPase to chymotryptic cleavage, *Biochem. Biophys. Res. Commun.,* 81, 439, 1978.
122. **Smith, J. B., Sternweis, P. C., and Heppel, L. A.,** Partial purification of active delta and epsilon subunits of the membrane ATPase from *Escherichia coli, J. Supramol. Struct.,* 3, 248, 1975.
123. **Nelson, N., Kanner, B. I., and Gutnick, D. L.,** Purification and properties of Mg^{2+}-Ca^{2+} adenosinetriphosphatase from *Escherichia coli, Proc. Natl. Acad. Sci. U.S.A.,* 71, 2720, 1974.
124. **Kobayashi, H. and Anraku, Y.,** Membrane-bound adenosine triphosphatase of *Escherichia coli.* Physico-chemical properties of the enzyme, *J. Biochem.,* 76, 1175, 1974.
125. **Abrams, A., Jensen, C., and Morris, D. H.,** Role of Mg^{2+} ions in the subunit structure and membrane binding properties of bacterial energy transducing ATPase, *Biochem. Biophys. Res. Commun.,* 69, 804, 1976.
126. **Sternweis, P. C. and Smith, J. B.,** Characterization of the purified membrane attachment (δ) subunit of the proton translocating adenosine triphosphatase from *Escherichia coli, Biochemistry,* 16, 4020, 1977.

127. **Bragg, P. D. and Hou, C.,** Binding of the Ca^{2+}, Mg^{2+} activated adenosine triphosphatase of *Escherichia coli* to phospholipid vesicles, *Can. J. Biochem.*, 56, 559, 1978.
128. **Futai, M., Sternweis, P. C., and Heppel, L. A.,** Purification and properties of reconstitutively active and inactive adenosinetriphosphatase from *Escherichia coli, Proc. Natl. Acad. Sci. U.S.A.*, 71, 2725, 1974.
129. **Smith, J. B. and Sternweis, P. C.,** Restoration of coupling factor activity to *Escherichia coli* ATPase missing the delta subunit, *Biochem. Biophys. Res. Commun.*, 62, 764, 1975.
130. **Hare, J. F.,** Purification and characterization of a dicyclohexylcarbodiimide-sensitive adenosine triphosphatase complex from membranes of *Escherichia coli, Biochem. Biophys. Res. Commun.*, 66, 1329, 1975.
131. **Altendorf, K. H. and Ziztmann, W.,** Identification of the DCCD-reactive protein of the energy transducing adenosine triphosphatase complex from *Escherichia coli, FEBS Lett.*, 59, 268, 1975.
132. **Altendorf, K., Kohl, B., Lukas, M., Muller, C. R., and Sandermann, H., Jr.,** Isolation and purification of bacterial membrane proteins by the use of organic solvents: the lactose permease and the carbodiimide-reactive protein of the adenosinetriphosphatase complex of *Escherichia coli, J. Supramolec. Struct.*, 6, 229, 1977.
133. **Cohen, N.-S., Lee, S., and Brodie, A.,** Purification and characteristics of hydrophobic membrane protein(s) required for DCCD sensitivity of ATPase in *Mycobacterium phlei, J. Supramolec. Struct.*, 8, 111, 1978.
134. **Kagawa, Y.,** Transport of nutrients by a thermophilic bacterium. Reconstitution of vesicles from crystalline ATPase or solubilized alanine carrier, *J. Cell. Physiol.*, 89, 569, 1976.
135. **Okamato, H., Sone, H., Hirata, H., Yoshida, M., and Kagawa, Y.,** Purified proton conductor in proton translocating adenosine triphosphatase of a thermophilic bacterium, *J. Biol. Chem.*, 252, 6125, 1977.
136. **Sone, N., Yoshida, M., Hirata, H., and Kagawa, Y.,** Resolution of the membrane moiety of the H^+-ATPase complex into two kinds of subunits, *Proc. Natl. Acad. Sci. U.S.A.*, 75, 4219, 1978.
137. **Sone, N., Yoshida, M., Hirata, H., and Kagawa, Y.,** Carbodiimide-binding protein of H^+-translocating ATPase and inhibition of H^+ conduction by dicyclohexylcarbodiimide, *J. Biochem.*, 85, 503, 1979.
138. **Sone, N., Ikeba, K. and Kagawa, Y.,** Inhibition of proton conduction by chemical modification of the membrane moiety of proton translocating ATPase, *FEBS Lett.*, 97, 61, 1979.
139. **Adler, L. W. and Rosen, B. P.,** Properties of *Escherichia coli* mutants with alterations in Mg^{2+}-adenosine triphosphatase, *J. Bacteriol.*, 128, 248, 1976.
140. **Bragg, P. D. and Hou, C.,** Purification and characterization of the inactive Ca^{2+}, Mg^{2+}-activated adenosine triphosphatase of the unc A mutant *Escherichia coli* AN 120, *Arch. Biochem. Biophys.*, 178, 486, 1977.
141. **Butlin, J. D., Cox, G. B., and Gibson, F.,** Oxidative phosphorylation in *Escherichia coli* K12. Mutations affecting magnesium ion- or calcium ion-stimulated adenosine triphosphatase, *Biochem. J.*, 124, 75, 1971.
142. **Kamazawa, H., Saito, S., and Futai, M.,** Coupling factor ATPase from *Escherichia coli.* An unc A mutant (unc 401) with defective α subunit, *J. Biochem.*, 84, 513, 1978.
143. **Senior, A. E., Downie, J. A., Cox, G. B., Gibson, F., Langman, L., and Foyle, D. R. H.,** The *unc A* gene codes for the α-subunit of the adenosine triphosphatase of *Escherichia coli.* Electrophoretic analysis of *unc A* mutant strains, *Biochem. J.*, 180, 103, 1979.
144. **Rosen, B. P.,** Restoration of active transport in an Mg^{2+}-adenosine triphosphatase-deficient mutant of *Escherichia coli, J. Bacteriol.*, 116, 1124, 1973.
145. **Rosen, B. P. and Adler, L. W.,** The maintenance of the energized membrane state and its relation to active transport in *Escherichia coli, Biochim. Biophys. Acta*, 387, 23, 1975.
146. **Rosen, B. P.,** β-Galactoside transport and proton movements in an adenosine triphosphatase-deficient mutant of *Escherichia coli, Biochem. Biophys. Res. Commun.*, 53, 1289, 1973.
147. **Fillingame, R. H., Knoebel, K., and Wopat, A. E.,** Method for isolation of *Escherichia coli* mutants with defects in the proton-translocating sector of the membrane adenosine triphosphatase complex, *J. Bacteriol.*, 136, 570, 1978.
148. **Rosen, B. P., Brey, R. N., and Hasan, S. M.,** Energy transcuction in *Escherichia coli:* New mutation affecting the F_0 portion of the ATP synthetase complex, *J. Bacteriol.*, 134, 1030, 1978.
149. **Hasan, S. M., Tsuchiya, T., and Rosen, B. P.,** Energy transduction in *Escherichia coli:* physiological and biochemical effects of mutation in the *unc B* locus, *J. Bacteriol.*, 133, 108, 1978.
150. **Butlin, J. D., Cox, G. B., and Gibson, F.,** Oxidative phosphorylation in *Escherichia coli* K12; The genetic and biochemical characterization of a strain carrying a mutation in the unc B gene, *Biochim. Biophys. Acta*, 292, 366, 1973.
151. **Simoni, R. D. and Shandel, A.,** Energy transduction in *Escherichia coli:* genetic alteration of a membrane polypeptide of the (Ca^{2+}, Mg^{2+}) ATPase complex, *J. Biol. Chem.*, 250, 9421, 1975.

152. **Friedl, P., Schmid, B. I., and Schairer, H. U.,** A mutant ATP synthetase of *Escherichia coli* with an altered sensitivity to N,N'-dicyclohexylcarbodiimide: characterization in native membranes and reconstituted proteoliposomes, *Eur. J. Biochem.*, 73, 461, 1977.
153. **Fillingame, R. H. and Wopat, A. E.,** Carbodiimide-resistant mutants of *Escherichia coli:* suppression of resistance to dicyclohexylcarbodiimide by growth on glucose or glycerol, *J. Bacteriol.*, 134, 687, 1978.
154. **Cox, G. B., Crane, F. L., Downie, J. A., and Radin, J.,** Different effects of inhibitors on two mutants of *E. coli* K12 affected in the F_0 portion of the ATPase complex, *Biochim. Biophys. Acta,* 462, 113, 1977.
155. **Gibson, F., Cox, G. B., Downie, J. A., and Radik, J.,** A mutant affecting a second component of the F_0 portion of the magnesium ion stimulated ATPase of *E. coli* K12. The unc C 424 allele, *Biochem. J.*, 164, 193, 1977.
156. **Senior, A. E., Fayle, D. R. H., Downie, J. A., Gibson, F., and Cox, G. B.,** Properties of membranes from mutant strains of *Escherichia coli* in which the β-subunit of the adenosine triphosphatase is abnormal, *Biochem. J.*, 180, 111, 1979.
157. **Fayle, D. R. H., Downie, J. A., Cox, G. B., Gibson, F., and Radik, J.,** Characterization of the mutant-unc D-gene product in a strain of *Escherichia coli* K12. An altered β-subunit of the magnesium ion-stimulated adenosine triphosphatase, *Biochem. J.*, 172, 523, 1978.
158. **Decker, S. J. and Lang, D. R.,** Mutants of *Bacillus megaterium* resistant to uncouplers of oxidative phosphorylation, *J. Biol. Chem.*, 252, 5936, 1977.
159. **Decker, S. J. and Lang, D. R.,** *Bacillus megaterium* mutant deficient in membrane-bound adenosine triphosphatase activity, *J. Bacteriol.*, 131, 98, 1977.
160. **Abrams, A., Smith, J. B., and Baron, C.,** Carbodiimide-resistant membrane adenosine triphosphatase in mutants of *Streptococcus faecalis.* I. Studies of the mechanism of resistance, *J. Biol. Chem.*, 247, 1484, 1972.
161. **Lieberman, M. A. and Hong, J.,** Changes in active transport, intracellular adenosine 5'-triphosphate levels, macromolecular syntheses, and glycolysis in an energy-uncoupled mutant of *Escherichia coli, J. Bacteriol.*, 125, 1024, 1976.
162. **Hong, J.,** An *ecf* mutation in *Escherichia coli* pleiotropically affecting energy coupling in active transport but not generation or maintenance of a membrane potential, *J. Biol. Chem.*, 252, 8582, 1977.
163. **Lieberman, M. A., Simon, M., and Hong, J.,** Characterization of *Escherichia coli* mutant incapable of maintaining a transmembrane potential, *J. Biol. Chem.*, 252, 4056, 1977.
164. **Epstein, W., Whitelaw, V., and Hesse, J.,** A K^- transport ATPase in *Escherichia coli, J. Biol. Chem.*, 253, 6666, 1978.
165. **Laimins, L. A., Rhonds, D. B., Altendorf, K., and Epstein, W.,** Identification of the structural components of an ATP-driven potassium transport system in *Escherichia coli, Proc. Natl. Acad. Sci. U.S.A.*, 75, 3216, 1978.
166. **Wieczorek, L. and Altendorf, K.,** Potassium transport in *Escherichia coli.* Evidence for a K^--transport adenosine-5'-triphosphatase, *FEBS Lett.*, 98, 233, 1979.
167. **Fisher, R. J. and Sanadi, D. R.,** Energy-linked nicotinamide adenine dinucleotide transhydrogenase in membrane particles from *Escherichia coli, Biochim. Biophys. Acta,* 245, 34, 1971.
168. **Hanson, R. L. and Gerolimatos, B.,** *E. coli* pyridine nucleotide transhydrogenase, purification and regulation, *J. Supramolec. Struct.*, Suppl. 1, 125, 1977.
169. **McFadden, B. J. and Fisher, R. R.,** Resolution and reconstitution of *Rhodospirillum rubrum* pyridine nucleotide transhydrogenase: localization of substrate binding sites, *Arch. Biochem. Biophys.*, 190, 820, 1978.
170. **Rydstrom, J.,** Energy-linked nicotinamide nucleotide transhydrogenases, *Biochim. Biophys. Acta,* 463, 155, 1977.
171. **Cox, G. B., Newton, N. A., Butlin, J. D., and Gibson, F.,** The energy-linked transhydrogenase reaction in respiratory mutants of *Eschericha coli* K12, *Biochem. J.*, 125, 489, 1971.
172. **Bragg, P. D. and Hou, C.,** Reconstitution of energy-dependent transhydrogenase in ATPase-negative mutants of *Escherichia coli, Biochem. Biophys. Res. Commun.*, 50, 729, 1973.
173. **Zahl, K. J., Rose, C., and Hanson, R. L.,** Isolation and partial characterization of a mutant of *Escherichia coli* lacking pyridine nucleotide transhydrogenase, *Arch. Biochem. Biophys.*, 190, 598, 1978.
174. **Macnab, R. M.,** Bacterial motility and chemotaxis: the molecular biology of a behavioral system, Crit. Rev. Biochem., 5, 291, 1978.
175. **Szmelcman, S. and Adler, J.,** Change in membrane potential during bacterial chemotaxis, *Proc. Natl. Acad. Sci. U.S.A.*, 73, 4387, 1976.
176. **DeJong, M. H. and van der Drift, C.,** Control of the chemotactic behavior of *Bacillus subtilis* cells, *Arch. Microbiol.*, 116, 1. 1978.

177. **Haury, J. F. and Wolk, C. P.,** Classes of *Anabeana variabilis* mutants with oxygen-sensitive nitrogenase activity, *J. Bacteriol.,* 136, 688, 1978.
178. **Oppenheim, J., Fisher, R. J., Wilson, P. W., and Marcus, L.,** Properties of a soluble nitrogenase in *Azotobacter, J. Bacteriol.,* 101, 292, 1970.
179. **Drozd, J. W., Tobb, R. S., and Postgate, J. R.,** A chemostat study of the effect of fixed nitrogen sources on nitrogen fixation, membranes, and free amino acids in *Azotobacter chroococcum, J. Gen. Microbiol.,* 73, 221, 1972.
180. **Yates, M. G. and Jones, C. W.,** Respiration and nitrogen fixation in *Azotobacter, Adv. Microbial Physiol.,* 11, 97, 1974.
181. **Jones, C. W., Brice, J. M., Wright, V., and Ackrell, P. A. C.,** Respiratory protection of nitrogenase in *Azotobacter vinelandii, FEBS Lett.,* 29, 77, 1973.
182. **Crane, P. K.,** Hypothesis for mechanism of intestinal active transport of sugars, *Fed. Proc.* Fed. Am. Soc. Exp. Biol., 21, 891, 1962.
183. **Misra, P. C. and Hofer, M.,** An energy-linked proton extrusion across the cell membrane of *Rhodotorula gracilis, FEBS Lett.,* 52, 95, 1975.
184. **Hauer, R. and Hofer, M.,** Evidence for interactions between the energy-dependent transport of sugars and the membrane potential in the yeast *Rhodotorula gracilis (Rhodosporidium toruloides), J. Membrane Biol.,* 43, 335, 1978.
185. **Hofer, M. and Misra, P. C.,** Evidence for a proton/sugar symport in the yeast *Rhodotorula gracilis (glutinis), Biochem. J.,* 172, 15, 1978.
186. **Cockburn, M., Earnshaw, P., and Eddy, A. A.,** The stoichiometry of the absorption of protons with phosphate and L-glutamate by yeasts of the genus *Saccharomyces, Biochem. J.,* 146, 705, 1975.
187. **Seaston, A., Carr, G., and Eddy, A. A.,** The concentration of glycine by preparations of the yeast *Saccharomyces carlsbergensis* depleted of adenosine triphosphate: effects of proton gradients and uncoupling agents, *Biochem. J.,* 154, 669, 1976.
188. **Komor, E. and Tanner, W.,** The determination of the membrane potential of *Chlorella vulgaris,* evidence for electrogenic sugar transport, *Eur. J. Biochem.,* 70, 197, 1976.
189. **Midgley, M.,** The transport of α-aminoisobutyrate into *Crithidia fasiculata, Biochem. J.,* 174, 191, 1978.
190. **Slayman, C. L.,** Electrical properties of *Neurospora crassa:* effects of external cations on the intracellular potential, *J. Gen. Physiol.,* 49, 69, 1965.
191. **Slayman, C. L.,** Electrical properties of *Neurospora crassa:* respiration and the intracellular potential, *J. Gen. Physiol.,* 49, 93, 1965.
192. **Slayman, C. L. and Slayman, C. W.,** Net uptake of potassium in *Neurospora:* exchange for sodium and hydrogen ions, *J. Gen. Physiol.,* 52, 424, 1968.
193. **Slayman, C. L., Long, W. S., and Lu, C. Y.,-H.,** The relationship between ATP and an electrogenic pump in the plasma membrane of *Neurospora crassa, J. Membrane Biol.,* 14, 305, 1973.
194. **Johnson, R. G., and Scarpa, A.,** Internal pH of isolated chromaffin vesicles, *J. Biol. Chem.,* 251, 2189, 1976.
195. **Casey, R. P., Njus, D., Radda, G. K., and Sehr, P. A.,** ATP-Evoked release of catecholamine by chromaffin granules: osmotic lysis as a consequence of H^+-translocation, *Biochem. J.,* 158, 583, 1976.
196. **Flatmark, T. and Ingebretsen, O. C.,** ATP-Dependent proton translocation in released chromaffin granule ghosts, *FEBS Lett.,* 78, 53, 1977.
197. **Pollard, H. B., Shindo, S., Creutz, C. E., Pazoles, C. J., and Cohen, J. S.,** Internal pH and state of ATP in adrenergic chromaffin granules determined by ^{31}P nuclear magnetic resonance spectroscopy, *J. Biol. Chem.,* 254, 1170, 1979.
198. **Pollard, H. B., Zinder, O., Hoffman, P. G., and Nikodejevic, O.,** Regulation of the transmembrane potential of isolated chromaffin granules by ATP, ATP analogs, and external pH, *J. Biol. Chem.,* 251, 4544, 1976.
199. **Johnson, R. G., Carlson, N. J., and Scarpa, A.,** ΔpH and catecholamine distribution in isolated chromaffin granules, *J. Biol. Chem.,* 253, 1512, 1978.
200. **Schuldiner, S., Fishkes, H., and Kanner, B. I.,** Role of a transmembrane pH gradient in epinephrine transport by chromaffin granule membrane vesicles, *Proc. Natl. Acad. Sci. USA,* 75, 3716, 1978.
201. **Holz, R. W.,** Evidence that catecholamine transport into chromaffin vesicles is coupled to vesicle membrane potential, *Proc. Natl. Acad. Sci. U.S.A.,* 75, 5190, 1978.
202. **Daniele, R. P. and Holian, S. K.,** A potassium ionophore (valinomycin) inhibits lymphocyte proliferation by its effects on the cell membrane, *Proc. Natl. Acad. Sci. USA,* 73, 3599, 1976.
203. **Deutsch, C. J., Holian, A., Holian, S. K., Daniele, R. P., and Wilson, D. F.,** Transmembrane electrical potential and pH gradients across human erythrocytes and human peripheral lymphocytes, *J. Gen. Physiol.,* 99, 79, 1979.

204. **Laris, P. C., Pershadsingh, H. A., and Johnstone, R. M.,** Monitoring membrane potentials in Ehrlich ascites tumor cells by means of a fluorescent dye, *Biochim. Biophys. Acta,* 436, 475, 1976.
205. **Grollman, E. F., Lee, G., Ambesi-Impiombato, F. S., Meldolesi, M. F., Alaj, S. M., Coon, H. G., Kaback, H. R., and Kohn, L. D.,** Effects of thyrotropin on the thyroid cell membrane: hyperpolarization induced by hormone-receptor interaction, *Proc. Natl. Acad. Sci. U.S.A.,* 74, 2352, 1977.

Chapter 3

IMMUNOLOGY OF THE BACTERIAL MEMBRANE

Peter Owen*

TABLE OF CONTENTS

* Abbreviations used: ADP, adenosine 5′-diphosphate; ATP, adenosine 5′-triphosphate; BANA, α-*N*-benzoyl-DL-arginine-β-napthylamide•HC1; CIE, crossed immunoelectrophoresis; deamino NADH, nicotinamide hypoxanthine dinucleotide, reduced form; EDTA, ethylenediaminetetraacetic acid; FMA, fast moving antigen; ID, immunodiffusion; LPS, lipopolysaccharide; NAD(H), nicotinamide adenine dinucleotide (reduced form); NADP(H), nicotinamide adenine dinucleotide phosphate (reduced form); Q-1, ubiquinone-1; SDH, succinate dehydrogenase; SDS, sodium dodecyl sulphate; TNBT, 2,2′,5,5′-tetra-*p*-nitrophenyl-3-,3′-(3,3′-dimethoxy-4,4′-diphenylene) ditetrazolium chloride; Tris, *tris* (hydroxymethyl) methylamine; UDP-galactose, uridine-5′-diphosphogalactose.

I. INTRODUCTION

The origins of modern immunochemistry can be traced back to 1929 and to the pioneering studies of Michael Heidelberger and his colleagues.[1,2] By applying the precise methods of quantitative analytical chemistry to the measurement of antigens and antibodies, Heidelberger and Kendall[3] established that precipitation of an antigen by its corresponding antibody could be treated as an ordinary chemical reaction, albeit a somewhat complex one. In the 1930s, largely through the work of Landsteiner, Avery, Goebel and others, the chemical basis underlying the specificity of serological reactions was firmly established, viz., that small regions or entities in macromolecules could act as antigenic determinants.[4,5] These developments allowed the structure of complex molecules to be probed in a surprisingly simple fashion by assessment of precipitation reactions conducted against a well selected stock of antisera.[6,7] Some of the first substances to be studied in this era were of bacterial origin. Notable examples are the pneumococcal capsular polysaccharides and the lipopolysaccharides (LPS) of *Salmonella*.[6,7] Our present detailed understanding of the structure and immunochemistry of these and other bacterial cell surface components is also due in large part to the more recent yet equally significant immunochemical and biochemical investigations of Westphal, Lüderitz, Staub and their colleagues.[8-11]

Important as these studies are, it should be noted that they are restricted largely to the major carbohydrate antigens of the cell surface and/or to components that are readily extracted and isolated from intact bacteria. Unfortunately, most components of the cytoplasmic (plasma) membrane do not fall into this category. As a consequence, detailed immunochemical analysis of bacterial plasma membranes and their components has had to await the introduction of techniques designed to remove or dissolve the rigid cell wall of Gram-positive bacteria,[12-16] and, in the case of Gram-negative microorganisms, the introduction of methods that allow separation of the inner and outer membrane systems.[17-19] Moreover, classical immunological methods proved to be of limited value in the analysis of particulate multicomponent systems such as membranes,[20] and use had to be made of gel precipitation techniques[21-24] in order to resolve individual membrane antigens.

The results of the first immunological studies of isolated microbial membranes appeared just over a decade ago and were not overly illuminating.[15,25-27] The agar gel precipitation methods in routine use at the time, viz., immunodiffusion[21,22] and immunoelectrophoresis[23] resolved only a handful of antigens for solubilized plasma membrane preparations.[25-27] Clearly, this neither reflects the known functional complexity of the plasma membrane[14-16,28-35] nor parallels the multiplicity of polypeptides resolved when similar membrane preparations are analysed by other techniques such as sodium dodecyl sulfate (SDS)-polyacrylamide gel electrophoresis.[36-38] Several possible reasons were proposed at the time to explain this apparent paradox. These included the poor immunogenic properties of membrane proteins,[25] possession by various membrane components of common antigenic determinants,[27,39,40] and possible denaturation of membrane antigens during solubilization by detergents.[41] Although none of these possibilities can be ruled out totally, it is now evident that the small number of immunoprecipitates detected in these early studies was largely a consequence of the practical limitations involved in clearly resolving a large number of precipitin bands in several square millimeters of agarose gel.

Two recent technical advances have served to popularize bacterial cell-surface immunology. The first was the introduction by Singer and his colleagues[42-44] of the ferritin-labeled antibody technique. This method allows the cellular location of antigens to be determined with precision, and has greatly advanced our understanding of cell surface architecture. Second, and perhaps more important, was the development of the two-dimensional (crossed) immunoelectrophoresis technique.[24,25] The pioneering work in this field was performed by Laurell[24,46-50] in 1965, and subsequent refinements and modifications of the method were largely the work of his colleagues in Scandinavian countries.[51-54] Initially, crossed immunoelectrophoresis was employed in the analysis of serum proteins[55] and it was not until the mid 1970s that its potential in the study of cell surface antigens began to be exploited.[56-58] It has since become evident that crossed immunoelectrophoresis, or CIE as it is now popularly known, is the immunological method of choice for analysis of bacterial membranes. Not only does the technique resolve the antigenic complexity of cell membranes with startling clarity but membrane antigens analyzed by this method usually retain sufficient intrinsic biological activity to permit functional characterization.[59-61] This contrasts markedly with other high resolution systems currently available for the analysis of membrane proteins, viz., one and two dimensional SDS-polyacrylamide gel electrophoresis.[62,63] Here, the intrinsic biological activity associated with individual polypeptides is usually destroyed by the denaturing effects of the anionic detergent.[64] An additional compelling feature of CIE relates to the quantitative nature of the method,[45,49,65] and to the ease and sensitivity with which removal of antibodies can be monitored during adsorption experiments designed to probe the antigenic architecture of membranes and surfaces. These aspects were first demonstrated by Johansson and Hjertén[57] in their elegant structural studies of mycoplasma cell membranes.

The principal aims of this review are to introduce in some detail the relatively new technique of CIE and to review the manner in which both CIE and the ferritin-labeled antibody methods have contributed to our current understanding of the antigenic composition and molecular structure of bacterial membranes. The scientific literature concerning the methodology of CIE is unneccessarily large, and is cluttered with many reports of refinements and modifications of, frankly, limited use. Consequently, no attempt is made to cover these in detail, and the interested reader is directed to several recent monographs[50-52] and reviews.[53,54] Instead, concentration will be placed on a few selected modifications of CIE which have regularly proved useful in the study of bacterial membranes.

Also included is a section dealing with aspects of the immunology and immunochemistry of several plasma membrane and outer membrane components, and of exported proteins. This particular section is restricted to coverage of immunogenic proteins and lipids, since the serology and chemistry of the main carbohydrate antigens of the cell surface, i.e., O-antigens (lipopolysaccharide) and polysaccharide K-antigens of Gram-negative bacteria, and lipoteichoic acids of Gram-positive organisms, have been the subject of many recent and comprehensive articles.[9-11,66-76] Polysaccharide and protein antigens of the Gram-positive cell wall and capsule[77-85] are also determined to be outside the scope of this review, as are flagellar (H-antigens) and fimbrial antigens.[69,86]

II. CROSSED IMMUNOELECTROPHORESIS

A. Principles

Crossed immunoelectrophoresis is a two dimensional technique. Initially, the antigen sample is subjected to electrophoresis in an agarose gel free of antibody. Electrophoresis is then continued in the second dimension into an agarose gel containing uniformally dispersed immunoglobulins. The whole operation is usually conducted at a pH (about 8.6) at which the antibody molecules remain virtually stationary or move slowly towards the cathode. Fortunately, under these conditions most membrane components migrate anodally and thus interact with their homologous antibody. Soluble immune complexes are formed at first, since antigen molecules are in excess. Further amounts of specific antibody bind on continued migration, yielding eventually an insoluble immunoprecipitate that is usually stable to continued electrophoresis and is unaffected by traverse of antibodies or of other antigens.[55,87] The overall effect for a complex mixture of antigens is a well-resolved two dimensional array of immunoprecipitates (Figure 1). The antigenic relationship between the various constituent immunogens, viz., total identity, partial identity, or nonidentity, may be established from a consideration of whether the corresponding immunoprecipitates fuse, partially fuse, or intersect, respectively. However, it should be borne in mind that, as a result of electrophoresis in the first direction, pertinent antigens may be separated to such an extent that it is impossible to discern any interaction between resultant immunoprecipitates.[88] The reader is referred to the well-illustrated texts by Axelsen and others[89-98] for a full discussion of the various reactions observed between antigens in the CIE system.

The area subtended by each immunoprecipitate arc is related to the concentration of antigen and antibody as follows:[49,55,65,87,99,100]

$$A = \frac{k_1 c}{B} \tag{1}$$

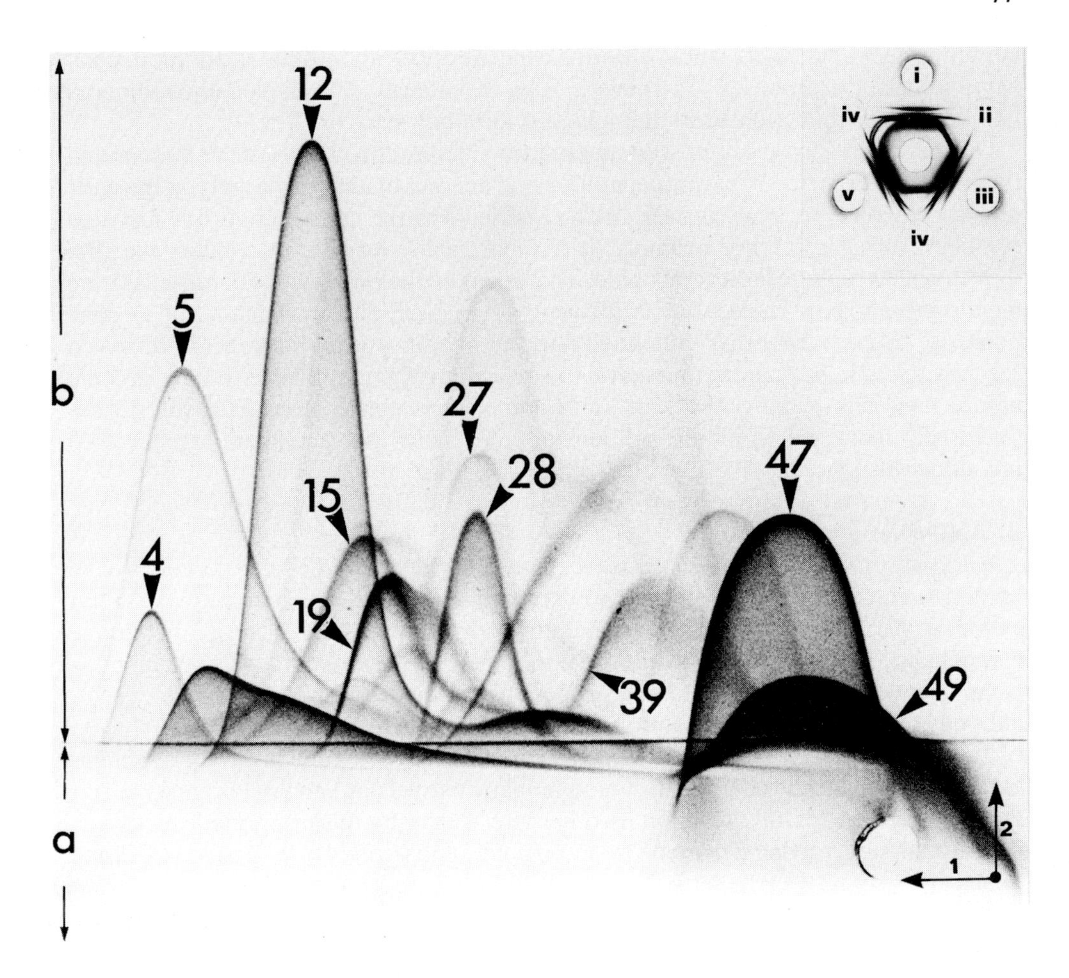

FIGURE 1. Crossed immunoelectrophoresis of a Triton® X-100-EDTA extract of membrane vesicles prepared from *E. coli* ML308-225. 67 μg of protein were subjected to electrophoresis into an antibody-free agarose gel (a) as indicated. Electrophoresis in the second direction was conducted into a gel (b) containing antivesicle immunoglobulin. Some of the salient immoprecipitates are identified numerically in order to allow comparison with the complete reference profile (*cf.* Figure 6) and various other Figures in the text. Many of the resolved antigens have been identified or partially characterized and details can be found in Table 1. The identity of some antigens, notably number 47, is still uncertain. The anode is to the left and top of the gel. The inset shows similar membrane extracts analyzed by immunodiffusion. Triton® X-100-EDTA extracts of *E. coli* ML308-225 membrane vesicles (111 μg; wells i, iii, and v), together with purified or partially purified preparations of Braun's lipoprotein (10 μg; well ii), lipopolysaccharide (10 μg; well iv) and matrix protein (36 μg; well vi) were tested against antivesicle immunoglobulins (central well). The clarity with which the vesicle antigens are resolved by CIE should be compared with the relatively poor resolution achieved by immunodiffusion. The immunoprecipitates resolved for preparations of matrix protein (well vi) have been shown by a variety of methods to result from contaminating lipoprotein and lipopolysaccharide. (Reprinted with permission from Owen, P. and Kaback, H. R., *Biochemistry,* 18, 1413, 1979. Copyright by the American Chemical Society.)

where A is the peak area subtended by immunoprecipitate, i, C is the amount of antigen i analysed, B is the amount of anti-i immunoglobulins present in the serum preparation, and k_1 is an area-loading constant that is dependent largely on the antigenic nature of i. This equation, which in simple language states that peak areas are proportional to corresponding antigen/antibody ratios, is extremely important and basic to an understanding of many aspects of CIE. For example, it explains why the CIE profile for complex antigen mixtures invariably contains immunoprecipitates of widely differing height (i.e., area) and/or intensity, since it would be anticipated that the individual immunogens would be present in differing concentrations and would, in addition, elicit varying antibody responses. As a general rule, the intensity of a given immunoprecipitate is a direct reflection of its homologous antibody titer.

One per cent agarose gels possess an exclusion limit for proteins of $\geq 10^8$ daltons and thus might be expected to permit separation and analysis of antigens largely on the basis of charge. However, analysis need not be restricted to the use of supporting matrices composed of agarose. For example, it is theoretically feasible to conduct the first direction electrophoresis in acrylamide. This opens up the possibilities of preseparating membrane antigens by isoelectric focusing or by SDS-polyacrylamide gel electrophoresis. These and other potentially useful modifications have been reviewed recently,[61] and it is apparent that certain technical problems still have to be overcome before they gain wider acceptance. This comment is particularly applicable to the potentially powerful combination technique of SDS-polyacrylamide gel-crossed immunoelectrophoresis.[101-103]

B. Antibodies and Antigens

Antibody production is not a precise science. Consequently, the remarks here will be limited to a few general comments. The host animals used most frequently are the rabbit and goat, and both appear to give good antibody titers against bacterial membrane components. However, the use of horse antibodies is to be discouraged since immune complexes involving horse γ-globulins appear to exhibit greater solubility in antigen or antibody excess.[55,61] A note of caution should also be introduced with respect to the use of rabbit immunoglobulins. Normal (preimmune) rabbit serum invariably contains antibodies to an acidic component found in many extracts of bacterial origin.[104-108] It is necessary to test for this reaction as otherwise it may be misinterpreted, especially during analysis involving the use of "monospecific" antiserum[108] (see Section II.C.2 and Figure 4).

CIE consumes relatively large amounts of antiserum. One standard 5 cm × 5 cm test plate usually requires about 0.5 to 1.0 m*l* of serum (or the equivalent volume of concentrated, purified IgG). This is roughly 100 times more than that commonly used in immunodiffusion tests, and is a consequence of having antibody dispersed evenly throughout the whole second dimension gel. Thus, for comprehensive studies, a pool of antiserum approaching a liter in volume is often needed. Acquiring a 1-*l* pool of rabbit serum is neither an inexpensive nor rapid exercise. It is also more appropriate to work from one large serum pool rather than from several smaller ones, since antibody titers can vary dramatically over the course of a prolonged immunization schedule. It is apparent from these practical considerations alone that some thought must be given to the nature of the initial immunogen. In addition, consideration should be given to the possibility that immunization with isolated membranes may not result in a detectable antibody response to important immunogens which are present in the membrane in relatively low concentrations. This appears to be the case for the penicillin-binding components of *Escherichia coli*[109] and may also explain the inability to detect by immunological procedures the *lac*-carrier protein in membrane vesicles from the same organism.[108] In instances such as these, it may be necessary either to amplify genetically the gene product in question or to relieve the immunological burden on the host animal by immunizing with a membrane fraction depleted or stripped of other superfluous antigens. This latter technique has been used to generate an antibody preparation which appears to interact with components of the D-glucose transport system of rat adipocyte plasma membranes.[110] There are also some indications that immunization with detergent-solubilized preparations enhances the antibody response to hydrophobic membrane proteins.[111]

A discussion of γ-globulin production would not be complete without a brief discussion of an extremely useful method for raising monospecific antisera. This technique involves immunization with discrete regions of immunoprecipitates cut from wet CIE immunoplates.[112,113] The excised region should be free of intersecting immunoprecipitates, a situation which can often be achieved by partial purification of the test antigen (Figure 9A). Immunization with excised immunoprecipitates is simple, extremely effective,[107] and would appear to be an ideal technique for raising monospecific antisera to membrane proteins which might be refractory to extensive purification.[114,115]

Membrane preparations are particulate in nature and thus require that they be solubilized prior to CIE analysis. Ideally, one should strive to achieve maximal solubilization with full retention of both immunological specificity and enzymic (or other biological) activity. This is not always an easy task. The choice of detergents and other methods for solubilization of membrane proteins, together with interference problems in protein determination, have been discussed in relation to immunological analysis elsewhere.[54,58,61] Suffice it to say that Triton® X-100 and Tween® 20 have proved to be the most popular detergents for CIE analysis of microbial membranes, although some workers have experimented with charged detergents such as cetyltrimethylammonium bromide[108] or sodium deoxycholate.[106,108,116-121] Sodium deoxycholate is claimed to give optimal resolution of mycoplasma membrane proteins.[116-121] It is important to remember, however, that amphiphilic membrane proteins, in contrast to hydrophilic proteins, bind significant amounts of detergents.[64] Consequently, their electrophoretic mobilities can be altered in a bidirectional manner if electrophoresis is conducted in the presence of a cationic detergent, such as cetyltrimethylammonium bromide, or an anionic detergent, such as deoxycholate.[122] This is known as a charge-shift effect,[122] and while it can undoubtedly be utilized to improve resolution of some membrane antigens,[121] it can be more usefully employed to determine which membrane antigens are amphiphilic and which are hydrophilic[108,122,123] (Figure 2).

Proteins of the outer membrane of Gram-negative bacteria are more difficult to solubilize with detergents than those of the inner plasma membrane.[124-127] Accordingly, it is often advisable to incorporate chelating agents such as ethylenediaminetetraacetic acid (EDTA) into extraction buffers for optimal solubilization of outer membrane components.[108,125,128] Chaotropic agents such as thiocyanates have also proved useful in the selective extraction of some gonococcal envelope antigens.[129]

Most of the extraction procedures outlined above result in only partial solubilization of the membrane. Indeed, in many instances less than 50% of the total membrane protein is amenable to extraction and analysis.[106,115,129,130] This is a serious problem, and there have been few systematic attempts to resolve it. Repeated extraction with the same detergent may be a solution in some,[128] but clearly not all, instances.[106,130] However, recent studies with *Micrococcus lysodeikticus* have shown that the Triton® X-100-insoluble membrane residue from this organism can be dispersed by controlled levels of SDS into a form which is amenable to analysis by CIE.[130] This method, which may have wide applicability, is discussed more fully in Section III.B.

C. Identification of Membrane Antigens

Crossed immunoelectrophoresis would be a technique of limited value if the resolved antigens could not be readily identified. Fortunately, many simple procedures have been devised for this purpose. By far the most useful, if the antigen in question happens to be an enzyme, is the technique of zymogram staining.[59-61] In the following paragraphs, some aspects of this approach and of three other popular methods for identifying immunoprecipitates in CIE reference profiles will be discussed. Other techniques are

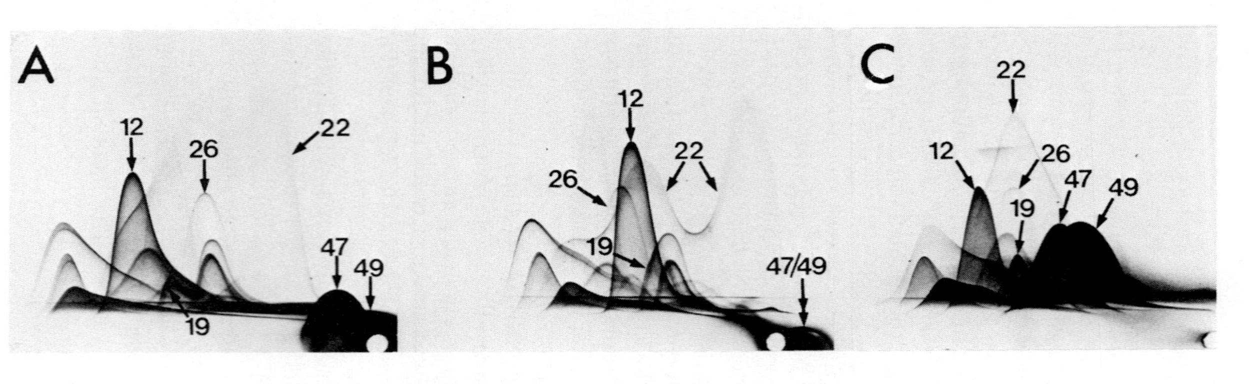

FIGURE 2. Characterization of membrane antigens by charge-shift CIE. A Triton® X-100-EDTA extract of membrane vesicles derived from *E. coli* ML308-225 (59 μg of protein) was electrophoresed in the first direction into an agarose gel containing Triton® X-100 alone (A), Triton® X-100 supplemented with 0.05% cetyltrimethylammonium bromide (B), or Triton® X-100 supplemented with 0.25% sodium deoxycholate (C). Electrophoresis in the second direction was conducted into gels containing antivesicle immunoglobulin and Triton® X-100 as sole detergent. Note that antigen numbers 26, 47, and 49 show obvious charge shifts in the presence of ionic detergents, while the electrophoretic mobilities of antigen numbers 12 and 19 (for example) are affected by cetyltrimethylammonium bromide only. Pronounced heterogeneity of immunoprecipitate number 22 in the presence of the cationic detergent is also evident (B). The diffuse staining reaction at the interface between the first dimension gel and the antibody gel in C is caused by deoxycholate-induced precipitation of serum protein, and impairs resolution of some antigens in that region of the gel. Antigens have been numbered in accordance with the reference profile shown in Figure 6. and the identities of those which display pronounced bidirectional charge shifts are indicated in Table 1. The anode is to the left and top of all gels. (Reprinted with permission from Owen, P. and Kaback, H. R., *Biochemistry,* 18, 1413, 1979. Copyright by the American Chemical Society.)

available, of course, and details of these can be found in several recent reviews.[51,52,54,58,61] In addition, the molecular weights of antigenic polypeptides can be readily obtained by analysis on SDS-polyacrylamide gels of excised regions of the corresponding immunoprecipitates.[131] In these instances, it is advisable to use radiolabeled antigen preparations.

1. Use of Purified Antigens

If one is fortunate enough to have access to a purified membrane antigen, identification can be established by coelectrophoresis with a standard preparation of solubilized membranes. It will be recalled that the peak areas of immunoprecipitates are directly (and solely) related to the antigen concentration if the amount of γ-globulin added to the second dimension gel is held constant (Equation 1). Consequently, the antigen in question may be identified since coelectrophoresis will cause a specific increase in the peak area of one immunoprecipitate compared to a reference control immunoplate.[108,120,128,132,133] This technique is illustrated in Figure 3 where it is used to identify Braun's lipoprotein[134] in the CIE profile of solubilized membrane preparations from *Escherichia coli.*[108]

A related method involves CIE of both purified sample and crude antigen mixture simultaneously from different antigen wells placed adjacently in such a way that diffusion for a period prior to electrophoresis permits the observation of reactions of identity in the final immunoprecipitin pattern. This method is termed tandem crossed immunoelectrophoresis[135] and has been used to effect the identification of the lipoyl dehydrogenase antigen (EC 1.6.4.3) in membrane vesicles of *E. coli* ML308-225 (Figure 12) and also to monitor D-lactate dehydrogenase (EC 1.1.1.27) in various mutants of the same organism (Figure 13). Tandem CIE is also extremely useful in demonstrating immunochemical relationships between antigens in different membrane preparations.[108]

2. Intermediate Gel Techniques

In intermediate gel techniques, an additional agarose strip is positioned between the preelectrophoresed antigen and the reference antibody gel.[93] Into this intermediate gel may be incorporated a wide range of substances with which one or more of the test antigens might be expected to interact. For example, use of an intermediate gel containing antiserum specific to one membrane antigen will result in partial or complete depression of the corresponding immunoprecipitate from the reference gel into the intermediate gel. In this manner, the immunoprecipitate corresponding to Braun's lipoprotein was clearly identified in CIE profiles of *E. coli* membranes.[108,128] CIE with intermediate gel, as this technique is aptly termed, is illustrated in Figure 4.

Intermediate gel techniques are not limited to the use of antiserum. Lectins, for example, may be used as affinity adsorbants in a variation of CIE known as crossed immunoaffinoelectrophoresis.[136] During electrophoresis in the second direction, polysaccharides and glycoproteins, which possess sugar residues in conformations allowing interaction with the lectin, form affinity precipitates in the intermediate gel and are generally recognized by loss or depression of corresponding immunoprecipitates from the reference antibody gel. The use of a panel of lectins with widely differing sugar specificities readily allows glycoprotein and polysaccharide antigens to be identified and partially characterized. This approach has been used successfully in the analysis of plasma membranes from *Micrococcus lysodeikticus* (see Section III.B)[88,132] and from *Acholeplasma laidlawii.*[137] It is also evident from four recent publications[138-141] that some antigens, notably the lipoteichoic acid of *Streptococcus faecalis*[140] and the

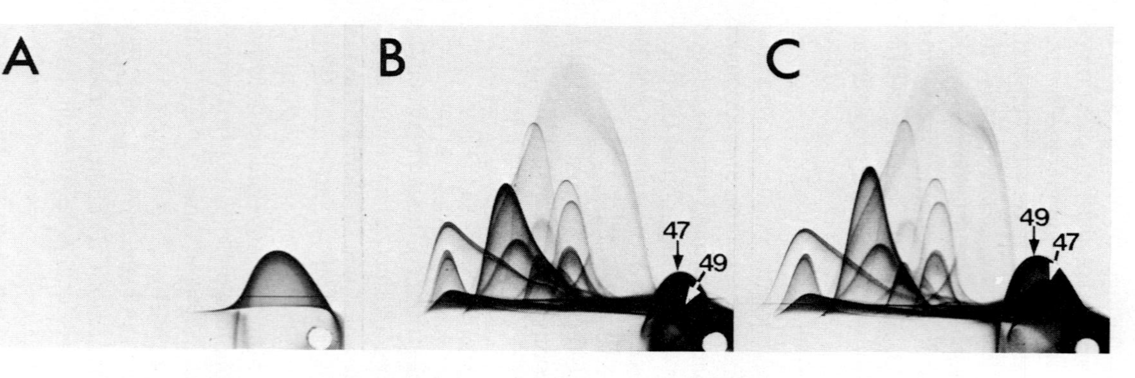

FIGURE 3. Identification by co-CIE of the immunoprecipitate corresponding to Braun's lipoprotein. A Triton® X-100-EDTA extract of membrane vesicles from *E. coli* ML308-225 (66 μg of protein) was electrophoresed in the CIE system with (C) or without (B) purified lipoprotein (5 μg). Lipoprotein alone (5 μg) was electrophoresed for comparison (A). For all immunoplates shown, electrophoresis in the second direction was conducted into gels containing antivesicle immunoglobulin. Only the area of immunoprecipitate number 49 increases following co-CIE with lipoprotein. (Unfortunately, the photographic process does not reproduce the clear definition between intersecting antigens numbers 47 and 49 that is evident in the original stained immunoplates). Immunoprecipitates have been numbered in accordance with the reference profile shown in Figure 6. The anode is to the left and top of all gels. (Reprinted with permission from Owen, P. and Kaback, H. R., *Biochemistry*, 18, 1413, 1979. Copyright by the American Chemical Society.)

lipomannan from *M. lysodeikticus*[138] may be conveniently analyzed and quantitated by direct interaction with lectins in an antibody-free variation of CIE known as crossed affinoelectrophoresis.[140,141]

3. Zymograms

One of the principle attractions of CIE is the fact that antigens generally appear to retain and express at least part of their intrinsic biological activity after interaction with antibody. Thus, it is usually possible to detect enzyme-containing immunoprecipitates by simple assay for the relevant catalytic activities. In practice, this can be achieved most conveniently by immersion of wet immunoplates in standard histochemical stains.[61,88,106-108,115,128,132] This simple method is surprisingly effective and has been responsible for detection of 14 different enzyme antigens in membrane preparations from *E. coli* alone[108,128] (see Section III.A). Some examples of enzyme-stained immunoplates or zymograms can be found in Figures 7, 10-13. The fact that enzyme antigens can be identified in this manner is often greeted with amazement if not scepticism. There appears to be a common misconception that most enzymes are totally inhibited by homologous antibody. This is not always the case, even in tube precipitin tests.[142] Furthermore, Harboe and Ingild have reported that the amount of antibody involved in immunoprecipitation in quantative immunoelectrophoresis is between two- and tenfold less than the equivalent amount of antibody required in some precipitin tests.[143] Thus, enzyme-containing immunoprecipitates in CIE may retain catalytic potential due partly to antigen excess.[61] In addition, immunoglobulins directed against determinants in the vicinity of active sites may not be the major antibody species involved in formation of CIE immunoprecipitates.[61] It should also be realized that enzyme antigens need only express a small percentage of their full catalytic potential in order to be detected by the very sensitive amplification methods of enzyme histochemistry. It is not intended to review here the theory and practice of enzyme staining reactions, since these aspects have been covered adequately elsewhere.[55,59-61,144,145] However, the Appendix lists zymogram stains which have proven useful in the author's laboratory for identification of enzyme antigens. In addition, the reader may find the recent comprehensive texts by Harris and Hopkinson[146,147] particularly useful in devising histochemical stains for new or relatively obscure enzymes.

Once the identity of an enzyme antigen has been established, it is then possible to probe some of its biochemical properties, such as its response to various metabolic inhibitors or its substrate and acceptor specificities, by monitoring relative rates of zymogram development. Although obviously a semiquantitative technique, this procedure has, on two different occasions, clearly differentiated two reduced nicotinamide adenine dinucleotide (NADH) dehydrogenases (EC 1.6.99.3) which were observed in a preparation of bacterial membranes[61,148] (see Sections III.A, III.B, and V.A.2). Analogous in vitro assays would not readily yield this degree of resolution.

Although enzyme staining is an extremely powerful technique, it does pose some problems with respect to interpretation. This is particularly true of immunoprecipitates which stain only faintly or slowly, since they may display apparent enzyme activity by virtue of nonspecific affinity for lipophilic substrates or coupling salts.[54] Another common problem is antigen entrapment.[60,61] Basically, this phenomenon results from nonspecific entrapment within part or all of a preformed immunoprecipitate of unrelated enzyme antigen molecules which, by reason of similar electrophoretic mobilities and differing antigen/antibody ratios, have to traverse the immunoprecipitate in question prior to formation of their own discrete precipitin arc. Artifacts of this type can often be recognized, since entrapment (and hence, staining) is restricted to those regions of immunoprecipitates which intersect the main enzyme-staining arc

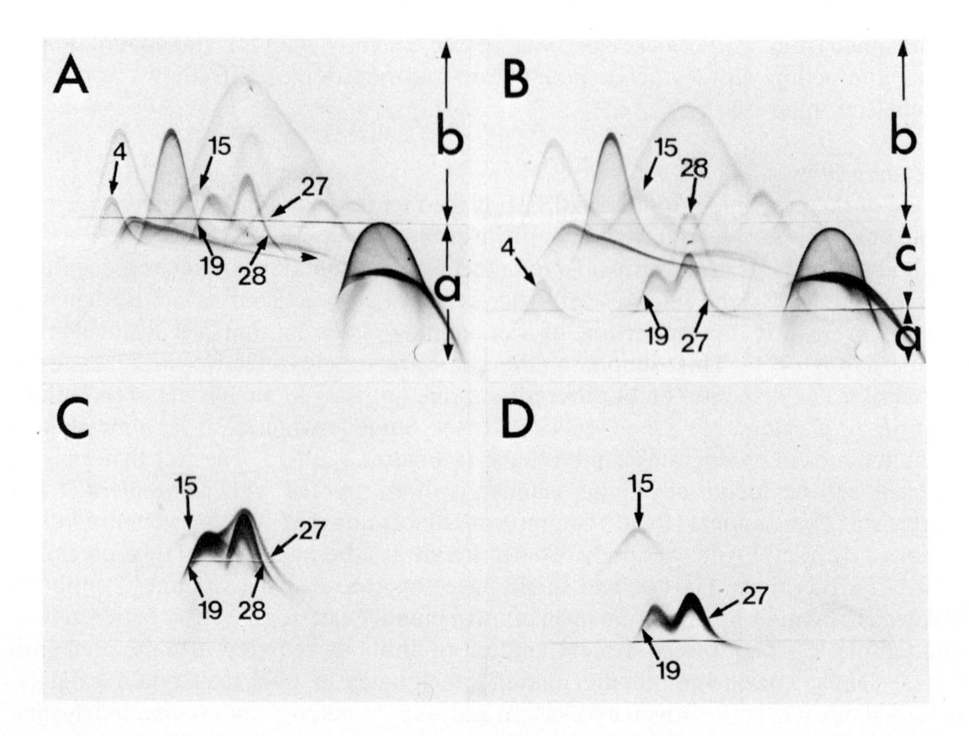

FIGURE 4. Resolution of NADH dehydrogenase-active immunoprecipitates by CIE with intermediate gel. Similar amounts (27 μg of protein) of a Triton® X-100-EDTA extract of *E. coli* ML308-225 membrane vesicles were analyzed in all instances. Gel regions marked a, b, and c contain agarose alone, antivesicle immunoglobulin, and antiserum to a partially purified membrane protein, respectively. Panel (A) represents the control immunoplate with no immunoglobulin in the intermediate gel, and (B) is the test immunoplate with serum in the intermediate gel. (C) and (D) represent immunoplates identical to those shown in (A) and (B), respectively, and have been stained by zymogram techniques for NADH dehydrogenase activity. At least four immunoprecipitates (numbers 15, 19/27, and 28) in (A) appear to stain for NADH dehydrogenase activity (C). However, in (B), antigen number 19/27 reacts strongly with antibody present in the intermediate gel, and the corresponding immunoprecipitate is depressed below number 28 which no longer stains for NADH dehydrogenase activity (D). This suggests that the activity demonstrated for antigen number 28 in (C) is an artifact of enzyme entrapment. Several other immunoprecipitates in (A) are depressed by the intermediate gel in (B). One of these (number 4) corresponds to an antigen which reacts with preimmune serum from a variety of sources. Antigens have been numbered in accordance with the reference profile shown in Figure 6. Anode is to the left and top of all gels.

(Figure 7). However, interpretation becomes more difficult if an enzyme-staining immunoprecipitate completely encloses another of apparently similar catalytic activity. One way to resolve this problem is to alter the relative positioning of the two precipitin arcs by CIE with intermediate gel (Figure 4).

The binding properties of some membrane antigens may also be used to aid in their identification. For example, the adrenaline receptors of rat liver plasma membranes can be detected in CIE by their ability to bind [^{14}C]-epinephrine.[149] The potential of this sort of approach to the identification of the various receptors and binding proteins of the bacterial envelope[34] should be obvious. Additionally, many important membrane metaloproteins[150-152] may be amenable to study following analysis of extracts obtained from cells grown in the presence of the appropriate isotope. Iron-containing antigens present in membranes of *E. coli*[153] and of *M. lysodeikticus* have been examined in this manner (see Sections III.A and III.B).

4. Immunoadsorption Procedures

It is not always possible to identify a particular immunogen by the procedures outlined in Sections II.C.1 to 3. Purified antigens and monospecific antisera may not be at hand, and the components under investigation may not be enzymes or receptors. In cases like these, use can sometimes be made of immunoadsorption procedures conducted with strains of microorganisms which specifically lack the gene product in question. This approach has been used recently in attempts to characterize the *lac*-carrier protein (the *lac* y gene product) in membrane vesicles derived from *E. coli.*[108] In these experiments, antiserum raised against membrane vesicles from *E. coli* ML308-225 ($i^- z^- y^+ a^+$) was adsorbed with increasing amounts of sonicated membrane vesicles derived from uninduced *E. coli* ML30 ($i^+ z^+ y^+ a^+$). Theoretically, this operation should result in progressive removal of antibodies to all vesicle immunogens except those coded for by the *lac* operon. It was then possible to analyze the adsorbed sera by CIE using constant amounts of antigen extract.[108] The results of these tests are depicted in Figure 5 (A-D), and it is notable that the areas of several immunoprecipitates are unaffected by adsorption. These must correspond to antigens which were unavailable to antibody during adsorption. (For a more detailed discussion of immunoadsorption see Section VI.A.) The obvious but possibly incorrect, interpretation is that they are all gene products of the *lac* operon. To be sure, the major immunoprecipitate is known to correspond to enzymatically inactive β-galactosidase[108] (EC 3.2.1.23; *E. coli* ML308-225 has a point mutation in the *lac* z gene). However, if antiserum is adsorbed instead with sonicated membrane vesicles derived from the homologous strain (*E. coli* ML308-225) and is then analyzed in an identical manner, it becomes apparent that several of the minor antigens are probably not products of the *lac* operon, since their corresponding immunoprecipitates behave in a similar fashion irrespective of whether adsorption is conducted with ML30 or ML308-225 vesicles (compare Figures 5 A-D with E-H). The reasons for this phenomenon may be various, e.g., poor expression of some antigens during adsorption, high antibody titers against minor antigenic components, or small variations in the antigenic composition of similar membrane preparations. In view of these considerations, it is advisable to treat identification based upon immunoadsorption experiments as tentative at best, and to confirm identity by other means.

III. ANALYSIS OF BACTERIAL MEMBRANES BY CROSSED IMMUNOELECTROPHORESIS

Several species and strains of microorganisms have been examined by quantitative immunoelectrophoresis. These include *Micrococcus,*[61,88,130,132,154-156] *Staphylococcus,*[157] *Bacillus,*[115,158] *Escherichia,*[108,128,153,159,160] *Salmonella,*[108] *Pseudomonas,*[104,161-165] *Proteus,*[166] *Serratia,*[167] *Bordetella,*[168,169] *Neisseria,*[61,106,107,129,170-172] *Haemophilus,*[173] *Rhodopseudomonas,*[174,174a] *Mycobacterium,*[97,175-181] *Chlamydia,*[182,183] mycoplasmas,[57,114,116-121,133,137,184-186] *Candida,*[187-191] *Actinomyces,*[192,193] and various other species of fungi.[194-200] The cross reactivities of *Pseudomonas,*[104] *Proteus,*[166] *Bordetella,*[169] and *Neisseria*[172] with many other bacterial genera have also been documented. CIE has also greatly revolutionized the analysis of the human antibody response to microbial infections. In this field, the major contributions have come from Axelsen and his colleagues[53,89,188,201-206] in studies of *Candida* infections and from Høiby and co-workers[105,162,207-213] in studies of *Ps. aeruginosa* infections in patients suffering from cystic fibrosis. Related studies have also been conducted to monitor the occurrence of precipitins in the sera of patients with respiratory tract infections,[211] whooping cough,[168] bacterial meningitis,[214] gonorrhoea,[215]

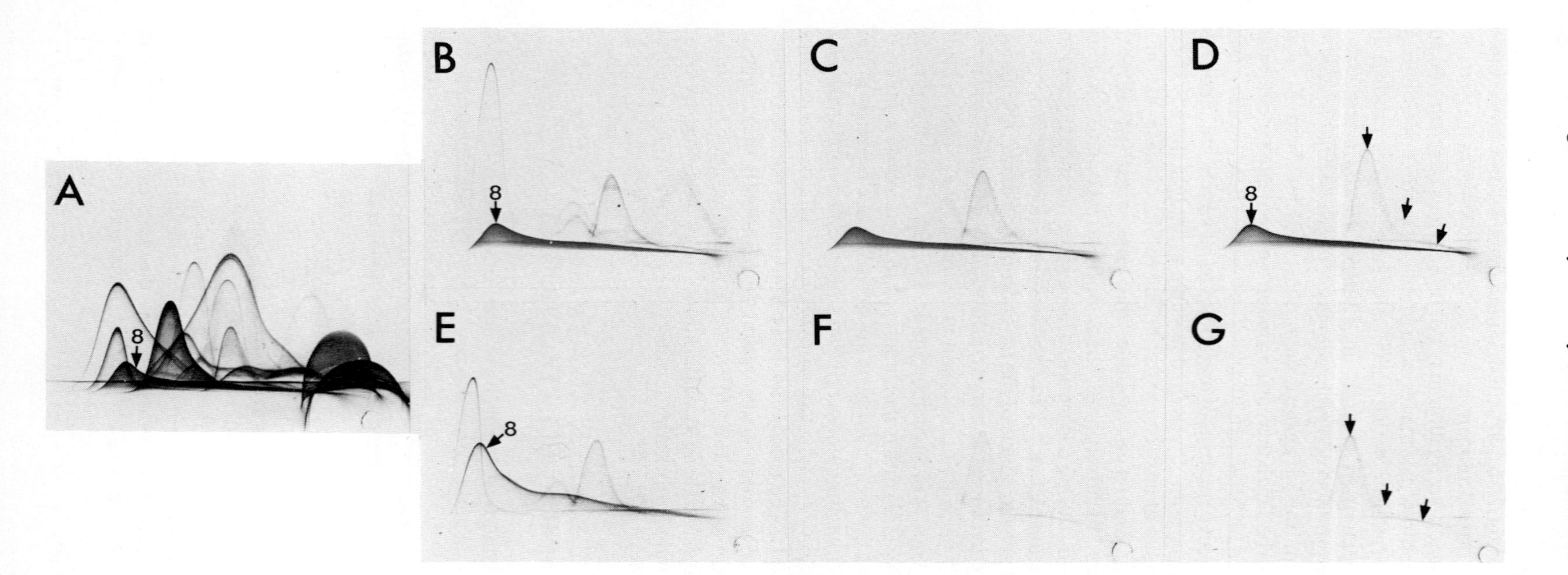

FIGURE 5. Presumptive identification of *lac* proteins by immunoadsorption procedures. Antiserum (63 mg of protein) to membrane vesicles from *E. coli* ML308-225 was adsorbed with 0 mg (A), 9.3 mg (B), 18.5 mg (C) and 27.8 mg (D) of sonicated ML30 membrane vesicle protein, or with 7.7 mg (E), 15.4 mg (F) and 23.1 mg (G) of sonicated ML308-225 membrane vesicle protein. Equivalent amounts of immunoglobulin fraction were incorporated into agarose gels and tested by CIE against a Triton® X-100-EDTA extract of membrane vesicles from *E. coli* ML308-225 (88 μg of protein). Note that the peak areas of most immunoprecipitates increase following adsorption with either ML30 or ML308-225 vesicles, thus indicating removal of antibodies to these antigens. However, the peak area of immunoprecipitate number 8 is differentially affected by adsorption with ML30 and ML308-225 vesicles. This precipitate corresponds to *lac* z (β-galactosidase). Note that the peak areas of several other immunoprecipitates (arrowed) appear to be virtually unaffected by adsorption with either type of vesicle preparation. It seems unlikely that these can correspond to *lac* proteins. Antigens have been numbered in accordance with the reference profile shown in Figure 6. Anode is to the left and top of all gels.

leprosy,[180,216-218] actinomycosis,[193] and chlamydial infections.[183] Unfortunately, in only a minority of these reports have defined membrane fractions been employed. In most intances, analysis has involved the use of whole cell lysates or sonicated cell suspensions, with few attempts[165,173,191,194,197] made to determine the nature or cellular origin of the numerous antigens detected. Thus, their relevance to the immunology of the membrane is unclear, at least at the present time—a situation, incidentally, which could be radically altered by a few simple experiments. The following remarks will thus be confined to CIE studies involving the use of isolated membrane preparations.

A. Membranes of *Escherichia coli*

The recent immunological investigations of Smyth et al.[128] and of Owen and Kaback[108,159,160] have provided a detailed picture of the antigenic composition of the membrane from this organism. In the former study, isolated inner and outer membranes from *E. coli* K12 were compared by CIE against antienvelope immunoglobulins.[128] A well-resolved immunoprecipitin profile comprising over 46 discrete antigens was observed for solubilized inner membranes. Ten antigens were clearly identified by zymograms, and were shown to correspond to the following important membrane-associated enzymes:[30,31,219-222] adenosine triphosphatase (ATPase; EC 3.6.1.3), succinate dehydrogenase (EC 1.3.99.1), NADH dehydrogenase, D-lactate dehydrogenase, 6-phosphogluconate dehydrogenase (EC 1.1.1.43), glutamate dehydrogenase (EC 1.4.1.4; two separate components), dihydroorotate dehydrogenase (EC 1.3.3.1), glycerol-3-phosphate dehydrogenase (EC 1.1.99.5), and a chymotrypsin-like protease (EC 3.4.21.4). The corresponding immunoprecipitate pattern for isolated outer membranes comprised over 25 discrete antigens and differed strikingly from that obtained with inner membranes. It consisted principally of a group of seven relatively slow moving antigens, two of which were identified as lipopolysaccharide and Braun's lipoprotein.[134] An additional precipitate possessed protease activity similar to that observed in the plasma membrane.[128] This last observation is interesting, since it is conceivable that the enzyme represents the "signal" protease which is involved in processing exported proteins.[223,224] None of the antigens of either the inner or outer membrane appeared to react with the lectins concanavalin A or wheat germ agglutinin, confirming the lack of glycoproteins in this organism. Nor did any correspond to established cytoplasmic or periplasmic marker enzymes such as glucose-6-phosphate dehydrogenase (EC 1.1.1.49) or alkaline phosphatase (EC 3.1.3.1). Furthermore, the CIE immunoprecipitate pattern observed for extracts of isolated cell envelopes was in good qualitative agreement with the summation of the patterns seen individually for inner and outer membranes.[128]

This study clearly demonstrated that antigens from both the inner and outer membranes of a model Gram-negative bacterium were readily amenable to analysis and identification by quantitative immunoelectrophoresis techniques. It also catalyzed a series of immunochemical studies on the composition and structure of membrane vesicles from *E. coli* ML308-225.[108,153,159,160]

Topologically sealed cytoplasmic membrane vesicles from *E. coli* ML308-225 have provided a useful model system for the study of active transport,[225,226] and it now seems clear that they catalyze the accumulation of many solutes by a respiration-dependent mechanism in which the chemiosmotic phenomena appear to play a central role.[227] Various lines of evidence[228] suggest that these vesicles consist predominantly of topologically sealed plasma membranes possessing the same orientation as the intact cell, and that they are essentially devoid of cytoplasmic constituents. However, there is by no means universal agreement on these points and doubts have been expressed about both the chemical nature and orientation of the preparations.[222,229-237] In an attempt to resolve this controversy, Owen and Kaback[108,159,160] undertook a comprehensive

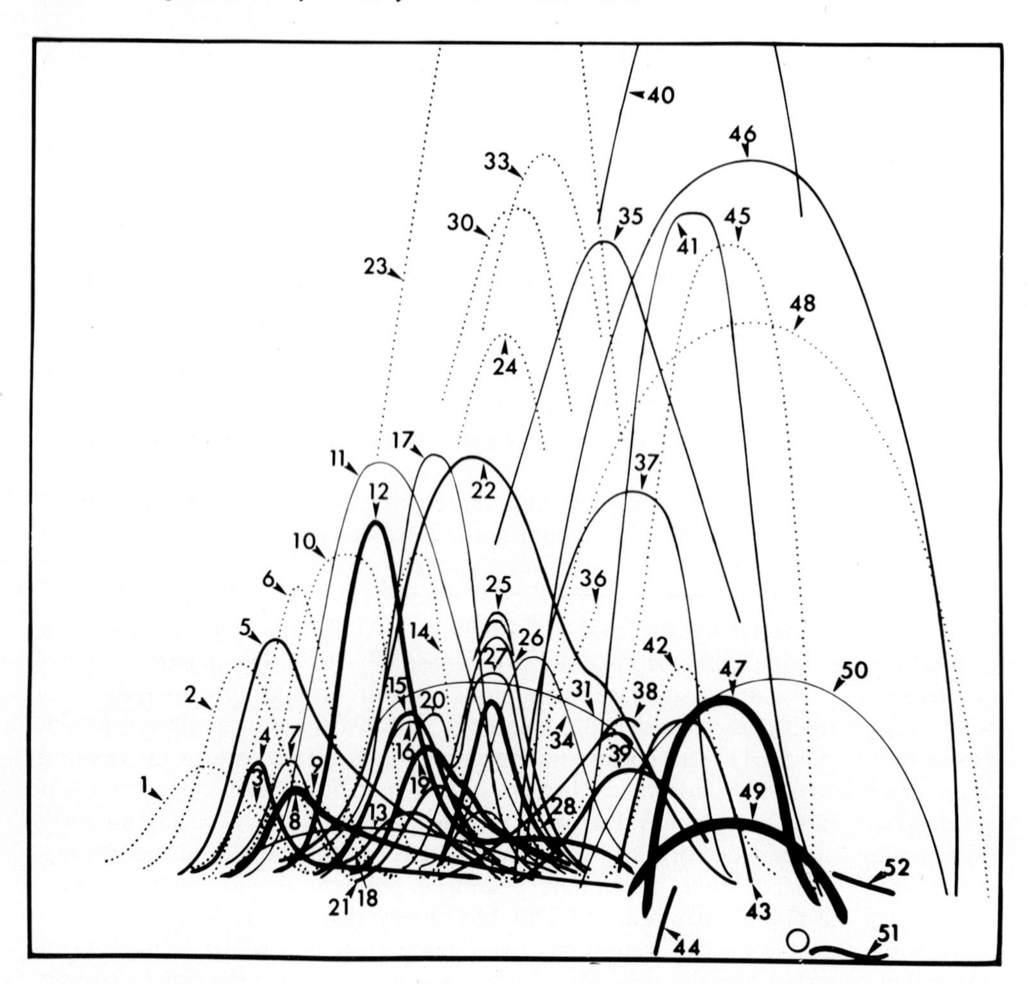

FIGURE 6. Schematic representation of the complete CIE pattern for the first Triton® X-100-EDTA extract of plasma membrane vesicles derived from *E. coli* ML308-225. The anode is to the left and top, and immunoprecipitates are numbered in order of decreasing electrophoretic mobility of the peak maxima. The line width is intended to indicate the intensities observed for the immunoprecipitates, dotted lines (. . . .) denoting precipitates that are detected only at relatively high concentrations of immunoglobulin. The lower portions of immunoprecipitate numbers 20, 25, and 35 and the upper parts of immunoprecipitate numbers 36 and 38 were often difficult to visualize and have been omitted. Some numbers (i.e., 29 and 32) and the bottoms of some immunoprecipitates (i.e., numbers 23,24,33, and 40) are omitted for clarity. Many of the weaker immunoprecipitates shown are not visualized on photographic reproduction (cf., Figure 1). The identities of antigen numbers 4,5,7,8,12,15,19/27,20,21,22,35,41,45,49,51 have been established and many more have been partially characterized (see Table 1). Note the reactions of partial identity between antigen numbers 19 and 27. (From Owen, P. and Kaback, H. R., *Proc. Natl. Acad. Sci. U.S.A.*, 75, 3148, 1978. With permission.)

immunological study of membrane vesicles prepared from *E. coli* ML308-225. Details of their CIE reference pattern[108] will be summarized in the following paragraphs and results of architectural analysis[159,160] in Section VI.A.3.

Over 50 discrete immunoprecipitates were resolved when solubilized membrane vesicles from *E. coli* ML308-225 were analyzed by CIE against antivesicle immunoglobulins[108] (see Figures 1 and 6). Seven immunoprecipitates possessed enzyme activity similar to that detected in the study of *E. coli* K12 plasma membrane[128] viz, ATPase, NADH dehydrogenase (two separate components; see also Section V.A.2),

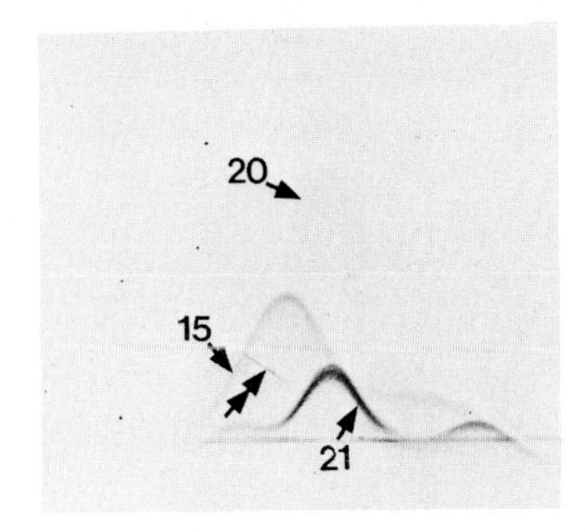

FIGURE 7. Characterization by zymogram of antigens possessing NADPH dehydrogenase activity. A Triton® X-100-EDTA extract (80 μg of protein) of *E. coli* ML308-225 membrane vesicles was analyzed by CIE against antivesicle immunoglobulin, and the wet immunoplate stained by zymogram techniques for NADPH dehydrogenase activity. Gels could be counterstained with Coomassie brilliant blue to confirm identity as indicated (cf. Figure 6). Part of an intersecting immunoprecipitate (number 5) is reinforced within the zymogram staining area (double headed arrow), a phenomenon which appears to be due to antigen entrapment. The well is delineated for clarity and the anode is to the left and top of the gel. (Reprinted with permission from Owen, P., Kaczorowski, G., and Kaback, H. R., *Biochemistry,* 19, 596, 1980. Copyright by the American Chemical Society.)

D-lactate dehydrogenase, 6-phosphogluconate dehydrogenase, glutamate dehydrogenase and dihydroorotate dehydrogenase. In addition, immunoprecipitates possessing polynucleotide phosphorylase[108] (EC 2.3.7.8) and nicotinamide adenine dinucleotide phosphate reduced form (NADPH) dehydrogenase[153] (EC 1.6.99.1; three separate antigens) were revealed (Figure 7). Components normally considered to partition in the cytoplasm, viz, β-galactosidase, and in the outer membrane, viz., lipopolysaccharide and Braun's lipoprotein, were also detected by coelectrophoresis and/or CIE with intermediate gel (Figure 3). However, results based upon both peak area estimations and analysis of cytoplasmic antigens suggests that the degree of contamination of the vesicle preparation with cytoplasm or outer membrane was minimal.[108] Some antiserum preparations were also shown to react with purified phosphatidyl serine synthetase[149] but exhaustive attempts to document an antibody interaction with the *lac* carrier protein were unsuccessful[108] (see, for example, Section II.C.4). Vesicle antigens were also probed by charge-shift CIE[123] to monitor their amphiphilic/hydrophilic properties (Figure 2) and by tandem CIE and co-CIE to assess their ability to cross react with vesicle immunogens from *Salmonella typhimurium*.[108] Of the identified antigens, glutamate dehydrogenase, dihydroorotate dehydrogenase, D-lactate dehydrogenase,

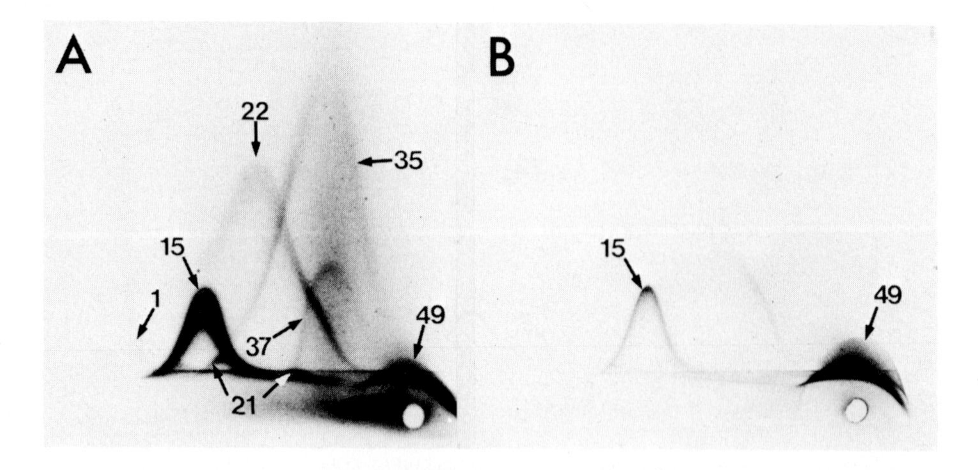

FIGURE 8. Resolution by CIE of the iron-containing antigens present in membrane vesicles prepared from *E. coli* ML308-225. A Triton® X-100-EDTA extract (95 μg of protein) of [^{59}Fe]-labeled membrane vesicles was analyzed by CIE against antivesicle immunoglobulin. Wet immunoplates were treated with distilled water (A) or extracted for nonheme iron[153] (B). Autoradiograms of both gels are presented. Gels could be stained with Coomassie brilliant blue to confirm identity as indicated (cf. Figure 6 and Table 1). Note that the main iron-containing antigen is number 15 (NADH dehydrogenase) and that most of the radioisotope in this and other antigens is removed following extraction of nonheme iron. The immunoprecipitate corresponding to Braun's lipoprotein (number 49) appears to possess tightly bound iron. The anode is to the left and top of both gels. (Reprinted with permission from Owen P., Kaczorowski, G., and Kaback, H. R., *Biochemistry,* 19, 596, 1980. Copyright by the American Chemical Society.)

and Braun's lipoprotein showed marked bidirectional alterations in electrophoretic mobility in the presence of anionic and cationic detergents, implying the presence of extensive hydrophobic regions in these molecules (Figure 2). Moreoever, ATPase, NADH dehydrogenase, 6-phosphogluconate dehydrogenase, and lipoprotein give reactions of full identity with components in *S. typhimurium* membrane vesicles.[108]

In a subsequent series of experiments, designed in part to analyze the redox chain components from this organism, Owen et al.[153] studied similar membrane vesicles obtained from *E. coli* cells grown in the presence of ^{59}Fe. Seven discrete [^{59}Fe] containing antigens were detected by autoradiography of CIE immunoplates and three of these were shown to correspond to the NADH dehydrogenase, NADPH dehydrogenase, and glutamate dehydrogenase antigens. In most instances, the iron seemed to be largely in the form of the nonheme derivative (Figure 8). It was speculated[153] that the immunoprecipitate possessing both ^{59}Fe and NADPH dehydrogenase activity may actually correspond to the complex hemoflavoprotein NADPH-sulfite reductase[238] (EC 1.8.1.2) which is known to manifest NADPH diaphorase activity.[239] The immunoprecipitate corresponding to Braun's lipoprotein also contained bound iron (Figure 8), an interesting and unexpected phenomenon which did not appear to be artifactual.[153] Since there have been no reports of bound iron in purified preparations of this well-characterized outer membrane component, it was suggested that the presence of ^{59}Fe might reflect an association of the lipoprotein molecule with iron-chelate receptors of the outer membrane.[34] Certainly, there is ample evidence for an in vivo association of lipoprotein with certain other molecules of the "porin" type,[240] and outer membrane preparations from *E. coli* are known to retain ferric-enterochelin binding activity following solubilization in Triton® X-100.[241]

These studies[108,128,153] have established the potential of CIE and its variants in immunochemical analysis of bacterial membranes, and have provided useful reference immunoprecipitin patterns for *E. coli* membranes in which over 20 antigens are identified or partially characterized. The properties of some of the *E. coli* ML308-225 vesicle immunogens are summarized in Table 1. The data clearly illustrate the abundance and diverse nature of membrane-associated antigens and provide the basis of a new approach to the analysis of microbial membranes and their components.

B. Membranes of *Micrococcus lysodeikticus*

For a number of years now, *M. lysodeikticus* has been used as a model bacterium for the study of structure and function relationship in biological membranes.[15,16,28-30] A great deal of the detailed information acrued over this period concerning chemical, biochemical, and immunological properties of the plasma membrane of this organism has emanated from the laboratory of Salton in New York. In 1975, Owen and Salton applied CIE to the analysis of *M. lysodeikticus* membranes, and were able to resolve 27 discrete antigens in tests against antimembrane serum.[89,132] Five of the major antigens were identified by zymograms and shown to correspond to the following enzymes: viz., ATPase, succinate dehydrogenase, NADH dehydrogenase (2 separate components), and malate dehydrogenase (EC 1.1.1.37). The two NADH dehydrogenase antigens are immunologically distinct and appear to possess different redox potentials,[61] whereas the succinate dehydrogenase is the major iron-containing antigen as judged from CIE of [^{59}Fe]-labeled plasma membranes.[154] Five other antigens probably possess α-D-mannopyranosyl or α-D-glucopyranosyl residues since they interact with concanavalin A in crossed immunoaffinoelectrophoresis experiments.[88,132] The major antigen in this latter group was identified as a succinylated lipomannan.[132] This membrane component is thought to be a lipoteichoic acid analogue for some micrococci.[38,242] An additional antigen appears to react with both soybean and ricin agglutinins, which have primary sugar specificities for *N*-acetyl-D-galactosaminyl[243] and D-galactose[244] residues, respectively. Despite extensive tests, the other antigens remain uncharacterized. It is relevant to note, however, that few of them appear to originate from the cytoplasm as judged by sensitive intermediate gel experiments conducted with anticytoplasm serum.[88] These and other properties are summarized in Table 2.

In a related series of experiments, Salton and Owen[155] analyzed by CIE the Triton® X-100-soluble antigens derived from "mesosomal" membrane vesicles of *M. lysodeikticus*. By comparison with corresponding profiles for plasma membranes, "mesosomal" membranes showed preferential enrichment of the succinylated lipomannan, and, furthermore, contained extremely small amounts of the ATPase and dehydrogenase antigens. It is notable that these data confirm the results of earlier chemical and biochemical analysis[38,245] and, moreoever, dispose of the argument that the enzymes are present in "mesosomal" vesicles in normal amounts but are inaccessible to substrate during direct enzyme analysis by virtue of physical constraints imposed by the right-side-out nature of the vesicles.

More recent immunological investigations on *M. lysodeikticus* have centered around the antigenic nature of the membrane residues remaining after extraction with Triton® X-100.[130] This is of obvious importance to a complete understanding of the antigenic character of the membrane because the Triton® X-100-insoluble residues are normally refractory to direct immunological analysis, and can account for over half of the total membrane protein. Initial studies using the ionic detergent, SDS, have been encouraging, and indicate that analysis of the residues is feasible, provided that dispersal is performed at carefully controlled protein/detergent ratios.[130] Under these conditions, two main immunoprecipitates are routinely detected by CIE (Figure 9).

Table 1
IDENTIFICATION AND PROPERTIES OF ANTIGENS DETECTED FOR MEMBRANE VESICLES OF *E. COLI* ML 308-225[a]

Antigen number[b]	Identity	Ferro-antigen[c]	Amphiphilic nature[d]	Reaction with *S. typhimurium* membrane vesicle antigens[e]	% Expression in intact vesicles[f]	Expression on cell surface[g]
1[h]	Unknown	+				
4	Common antigen[i]			+		–
5	Polynucleotide phosphorylase			+	9	–
7	6-Phosphogluconate dehydrogenase			+	5	–
8	β-Galactosidase				5	–
12	ATPase			+	11	–
15	NADH dehydrogenase	+		+	2	–
19/27	Dihydrolipoyl dehydrogenase				2	–
20[h]	NADPH dehydrogenase					
21	NADPH dehydrogenase[j]	+			minimal	–
22	Succinate dehydrogenase	+[k]	+	+		–
22.1[l]	Unknown					+
26	Unknown		+		5	
27	Unknown					–
28	Unknown				2	–
35	Glutamate dehydrogenase	+	+			
37	Unknown	+			minimal	–
38	Unknown					+
39	Unknown			±	82	+
41	Dihydroorotate dehydrogenase		+		6	–
43	Unknown				105	+
45	D-Lactate dehydrogenase		+		10	–
47	Unknown		+		100	–
49	Braun's lipoprotein	+	+	+	100	–
51	Lipopolysaccharide					+
13,16	Unknown					–
18,30,	Unknown					–
44	Unknown					–

[a] Data have been compiled from References 108, 153, 159, 160, and 295.
[b] Antigen number refers to the immunoprecipitates shown in Figure 6.
[c] Antigens identified from an analysis of [^{59}Fe]-labeled membrane vesicles.
[d] Amphiphilic nature denoted (+) indicates that the antigen in question exhibits bidirectional alteration in electrophoretic mobility during change-shift CIE experiments.
[e] Full (+) or partial (±) cross-reactivity is indicated.
[f] Values computed from graphs similar to those displayed in Figure 16.
[g] Nonexpression (−) or expression (+) based upon the results of progressive adsorption experiments conducted with whole cells.
[h] Correlation is only tentative due to faint nature of immunoprecipitin arc.
[i] Presumptive identity; see Reference 106.
[j] The possibility that this antigen represents NADPH-sulfite reductase cannot be ruled out.
[k] Immunoprecipitate is heterogeneous and does not label evenly.
[l] Antigen not shown in Figure 6.

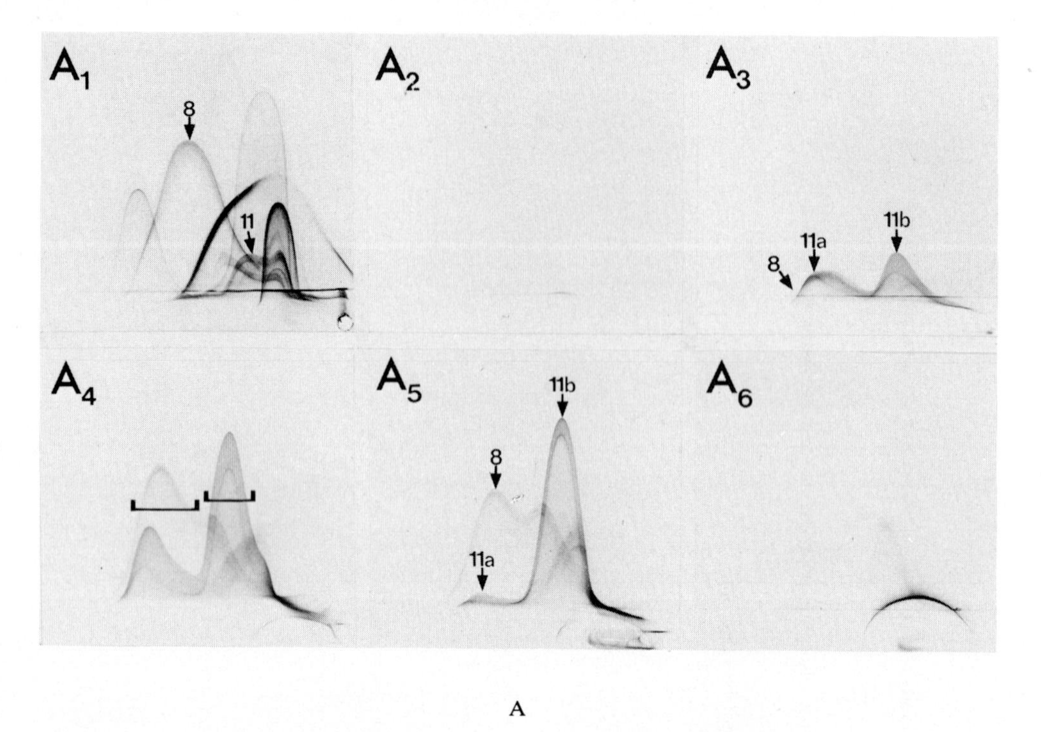

A

FIGURE 9. Immunochemical analysis of Triton® X-100-insoluble residues from membranes of *M. lysodeikticus*. Plasma membranes of *M. lysodeikticus* were extracted exhaustively with Triton® X-100 and the insoluble residue (final concentration 8.5 mg of protein per m*l*) treated with SDS at varying concentrations. Supernatant fractions obtained following centrifugation at 48,000 × g were analyzed by CIE (A_2-A_6) and for ATPase (B) and for succinate dehydrogenase (C). A_2-A_6 represents the profiles obtained by CIE for soluble extracts obtained following treatment of Triton® X-100-insoluble membrane residues with SDS at final concentrations of 0.01%, 0.05%, 0.167%, 0.25%, and 0.5%, respectively. Similar volumes (4 μ*l*) of SDS extracts were analyzed against antimembrane serum in all instances. The immunoprecipitin profile obtained following CIE of a Triton® X-100 extract (88 μg of protein) of *M. lysodeikticus* membranes is shown for comparison in panel A_1. Immunoprecipitates possessing ATPase (number 11) and succinate dehydrogenase (number 8) activity have been indicated in several of the panels (cf. Table 2). Brackets in panel A_4 indicate regions of precipitates number 8 and 11 which can be excised and used as immunogens to generate serum specific for succinate dehydrogenase and ATPase, respectively. Anode is to the left and top of all gels. Panels B and C: SDS extracts were analyzed colorimetrically for protein (○) and for enzyme activity (■) and by CIE in order to quantitate by peak area estimations the amount of enzyme antigen (●) present in each extract. Protein (○) and enzyme activity (■) are expressed as a percentage of the amounts observed in untreated Triton® X-100-insoluble residues. Similar calculations are not feasible for the estimation of enzyme antigen, since the Triton® X-100-insoluble membrane residues are not amenable to direct CIE analysis. Consequently, values in this instance (●) are quoted as a percentage of the maximal amount of antigen released. The specific activity of ATPase and succinate dehydrogenase in the Triton® X-100-insoluble residue was 1.61 μmoles ATP hydrolyzed/10 min/mg of protein and 0.11 μmole of succinate oxidized/min/mg of protein, respectively. Inset in (B) shows the differential effect of SDS on the peak area of the two components of the ATPase immunoprecipitate (i.e., numbers 11a and 11b in A). Note that the Triton® X-100-insoluble residue can be dispersed by SDS into a form which is amenable to CIE analysis, and which gives no obvious distortion of the immunoprecipitin profile. Note also that the spectrum of immunoprecipitates is relatively simple (e.g., A_5) in comparison to the Triton® X-100 extract (A_1), and consists of two main antigens, viz., ATPase and succinate dehydrogenase. Maximal release of the two enzyme antigens (●: B and C) is achieved at different concentrations of SDS, viz., 0.2% for succinate dehydrogenase and 0.13% for ATPase. In both cases, release is accompanied with retention of full or enhanced catalytic potential (■). Note that component 11a of the heterogeneous ATPase antigen is more susceptible to SDS than component 11b (A_3-A_5 and Inset of B). Both ATPase and succinate dehydrogenase lose immunological specificity and catalytic potential and become denatured at concentrations of SDS above 0.25% (Reprinted with permission from Owen, P. and Doherty, H., *J. Bacteriol.*, 140, 881, 1979. Copyright by the American Society for Microbiology.)

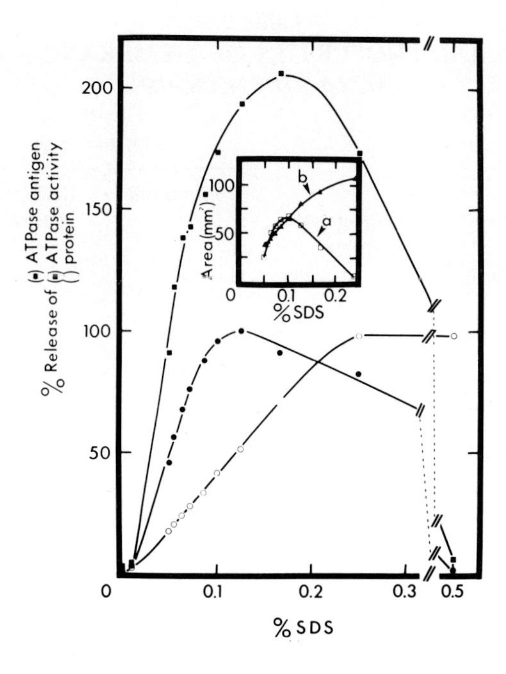

FIGURE 9B

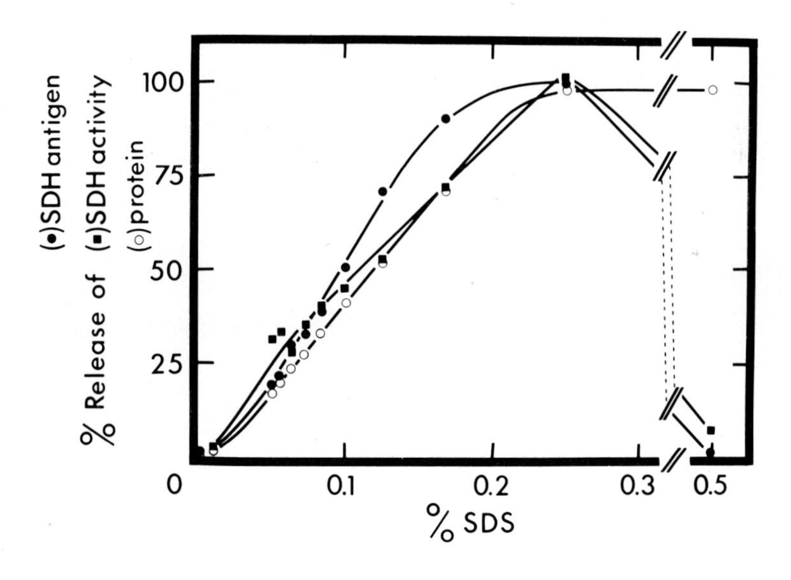

FIGURE 9C

These have been positively identified as ATPase and succinate dehydrogenase. Optimal release of both antigens from the Triton® X-100-insoluble residues can be monitored by peak area estimations, and occurs at protein/detergent ratios of 6.8/1 and 4.3/1 for ATPase and succinate dehydrogenase, respectively. Comparison of catalytic activities in SDS extracts and in untreated control Triton® X-100-insoluble residues reveals that both enzyme antigens are released with full (or enhanced) catalytic potential at or below concentrations of SDS required to affect maximal solubilization of

Table 2
IDENTIFICATION AND PROPERTIES OF MEMBRANE ANTIGENS FROM *M. LYSODEIKTICUS*[a]

Antigen number[b]	Identity	Ferro-antigen[c]	Reaction[d] with Concanavalin A	Reaction[d] with Ricin/soybean agglutinins	Expression on protoplast/cell surface[e]
4	Partial identity with number 17		±		+
5	Unknown		±		+
7	Unknown[f]		+		+
8	Succinate dehydrogenase	+			−
10	NADH dehydrogenase	+			−
11	ATPase				−
13	NADH dehydrogenase				−
14	Unknown[f]			+	−
15	Unknown[f]		+		+
17	Unknown[f]		+		+
18	Succinylated mannan		+		+
21	Malate dehydrogenase				−
23	Unknown[f]		+		+
1,2,3,	Unknown				+
12,24	Unknown				+
6,9,16	Unknown				−
19,20	Unknown				−
22,25	Unknown				−
27	Unknown				−

[a] Data have been compiled from References 88, 132, and 154.
[b] Antigen numbering system is that found in Reference 88.
[c] Antigens identified from an analysis of [^{59}Fe]-labeled membranes.
[d] (+) and (±) denote full adsorption and partial adsorption, respectively, in crossed immunoaffinoelectrophoresis experiments.
[e] Nonexpression (−) or expression (+) judged from results of progressive immunoadsorption experiments. All antigens denoted (−) are known to be expressed solely on the cytoplasmic face of the membrane.
[f] These components are probably glycoproteins.

the enzyme in question. Denaturation is evident only at protein/detergent ratios lower than about 3.4/1[130] (Figure 9)

The above studies establish the feasibility of analyzing the full complement of membrane antigens by gel precipitation methods. Two aspects of the study are somewhat surprising, however. The first relates to the simplicity of the precipitin profile observed for the Triton® X-100-insoluble residue, and the second, to the fact that the ATPase and succinate dehydrogenase antigens appear to be identical to similar antigens found in the Triton® X-100-soluble fraction of the membrane. Although the precise reasons for these phenomena are not clear at the present time, it could be speculated that they reflect in part a strong cohesive interaction between the catalytic moities and other functionally related components in the membrane (see Section IV).

C. Membranes of *Bacillus subtilis*

Bacillus subtilis and a number of related succinate dehyrogenase mutants have been analyzed immunochemically by Rutberg and his co-workers.[115,158,246] It seems that the membranes of this organism give a much simpler CIE immunoprecipitin pattern than those observed for membranes of either *E. coli*[108,128] or *M. lysodeikticus*.[88,132] This may be related in part to the low solubility of *B. subtilis* membranes in Triton® X-100.[115] Succinate dehydrogenase is one of the antigens resolved by CIE, however, and precipitate-excision techniques[112] have been employed to generate succinate dehydrogenase-specific antiserum.[115] This serum has been used to show that nine *cit F* mutants, which possess very low levels of membrane-associated dehydrogenase activity, lack the succinate dehydrogenase antigen in the membrane. Significantly, five of the mutants appeared to contain an antigenically competent, soluble form of the enzyme in the cytoplasm, suggesting that they are defective in a subunit which is normally involved in the interaction of the catalytic moiety with another membrane protein.[115,158,246]

D. Envelopes of *Neisseria gonorrhoeae*

Concerted efforts have been made over the past decade to develop sensitive serodiagnostic tests for gonococcal infections. Most have been directed towards evaluating the potential of antigenic extracts of the organism in screening for infection. In many instances, the nature of such preparations has been ill defined. Several tests also suffer from an unacceptable level of false positive reactions.[247-251] Efforts to resolve some of these difficulties have been made in careful immunochemical studies of the envelopes of *Neisseria gonorrhoeae*.[106,107,129,215]

The initial studies revealed that envelopes from this organism could be resolved satisfactorily by CIE into about 30 immunoprecipitates, providing agarose of low endoosmotic flow properties was used. The pattern of immunoprecipitates was highly reproducible and did not appear to be grossly affected by the method of cell disruption. Two of the resolved antigens possessed NADH dehydrogenase activity, and at least one other possessed D-lactate dehydrogenase activity.[106,107] None of the other components have been characterized, however. More recent studies have been directed towards identifying which of the main envelope-associated antigens are recognized by patients' immune systems, and towards resolution of antigen(s) responsible for the troublesome false positive reactions. Potential sources of false positive reactions have been identified in CIE experiments conducted with serum raised to apathogenic neisseriae such as *N. lactamis, N. perflava,* and *N. sicca.*[107] Furthermore, a very acidic antigen called "fast moving antigen" (FMA), which is present in both gonococcal envelopes and cytoplasm,[106,215] may also be implicated, since CIE with intermediate gel has revealed that about 60% of serum samples from "normal" healthy individuals contain anti-FMA immunoglobulins, compared with 100% for patients with gonococcal infections.[215] Thus, a likely source of false positive reactions in the serodiagnosis of gonorrhoeae seems to have been identified as FMA. Its chemical characterization and its relationship to "antigen a" described by Danielsson et al.[252] and the so called "common protein antigen", which is responsible for cross-reactions between a wide range of Gram-negative bacteria,[166] will be awaited with interest.

In contradistinction to the broad serum reactivity against FMA,[106,107,215] a reaction of potential serodiagnostic value has been observed for a major antigen present in barium thiocyanate extracts[129] of gonococcal envelopes.[215] This component reacts consistently (92%) in intermediate gel experiments with the serum of patients with *N. gonorrhoeae* infections. Similar reactions are observed at very low frequency (8%) in "normal" serum samples.[215] It will be of interest to learn whether this antigen corresponds to one of the major outer membrane proteins that are presently under extensive investigation in other laboratories (see Section V.C.2).

In other studies, Smith and co-workers[171,172] have suggested that a poorly defined immunoprecipitate found in CIE profiles of surface washes of in vivo, but not in vitro, grown gonococci corresponds to a type specific antigen synthesized by the cell during adaptation to growth in vivo. A vaccine prepared from the excised immunoprecipitate failed to protect against infection in the guinea pig chamber model, although a synergistic effect was reported when similar preparations were combined with formalin-killed gonococci lacking the type specific immunogen.[172] The type specific antigen appears to be proteinaceous in nature since it is affected by pronase,[172] but more detailed information regarding its chemical nature is lacking. However, the fact that it is susceptible to Triton® X-100[172] is reminiscent of the earlier immunological studies of Johnson et al.[253] in which it was shown that a major serotyping antigen from this organism could be dissociated into several[107] antigenic components following treatment with nonionic detergents. Thus, the type-specific antigen of Smith and others[171,172] may in fact represent a complex association of more than one surface component. The morphology of the corresponding immunoprecipitate is not incompatible with such a suggestion.

E. Mycoplasma Membranes

Membranes of mycoplasmas have been analyzed immunochemically by a number of workers,[26,57,114,116-121,137,185,254,255] notably by Hjertén and his colleagues in Uppsala. Indeed, the report by Johannson and Hjertén[57] in 1974 of the localization of Tween® 20-soluble membrane proteins of *Acholeplasma laidlawii* represents the first application of the CIE technique to the study of microbial membranes, and remains a landmark in the analysis of membrane structure and immunology. CIE of Tween® 20-soluble extracts of *A. laidlawii* membranes routinely resolves about 15 to 20 immunoprecipitates.[57] Detergents other than Tween® 20 have been tested,[116-121] but whereas deoxycholate does appear to enhance resolution of membrane antigens for some other mycoplasmas, e.g., *Spiroplasma citri*[119] and *M. arginini*,[121] it does little to improve on the clear spectrum of immunoprecipitates obtained for *A. laidlawii* following CIE in the presence of the nonionic detergents (i.e., Tween® 20[57,137,185] or Triton® X-100[119]). Several of the major proteins from *A. laidlawii* have been isolated and partially purified[57,137,185,256] and their identities with antigens observed in CIE profiles of Tween® 20[57] and deoxycholate–[120] solubilized membranes have been established. Furthermore, their relationships with bands observed following SDS-polyacrylamide gel electrophoresis of membrane proteins of this organism[257] have been determined, as have their probable molecular weights and some of their biochemical properties.[57,137,185,256,258] A summary of these data can be found in Table 3.

The membrane antigens of *M. arginini*[114,121] and *S. citri*[116,119] have also been analyzed by CIE, but in neither case has a profile of immunoprecipitates been established which is comparable in clarity to that obtained for *A. laidlawii.* The CIE profiles of both are characterized by the presence of an exceptionally strong immunoprecipitate which, in the case of *M. arginini,* shows considerable heterogeneity. From the reactions in charge-shift experiments[121] and against monospecific antisera,[114] it seems likely that this particular antigen from *M. arginini* is a complex of several membrane components. The principal immunoprecipitate observed for *S. citri*[116,119] probably corresponds to the main membrane protein of this organism, spiralin. This component has recently been purified[118] and shown to be a polypeptide of molecular weight 26,000 daltons accounting for 22% of the total membrane protein. Its amino acid composition has been determined and is characterized by a very low content of methionine, histidine, glycine, leucine, and aromatic amino acids, together with a relatively high content of threonine, alanine, valine, and polar amino acids (aspartic acid, glutamic acid, and leucine).[118]

The cross reactivities of the membrane antigens of several species and strains of the *Mycoplasmatales* have also been determined by quantitative immunoelectrophoretic techniques.[114,117,120]

F. Membranes and Chromatophores of *Rhodopseudomonas sphaeroides*

The membrane system of the photosynthetic microorganism, *R. sphaeroides,* has been analyzed recently by two groups of workers.[174,174a] Collins et al.[174a] have detected 31

Table 3
PROPERTIES OF SOME MEMBRANE ANTIGENS FROM *A. LAIDLAWII*[a]

Antigen number in different analytical systems: CIE in Tween®20	CIE in Deoxycholate	SDS-gel electrophoresis	Apparent molecular weight[b] (daltons)	Biochemical properties[c]	Expression on cell surface[d]
t_{1a}	n.d.[e]	D_{12}	140,000[f]	Glycoprotein possessing glucosamine and galactosamine; amino acid composition known; polarity 45%	±
t_{1b}	n.d.[e]	—	110,000[g]	Protein interacting with *Vicia ervilia* lectin[h]	±
T_2	5	D_6	95,000[i] 52,000	Protein of known amino acid composition; polarity 45%; sedimentation coefficient 6.8S	+
T_3	8	D_{11}	110,000	Protein of known amino acid composition; polarity 46%; sedimentation coefficient 4.6S	−
T_{4a}	4	D_5	34,000	Flavoprotein of known amino acid composition; polarity 45%; sedimentation coefficient 4.6S; binds 80 moles of Triton® X-100 per mole	−
T_{4b}	6	D_6	52,000	Protein of known amino acid composition; polarity 48%; sedimentation coefficient 3.4S	−

[a] Data have been compiled from References 57, 120, 137, 185, 256—258, 537, and 538.
[b] Except where indicated, molecular weight values are based upon mobilities in SDS-polyacrylamide gels.
[c] Polarities are calculated from the amino acid composition by the method of Capaldi and Vanderkooi.[539]
[d] Results are based upon progressive immunoadsorption experiments conducted with intact and lysed cells; (+) and (−) indicate expression on the cell surface and cytoplasmic face of the membrane, respectively, and (±) indicates that expression on the cell surface depends on the phase of growth.
[e] Not determined; t_{1a} and t_{1b} have not been conclusively identified in deoxycholate-CIE reference pattern.
[f] Value based on amino acid analysis.
[g] Fraction t_1, which contains antigens t_{1a} and t_{1b}, can be resolved into two species of molecular weights 110,000 and 140,000 daltons; definite correlation is lacking.
[h] This lectin has primary sugar specificities similar to those of concanavalin A.[540]
[i] Molecular weight increases from 52,000 to 95,000 daltons after boiling in SDS.

antigens following CIE analysis of the intracytoplasmic membranes (chromatophores), and have established that the major immunoprecipitate probably corresponds to the light-harvesting bacteriochlorophyll *a*·protein complex. A second immunoprecipitate was shown by CIE with intermediate gel to contain the photochemical reaction center, and two others were correlated by zymograms with NADH dehydrogenase and D-lactate dehydrogenase enzymes. Attempts to detect the ATPase or succinate dehydrogenase antigens were unsuccessful.[174a] In an independent study, Konings and his colleagues[174] have compared the plasma membranes and chromatophores of *R. sphaeroides* by quantitative immunoelectrophoresis. Both membrane preparations give complex arrays of over 50 immunoprecipitates in tests against homologous antiserum. The CIE profiles appear to be qualitatively similar, although some quantitative differences are evident. However, it is difficult to assess whether these reflect differences in the antigenic character of the two membrane preparations, or differences in the efficiencies of extraction (estimated at between 60 to 80%) or in antibody titers. (Similar considerations might also explain in part the dissimilarities between the CIE profiles obtained for ostensibly the same membrane preparation by two different groups of workers.[174,174a]) Unlike Collins et al.,[174a] Konings and his co-workers[174] have been able to identify both the ATPase and succinate dehydrogenase antigens, and have further revealed, for both the plasma and chromatophore membrane preparations, the identities of the malate dehydrogenase antigen and of 4 antigens apparently possessing NADH dehydrogenase activity (*cf.* the single NADH dehydrogenase antigen detected by Collins et al.[174a]). The orientation of the membrane preparations has also been assessed by immunoadsorption experiments.[174] (see Section VI.A.5).

IV. ANTIGENS OR ANTIGEN COMPLEXES?

Analysis of bacterial membranes by CIE involves the study of membrane immunogens in detergent solution, and although it is justifiable to identify a resolved immunoprecipitate on the basis of the inherent characteristics of a component antigen, e.g., catalytic activity or sugar moieties, it could be incorrect to assume that additional membrane components are not contained therein. For example, it is conceivable that some immunoprecipitates may represent a complex association of membrane components which perform coupled biological function and which resist dissociation by virtue of strong cohesive interactions in localized membrane microenvironments. This point was brought into sharp focus by the somewhat controversial studies of Blomberg and his colleagues.[149,259-261] These investigations, the results of which have been critically reviewed elsewhere,[61] concerned the CIE analysis of certain mammalian biomembranes. In short, numerous immunoprecipitates were detected, each of which appeared to possess several different enzyme activities as judged by zymograms. Some of them also seemed to contain phospholipids and epinephrine receptors. The authors termed these antigens "multienzyme complexes" and suggested that they reflected the organization of the individual components in the native membrane structure.[149,259-261] Although some aspects of these studies have been criticized,[61] the conclusions have a certain appeal, and there is now a growing volume of experimental evidence from the study of both mammalian[131,262] and microbial membranes to suggest that this situation does occur in some instances. Pertinent evidence obtained from investigations of microbial membranes is discussed below.

The reduction of nitrate in *E. coli* is catalyzed under anaerobic conditions by a series of membrane-bound components including formate dehydrogenase, cytochrome b_1 and nitrate reductase.[263] The nitrate reductase moiety has been extensively studied by MacGregor and colleagues, and a method for its purification has been devised which initially involves release from the membrane by activation of a membrane-bound protease.[264,265] This soluble form of nitrate reductase possesses two subunits, A and B, of molecular weights 142,000 and 60,000 daltons, respectively.[265] When antiserum raised to the purified enzyme was used to induce precipitation of nitrate reductase from Triton® X-100-solubilized plasma membranes, it was found that the immune complex contained, in addition to the subunits A and B, a third component which was characterized as the 19,500 dalton apoprotein of cytochrome b.[266] The Triton® X-100-soluble nitrate reductase complex had a molecular weight of about 500,000 daltons and contained subunits A and B and cytochrome b in the molar ratio 2:2:4.[266,267] Studies with *hem A* mutants which were unable to synthesize the heme precursor δ-aminolevulinic acid further suggested that cytochrome b was involved in the attachment and integration of the enzyme into the membrane.[268] Thus, antibodies directed solely against the A and B subunits of nitrate reductase are able to coprecipitate from Triton® X-100 solubilized membranes a third component (the cytochrome b) with which the enzyme is closely associated in vivo.

A very similar situation was observed recently during studies of succinate dehydrogenase in *B. subtilis*.[115,158,246] (See also Section III.C.) In this instance, three polypeptides of molecular weights 65,000, 28,000 and 19,000 daltons were precipitated in equimolar quantities from Triton® X-100-solubilized membranes by specific antisuccinate dehydrogenase serum. The component of highest molecular weight was a flavoprotein and various lines of evidence indicated that the smallest polypeptide was probably a cytochrome b.[115,158] Other data, based upon the analyses of *cit F* and 5-aminolevulinic acid-requiring mutants, suggested that succinate dehydrogenase was bound to the membrane via an interaction between the 28,000 dalton subunit and cytochrome b.[158,246] Therefore, it seems probable that the succinate dehydrogenase-active immunoprecipitate observed following CIE of *B. subtilis* membranes contains all three components, although this has not been shown directly as yet.

Further evidence to support the contention that some membrane components can retain their close *in vivo* association during solubilization and CIE comes from studies of *Micrococcus lysodeikticus* membranes.[61,132] Partially purified F_1·ATPase from *M. lysodeikticus* was shown by Owen and Salton[132] to have approximately twice the electrophoretic mobility of the ATPase antigen detected in Triton® X-100-solubilized membranes of this organism. Both stained for ATPase activity by zymograms (Figure 10) and were clearly related immunologically since they gave lines of identity in tandem CIE experiments.[61,132] It was suggested at the time that the Triton® X-100-extractable ATPase antigen might represent the "native enzyme complex coupled with other components of the membrane energizing system."[132] Very similar results were reported 3 years later by Schmitt et al.[269] In this study,[269] a preparation of detergent-solubilized F_0F_1·ATPase from *M. lysodeikticus* was shown to have an electrophoretic mobility comparable to that reported earlier for ATPase in Triton® X-100-solubilized membranes from this organism.[132] It thus seems reasonable to suggest that the Triton® X-100-extractable ATPase antigen represents an integrated form of ATPase (e.g., F_0F_1·ATPase), and that ATPase-active antigens of greater electrophoretic mobilities represent degenerate forms of this complex which have lost some of the components or subunits which are not themselves involved in the primary catalytic event (i.e., the hydrolysis of ATP).

The presence of antigen complexes has also been invoked to explain the association of iron with the precipitate corresponding to Braun's lipoprotein in the CIE profile of *E. coli* membranes[153] (see Section III.A), the simplicity of the CIE profile obtained for Triton® X-100-insoluble membrane residues of *M. lysodeikticus*[130] (see Section III.B), and the complex CIE profiles consistently displayed by some immunoprecipitates.[114,132]

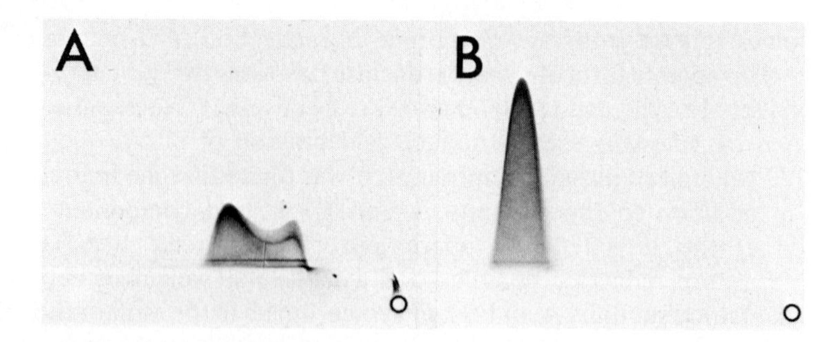

FIGURE 10. Electrophoretic properties of *M. lysodeikticus* ATPase. A Triton® X-100 extract (48 μg of protein) of *M. lysodeikticus* membranes (A) and a purified preparation of *M. lysodeikticus* F_1. ATPase (5.6 μg of protein; B) were analyzed by CIE against antimembrane serum. Both gels were stained by zymogram techniques to highlight the ATPase antigen. Wells have been delineated for clarity. Note that the electrophoretic mobility of purified F_1. ATPase is considerably greater than that of the ATPase complex present in Triton® X-100 extracts. Anode is to the left and top of both gels.

However, most of the evidence accumulated to date is indirect, and additional data will be required before firm conclusions can be drawn about the nature of many detergent-solubilized membrane antigens. Analysis of the radiolabeled membrane components interacting with antiserum raised to excised CIE immunoprecipitates[112] would appear to be one of the more productive approaches to this problem, and would, in addition, provide invaluable information about membrane ultrastructure. In the light of the above considerations and the detailed information regarding the monomeric nature of many of the *A. laidlawii* membrane antigens (Table 3), it seems that both individual antigens and antigen complexes are likely to be present in bacterial membranes that have been solubilized by nonionic detergents.

V. IMMUNOCHEMICAL PROPERTIES OF SOME MEMBRANE COMPONENTS

Prior to a discussion of the antigenic architecture of microbial membranes (Section VI) it is appropriate to review the immunochemical properties of some of the better characterized microbial membrane components. It is not intended, however, to cover the chemistry and serology of membrane-associated carbohydrate antigens, i.e., lipoteichoic acids and lipopolysaccharides, since they have been discussed in adequate detail elsewhere.[9-11,66-76] Nor is the intention to provide a comprehensive treatise on the immunochemical properties of all the remaining membrane components. Instead, coverage will be largely restricted to a few selected membrane components which have been isolated and characterized, and it is hoped that this approach will impress the reader with the variety and detail of information that immunochemical analysis can offer.

A. Plasma Membrane Proteins

1. ATPase

ATPase is one of the more extensively characterized membrane-associated enzymes.[30,152,270] It is a complex molecule comprising a soluble catalytic moiety (F_1) and an insoluble portion (F_0), which is an integral part of the membrane and which is proposed to act as a proton channel. Analysis of F_1·ATPase from a variety of microorganisms suggests that most are remarkably similar in subunit composition. Five polypeptide chains are usually observed ($\alpha,\beta,\gamma,\delta,\epsilon$), with molecular weights ranging from 60,000 to 10,000 daltons. Although the subunit stoichiometry is still controversial, it is generally accepted that α and β are essential for ATP hydrolysis, and that δ and ϵ are involved, at least in part, in binding $\alpha\beta\gamma$ to F_0.[152,271,272]

The ATPases from a number of bacterial sources, notably *M. lysodeikticus*[27,39,41,61,269,273,274] and *E. coli*,[275-280] have been analyzed immunochemically. Antiserum raised to F_1.ATPase from *M. lysodeikticus* allows quantitation of the homologous enzyme by radial immunodiffusion[41] and rocket immunoelectrophoresis,[61] and reacts with membrane-bound ATPase[274] as well as with the form of the enzyme obtained following solubilization of the membrane in Triton® X-100.[61,269] Similar observations have also been documented for the ATPases of *E. coli*[275,276] and *Rhodospirillum rubrum*.[281] In most instances, anti-ATPase immunoglobulins almost totally inhibit ATP hydrolysis, irrespective of whether the enzyme is in the soluble or membrane-associated form.[41,274,275,281,282] However, there is some evidence that larger quantities of specific antiserum are necessary to achieve inhibition in the case of the membrane-bound enzyme, a feature which may indicate that some determinants are masked in this form of the complex.[270] ATPases isolated from *M. lysodeikticus*[41] and *R. rubrum*[281] are inhibited in a noncompetitive fashion by homologous antibody, suggesting that antibody molecules are physically excluding substrate from the catalytic site rather than directly binding to the active centers of the enzymes. This contrasts with the situation reported for the ATPase from *Proteus*.[283] In this instance, immunoglobulins appear to be interfering with ATP hydrolysis by inducing a conformational change at the active site.

The proposal that the α and β subunits of $F_1 \cdot$ATPase play a central obligatory role in oxidative phosphorylation and ATP-dependent energy coupling reactions receives support from the immunochemical studies of Gutnick and his colleagues[276,277] who observed that antisera prepared against the (α + β) subunits of *E. coli* completely inhibited not only ATP hydrolysis, but also ATP-P_i exchange and ATP-linked transhydrogenation. Notably, no inhibition of respiration-linked transhydrogenation was observed in these experiments[276,277] (see also References 275 and 278). Antibodies raised to individual subunits have also been used in inhibition studies but the results are less conclusion. For example, Kanner et al.[277] found that the ATPase from *E. coli* could be inhibited by anti-α immunoglobulins and to a lesser extent by anti-β and anti-γ antibodies, whereas Mollinedo and his colleagues[283a] showed that only anti-β immunoglobulins significantly inhibited *M. lysodeikticus* ATPase. In further contrast, both anti-α and anti-β but not anti-γ, anti-δ, or anti-ϵ antibodies have been shown to inhibit ATP hydrolysis and ATP-P_i exchange for the enzyme isolated from the thermophilic bacterium PS3[283b]. It seems probably that these discrepancies may result, at least in part, from a loss of native conformation following isolation and purification of the individual subunits.

The F_0 portion of the proton-translocating ATPase is not as well characterized as the F_1 complex. However, antiserum raised to the F_0 moiety from the thermophilic bacterium PS3 has been shown to inhibit proton conduction in, and F_1 binding to, vesicles reconstituted from purified F_0 and phospholipids. It was suggested that a hydrophilic site, which is essential for both proton translocation and F_1 binding, is exposed on F_0 at the outer surface of the reconstituted vesicles.[284] This is not entirely compatible with a variety of immunochemical and biochemical data which have clearly shown that $F_1 \cdot$ATPase is located solely on the inner surface of the plasma membrane[132,159,274,285] (see Sections VI.A.2 and 3 and V.I.B.1.a). It seems probable that F_0 is oriented in these reconstituted vesicles in the direction opposite to that which exists in vivo.

An alternative approach to the study of bacterial ATPase has involved dissociation of the enzyme into, or reconstitution of active ATPase from, isolated subunits. A number of years ago it was shown that treatment of *M. lysodeikticus* ATPase with SDS or guanidinium hydrochloride resulted in a total loss of reactivity against anti-ATPase.[41] Curiously SDS-treated ATPase did appear to react with antiserum raised to sonicated membranes of this organism,[27] and it was suggested that sonication or dissociation with SDS revealed an additional antigenic site on the enzyme.[27,41] Whether this is a product of dissociation or of denaturation appears to be an open question, however. A clearer picture of the dissociation events has emerged from CIE experiments using antiserum raised to *M. lysodeikticus* ATPase preparations that have been fully dissociated by treatment with urea, dithiothreitol, and iodoacetamide.[61] In these experiments, it was clearly demonstrated that F_1·ATPase could be quantitatively converted to antigens of reduced electrophoretic mobilities by concentrations of SDS which did not denature other membrane immunogens present in the preparation. Unfortunately, the antigenic products of dissociation were not identified in terms of individual subunits.[61] This has been achieved to some extent for the ATPase from *E. coli*.[279] Vogel and Steinhart[279] have dissociated the F_1·ATPase by freezing in salt solutions, and have split the enzyme complex into two main entities, neither of which possesses any catalytic activity. One is a complex of the α, γ, and ϵ subunits and the other corresponds to the β subunit. Both have reduced electrophoretic mobilities when compared with the native enzyme and both give lines of partial identity with F_1·ATPase in gel diffusion tests. No serological relationship could be demonstrated between the ($\alpha\gamma\epsilon$) complex and the β subunit, however. These results emphasize the difference in primary structure that must exist between individual subunits[279] (see also References 61 and 280) and also demonstrate that antigenic determinants on the β subunit are expressed in the intact F_1·ATPase.

ATPase has also been used by Whiteside and co-workers[41,273] in serotaxonomic studies of the *Micrococcaceae*. The serological cross-reactivities of ATPases from various microorganisms were assessed by gel diffusion and enzyme inhibition experiments conducted with antiserum to *M. lysodeikticus* ATPase. The results were shown to correlate closely with a series of independent biochemical parameters, which included phospholipid and fatty acid composition, G + C contents and cell wall structure. It was concluded that the taxonomic relationship between the various microorganisms is reflected in the degree of similarity of their membrane ATPases.[273]

2. NADH Dehydrogenase

The biochemical properties of NADH dehydrogenase from *E. coli* have been studied by numerous investigators.[286-293] However, many of the results are contradictory and there is also conflicting evidence as to whether the enzyme is proton translocating or not.[152] It is clear that much of the present confusion surrounding the status of this key enzyme relates to the facts that it is refractory to extensive purification and that analytical systems such as polyacrylamide gel electrophoresis do not allow clear resolution of the catalytic species. Initial CIE studies using either Triton® X-100-solubilized membranes from *E. coli* K12[128] or partially purified preparations of the enzyme[292] were also dissappointing. Although a NADH dehydrogenase-active immunoprecipitate was detected, it was ill-defined and diffuse in nature. There was also some evidence to suggest that the enzyme was associating with outer membrane components, possibly lipopolysaccharide.[294]

A somewhat more illuminating picture was obtained following CIE of membrane vesicles of *E. coli* ML308-225.[108,153,159] Two immunoprecipitates possessing NADH dehydrogenase activity (numbers 15 and 19/27 in Figure 6) were clearly defined in tests against antivesicle immunoglobulins.[108,159] The two enzymes were antigenically unrelated, although one (number 19/27), which fractionated in both the envelope and cytoplasm fractions of the cell,[108] appeared to show partial identity with another vesicle antigen (Figure 6). The second NADH dehydrogenase (antigen number 15) partitioned almost exclusively in the membrane fraction[108] and, unlike antigen number 19/27, possessed nonheme iron[153] (Figure 8). The ability to clearly resolve the two functional NADH dehydrogenases in CIE test plates allowed some of the biochemical properties of the two enzymes to be directly compared.[295] Several interesting and distinctive features have emerged. First, the iron-containing enzyme (antigen number 15) is rapidly inactivated by 100 μM substrate in the absence of electron acceptors (see Figure $11A_3$). Second, the other NADH dehydrogenase (number 19/27) is partially inhibited by 10 m*M* adenine monophosphate (*cf.* Reference 291) and is exquisitively sensitive to low concentrations of guanidinium hydrochloride (see Figure $11A_4$). Furthermore, antigen number 15 can oxidize the NADH analog, nicotinamide hypoxanthine dinucleotide reduced form (deamino NADH), as effectively as NADH, but antigen number 19/27 cannot (see Figure $11A_1$). NADH and deamino NADH differ structurally at the 6-position of the purine ring, i.e., at a site removed from the nicotinamide moiety. Consequently, they have indentical redox potentials, and it is therefore difficult to explain the inability of antigen number 19/27 to oxidize the NADH analog in terms other than those in which the 6-amino nitrogen plays an obligate role in binding. In other words, the catalytic sites of the two enzymes are different. The fact that the two enzymes have different substrate specificities is also important in a practical sense in that it allows the properties of NADH dehydrogenase number 15 to be selectively monitored in membrane preparations which contain both enzymes. Thus, by using either NADH or deamino-NADH as substrates in transport experiments, it has been possible to show that antigen number 19/27, and not number 15, is responsible for ubiquinone-mediated NADH-dependent uptake of proline by membrane vesicles of *E. coli* ML308-225[295] (see Figure 11B and C); it will be remembered that Stroobant and Kaback[228] showed that transport systems in these vesicles were coupled very poorly to the oxidation of NADH in the absence of exogenously added ubiquinone-1). These results may explain the earlier observation that quanidinium hydrochloride totally impairs NADH-mediated transport in membrane vesicles[228] in a fashion which is not reversed by restoring the proton permeability of the membrane.[296] It seems likely that the major functional enzyme in this system is denatured by the chaotropic agent (see Figure $11A_4$). The conclusion that antigen number 19/27 is the enzyme primarily involved in the ubiquinone-mediated NADH-dependent transport phenomena in membrane vesicles of *E. coli* ML308-225 is surprising, since it might be expected that the membrane-associated iron-containing NADH dehydrogenase (antigen number 15) would be the major proton-translocating species. The situation is made all the more intriguing by the positive identification of antigen number 19/27 as dihydrolipoyl dehydrogenase in tandem CIE experiments conducted with the purified enzyme (Figure 12). Dihydrolipoyl dehydrogenase is an integral component of two multienzyme complexes in *E. coli*, viz., the pyruvate dehydrogenase and α-ketoglutarate dehydrogenase enzyme systems, and a well-documented feature is its ability to display NADH dehydrogenase activity,[297,298] (i.e., it can oxidize NADH at the expense of artificial electron acceptors, such as tetrazolium salts and quinones, which bear little structural resemblance to the natural acceptor, the lipoic acid moiety).

Thus, the combined application of immunochemical and biochemical techniques to the study of NADH oxidation in *E. coli* membrane vesicles has clearly shown that ubiquinone-mediated NADH-dependent transport[228] is largely a function of dihydrolipoyl dehydrogenase and not of a respiratory chain NADH dehydrogenase. Further experimentation, possibly involving the use of *lpd* mutants which lack functional

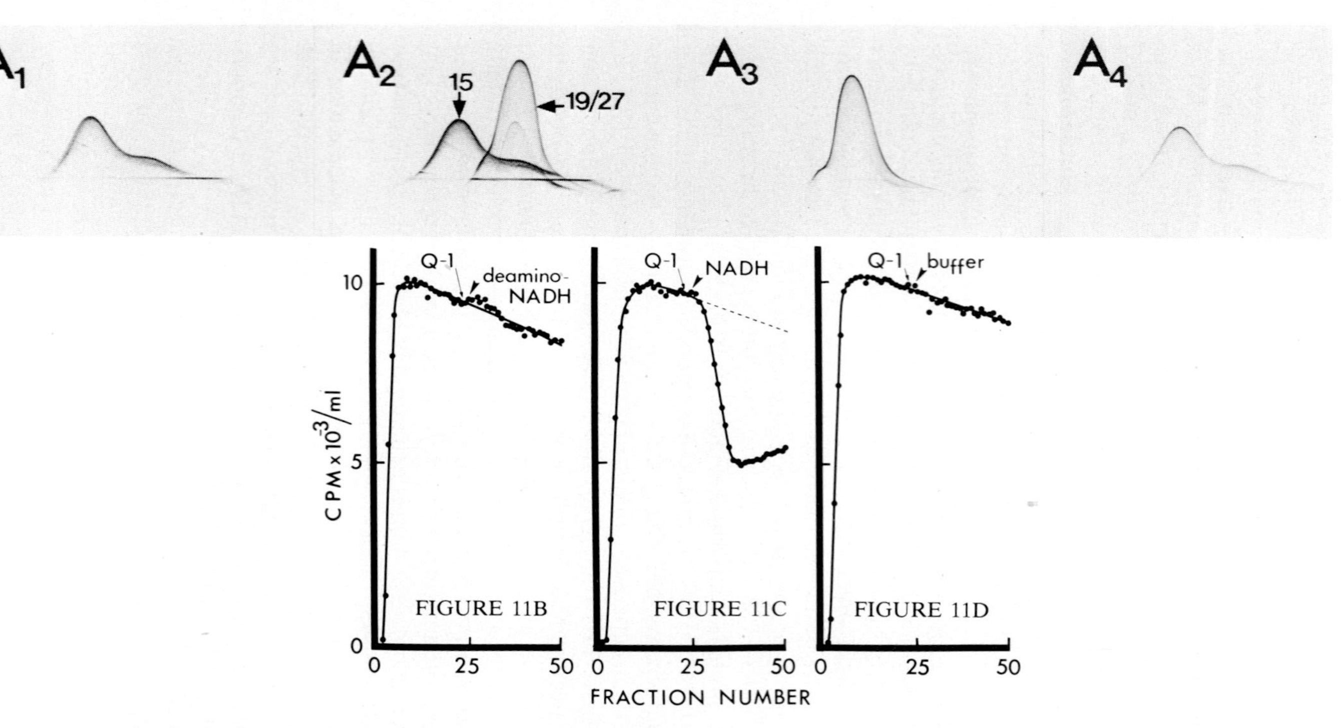

FIGURE 11. Resolution and identification of the antigen catalyzing ubiquinone-mediated NADH-dependent transport in membrane vesicles of *E. coli* ML308-225. (A) Triton® X-100-EDTA extracts (80 μg of protein) of plasma membrane vesicles were analyzed by CIE against antivesicle immunoglobulin. Wet gels were stained for deamino-NADH dehydrogenase (A_1), NADH dehydrogenase (A_2), and for NADH dehydrogenase after preincubation (10 min) with 100 μ*M* NADH (A_3) or 0.25 *M* guanidinium hydrochloride (A_4). The system used to number immunoprecipitates follows that shown in Figure 6. Anode is to the left and top of all gels. (B-D) The uptake of [^{14}C-] proline (250 Ci/mole) by membrane vesicles (4.8 mg of protein) was monitored by flow dialysis.[227] Ubiquinone-1 (Q-1), NADH, and deamino-NADH were added to the reaction mixture as indicated and to final concentrations of 80 μ*M*, 1m*M*, and 1 m*M*, respectively. Note that two immunologically unrelated antigens (number 15 and 19/27) are capable of oxidising NADH (A_2), and that only one of these (number 19/27) has substrate specificity (compare A_1 and A_2) similar to that evidenced during ubiquinone-mediated proline uptake by membrane vesicles (compare B and C). Vesicles do not transport proline to a significant extent in the absence of either NADH (D) or ubiquinone-1 (not shown). (Reprinted with permission from Owen, P. and Kaback, H. R., *FEMS Microbiol. Lett.*, 7, 345, 1980. Copyright by the Federation of European Microbiological Societies.)

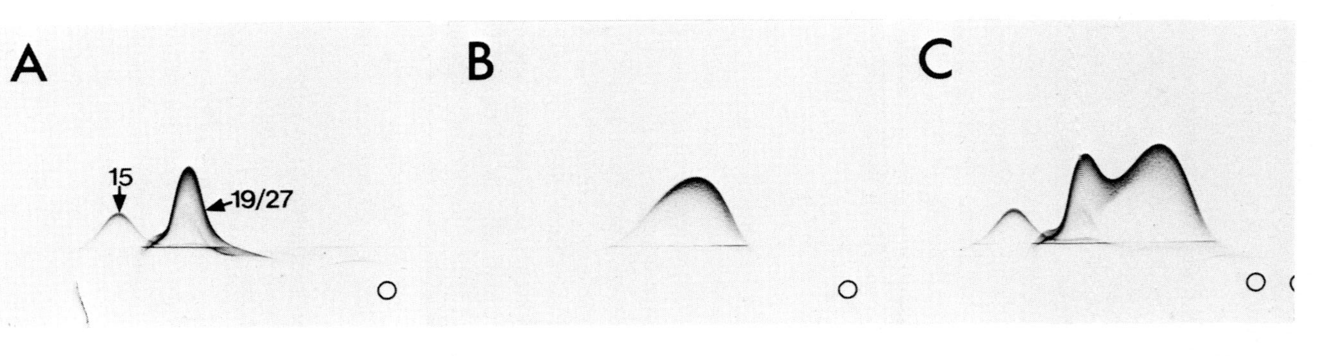

FIGURE 12. Identification of antigen number 19/27 by tandem CIE. A Triton® X-100-EDTA extract (44 μg of protein) of *E. coli* ML308-225 membrane vesicles (A) and purified dihydrolipoyl dehydrogenase from *E. coli* B (0.4 μg of protein; B) were analyzed by CIE against anti-vesicle immunoglobulin. (C) represents the corresponding tandem-CIE immunoplate in which the Triton® X-100-EDTA vesicle extract and the dihydrolipoyl dehydrogenase were analyzed simultaneously by electrophoresis from the front and rear wells, respectively. Wet immunoplates were finally stained by zymogram techniques for NADH dehydrogenase activity. Note that the immunoprecipitate corresponding to dihydrolipoyl dehydrogenase stains for NADH dehydrogenase activity (B), and fuses completely with immunoprecipitate number 19/27 (A) in tandem CIE experiments (C). The system used to number immunoprecipitates conforms with that shown in Figure 6. The positions of all wells have been delineated for clarity and the anode is to the left and top of all gels. (Reprinted with permission from Owen, P. and Kaback, H. R., *FEMS Microbiol. Lett.*, 7, 345, 1980. Copyright by the Federation of European Microbiological Societies.)

dihydrolipoyl dehydrogenases,[229] will be necessary before the full physiological significance of these observations is realized.

3. D-Lactate Dehydrogenase

Active transport of many solutes in *E. coli* ML308-225 membrane vesicles is coupled most effectively to the oxidation of D-lactate, a reaction which is catalyzed by a membrane-bound flavin-linked D-lactate dehydrogenase.[225,226] This enzyme has been purified to homogeneity by Kaback and co-workers[219,300] and many of its biochemical properties have been studied, including its reaction with specific antibody.[301] Anti-D-lactate dehydrogenase immunoglobulins are highly specific for the homologous flavin-linked enzyme and do not react with either soluble, pyridine nucleotide-dependent D-lactate dehydrogenase or membrane-bound flavin-linked L-lactate dehydrogenase, both of which catalyze very similar reactions. Antibody-induced inhibition of enzyme activity occurs in a manner that is dependent on both time and temperature and is maximal (95% inhibition) after about 6 hr at 37°C.[301] A radioimmunoassay utilizing D-lactate dehydrogenase labeled with either ^{125}I or the suicide substrate, [1 – ^{14}C]-hydroxybutynoic acid,[300] has been developed by Short et al.,[301] and has been used to show that membrane vesicles prepared from a strain of *E. coli* which lacks D-lactate dehydrogenase activity (*E. coli* ML 308-225dld-3) contain a catalytically inactive form of the enzyme which shares some determinants in common with the native protein. These results are confirmed and extended in Figure 13 where CIE immunoplates displaying tandem and co-CIE experiments have been stained by zymogram techniques for D-lactate dehydrogenase activity. Clearly, membranes from the wild type organism possess a D-lactate dehydrogenase-active antigen (Figure 13A) whereas those from *E. coli* ML308-225dld-3 do not (Figure 13B). Although not shown, it is also notable that an immunoprecipitate corresponding to the enzymatically active enzyme, but not the mutant enzyme, can be detected by protein-staining methods. Nevertheless, an antigenically related species must be present in dld-3 membranes since the profile of the D-lactate dehydrogenase-active immunoprecipiate (1) increases in peak area following co-CIE experiments (Figure 13C) and (2) is altered dramatically in tandem CIE experiments (Figure 13D). These results suggest that soluble immune complexes are formed by the interaction of the inactive dld-3 antigen with antibodies to the wild type enzyme. It seems probable that the dld-3 mutation induces a major structural change in the D-lactate dehydrogenase molecule, resulting in the loss of all but one of the determinants originally expressed thereon.[148]

Anti-D-lactate dehydrogenase immunoglobulins have also been utilized by a number of workers to probe the accessibility of D-lactate dehydrogenase in, and, by implication, the structure of, isolated bacterial membrane vesicles.[234,237,302] Short et al.[302] found that specific antibody had little effect on D-lactate dehydrogenase activity, D-lactate oxidation or D-lactate dependent-lactose transport in membrane vesicles derived by osmotic lysis[303] from *E. coli* ML308-225, observations which would imply that D-lactate dehydrogenase was located on the inner surface of the vesicles. In contrast, lactate oxidation and D-lactate-dependent transport could be blocked readily by specific antiserum in *E. coli* ML308-225dld-3 vesicles reconstituted with active D-lactate dehydrogenase enzyme.[302] In this instance, the D-lactate dehydrogenase appears to be located preferentially on the outer surface of the (reconstituted) vesicles. Futai and Tanaka[234] have confirmed much of this data and have further shown that D-lactate dehydrogenase and D-lactate-dependent oxygen uptake are completely inhibited by anti-D-lactate dehydrogenase antibody in membrane vesicles prepared with the aid of a French pressure cell. These and other data suggest that lysis of *E. coli* in the French press results in vesicles with an inverted topology,[234] and may explain, in part, the observation of Yamato et al.[237] that 60% of the D-lactate oxidase activity in their vesicle

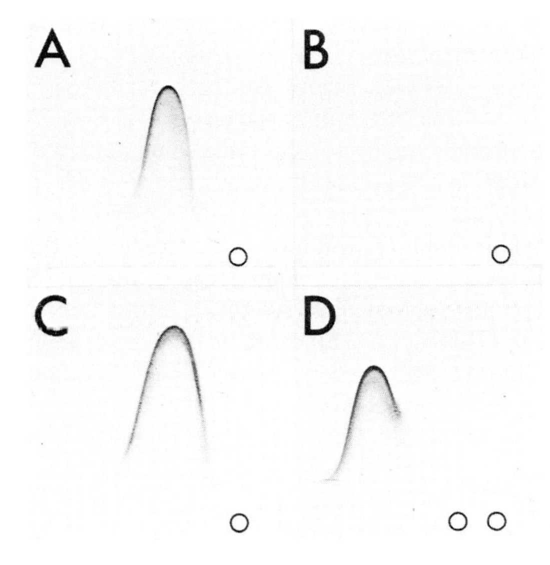

FIGURE 13. Analysis of D-lactate dehydrogenase by CIE and zymogram techniques. Triton® X-100-EDTA extracts of membrane vesicles from *E. coli* ML308-225 (5.3 μg of protein) and from *E. coli* ML308-225dld-3 (38 μg of protein) were anlayzed by CIE (A and B, respectively) and by co-CIE (C) and tandem CIE (D) against antivesicle immunoglobulin. In the tandem CIE experiments (D), the dld-3 extract occupied the rear well. Wet immunoplates were stained by zymogram techniques for D-lactate dehydrogenase activity. Note that the immunoprecipitate in (A) increases in area following co-CIE (C) even though no enzyme-active or protein-staining immunoprecipitate is evident for the dld-3 extract (B). Note also the break in the profile of immunoprecipitate observed following tandem CIE (D). The wells have been delineated for clarity and the anode is to the left and top of all gels.

preparations was inhibited by specific antibody. Certainly, there is now a considerable body of evidence to suggest that D-lactate dehydrogenase is largely inaccessible to homologous antibody in membrane vesicles prepared from *E. coli* by osmotic lysis[108,160,234,302] (see also Section VI.A.3).

4. *D-Alanine Carboxypeptidase*

Many bacteria possess in their membranes proteins that are capable of binding penicillins and related β-lactam antibiotics.[304,305] The most abundant of these, a D-alanine carboxypeptidase involved in peptidoglycan biosynthesis, has been studied extensively by Ghuysen.[305] In a series of careful experiments, it was shown that antiserum to the DD-carboxypeptidase from *Streptomyces* R61 inhibited about 80 to 85% of the DD-carboxypeptidase and transpeptidase activities, as well as the penicillin-binding capabilities of the homologous purified enzyme.[306] The same antiserum also inhibited the soluble DD-carboxypeptidase from *Streptomyces* K15 and *S. rimosus,* suggesting certain structural similarities between these enzymes. In contrast, the corresponding enzymes from *Actinomadura* R39 and *S. albus* were unaffected. It should be noted that the above data pertain to the interaction of immunoglobulins with soluble forms of DD-carboxypeptidase. Strikingly different results were obtained for the membrane-bound enzymes. Antiserum to the R61 enzyme had relatively little or no effect on the

DD-carboxypeptidase or transpeptidase activity expressed by membranes isolated from *Streptomyces* R61, K15 or *rimosus*.[306] These results are in marked contrast to the situation for bacterial ATPase, where antibody can readily inhibit both the membrane-associated and soluble form of the enzyme[41,274,275,281,282] (see Section V.A.1). DD-carboxypeptidase probably does not share the same degree of surface exposure as F_1·ATPase, and is buried in a region of the membrane where its antigenic sites are less accessible to antibody.

Several other penicillin-binding proteins can be detected in the membranes of most microorganisms.[304] However, these do not appear to be related immunologically to the D-alanine carboxypeptidase, at least in *B. subtilis*.[307] Another class of bacterial products which also share the ability to bind penicillins and cephalosporins are the β-lactamases.[308] These too appear to be structurally and immunologically unrelated to the other penicillin-binding proteins.[309]

B. Exported Proteins

1. General Comments

Proteins undergo a variety of conformational changes from initial synthesis on ribosomes to ultimate location in the cell envelope. A hypothesis which is gaining rapid acceptance suggests that exported proteins are initially synthesized with an additional hydrophobic sequence of amino acids attached to the N termini.[223-224] It is proposed that these leader or "signal" sequences function to bind polysomes to the membrane, thus allowing the nascent proteins to be vectorially discharged through hydrophilic pores. The signal sequences are removed by proteolysis as the nascent polypeptides traverse the membrane, and the extruded proteins are then free to adopt their native conformations.[223,224] Several important aspects of this hypothesis have been confirmed with the aid of specific antibody prepared against purified exported proteins.

At least two classes of exported proteins may be recognised in a typical Gram-negative organism, viz., those of the outer membrane and those of the periplasmic space. Proteins of both classes appear to be synthesized on membrane-bound polysomes as judged from the results of immunoprecipitation experiments.[310-313] These include the λ-receptor of the outer membrane, and three periplasmic proteins, viz., the maltose-binding protein, the arabinose-binding protein, and alkaline phosphatase. On the other hand, soluble cytoplasmic proteins, such as elongation factor Tu,[311,312] are synthesized on free ribosomes. By analysis of precipitated material obtained following addition of monospecific antisera to the products of either in vitro protein synthesis or of synthesis by toluene-treated cells, it has been firmly established that lipoprotein,[314,315] the λ-receptor,[312] alkaline phosphatase,[316] the maltose- and arabinose-binding proteins,[312] and penicillinase,[317] but not cytoplasmically located elongation factor Tu,[312] are synthesized in forms a few thousand daltons longer than the authentic proteins. These results lend strong support to the "signal" hypothesis[223,224] and emphasize an important use of specific antisera, i.e., in the selective resolution of related or cross-reacting antigens. The results also suggest that a hydrophobic N-terminal extension of between 15 to 30 amino acids is probably insufficient, in many instances, to effect a radical change in protein conformation. Of course, this does not imply that exported proteins are extruded from the membrane in a completely folded conformation. Ample evidence exists to the contrary,[318,319] some from immunological studies of alkaline phosphatase.

2. Alkaline Phosphatase

Active alkaline phosphatase has a molecular weight of 86,000 daltons and contains two identical subunits, as well as four tightly bound Zn^{2+} ions per mole and phosphate molecules. Although it is normally considered to be a periplasmic enzyme, its inclusion here is relevant for a number of reasons. First, during the initial stages of synthesis the nascent polypeptides are membrane associated.[310,313] Second, the immunochemistry of the enzyme has been studied in detail, notably in the laboratories of Schlesinger[310,320-329] and of Lazdunski.[313,330-333] Finally, an appreciation of the conformational changes experienced by a dimeric periplasmic enzyme from synthesis to final assembly may be instructive when considering possible mechanisms of, and experimental procedures for probing, integration, and correct spatial positioning of membrane-bound proteins which perform coupled biological function.

A singularly effective approach to the study of alkaline phosphatase has involved the generation and purification of antibody populations which have primary specificities for different conformational forms of the molecule. This has been achieved by immunization with denatured forms of the subunit, the active dimer, and with inactive Zn^{2+}-deficient alkaline phosphatase molecules which exist in monomer-dimer equilibrium.[322,331,332] The conformationally specific immunoglobulins have been applied to the analysis of the wild type enzyme,[320,322,331] the Zn^{2+}-deficient apoenzyme,[321,329,332] inactive alkaline phosphatase molecules containing amino acid analog,[324,326-328] or having defects in their structural genes,[323] as well as to the study of alkaline phosphatase during initial synthesis on polysomes,[310,316] and during secretion by spheroplasts.[325,333] Much detailed information has also been derived from the quantitative study of the interaction between various molecular forms of the enzyme and ^{125}I-labeled γ-globulins.[330-332] From these data the following picture has emerged.

Alkaline phosphatase is synthesized preferentially on membrane-associated ribosomes, and the nascent monomer chains appear to be extruded from the membrane initially in an extended conformation, since they are capable of reacting with antisera directed against denatured subunits.[310,313] At what stage during the extrusion process the nascent chain adopts or begins to adopt its native conformation is unclear. In any event, it has been estimated that an uncoiled denatured monomer molecule possesses two antigenic determinants, only one of which is eventually expressed in the active dimer.[331] This explains the lack of cross-reactivity observed between denatured monomers and active alkaline phosphatase preparations in gel precipitation tests.[322,331] During the process of refolding one of the two antigenic determinants on the extended form of the monomer is lost.[331] Folded monomers are still capable of interacting with antiserum raised to denatured subunits by virtue of the remaining determinant, and bound immunoglobulin molecules can prevent the next stage in the assembly process, namely that of association of the two folded subunits to form the apoenzyme dimer.[322] Distinct changes in conformation occur during dimerization. These involve masking of the remaining antigenic site on each monomer and, in addition, the expression of two new conformational determinants, each of which can be recognized by antiserum to the native enzyme.[331] Replacement of histidine, proline, and arginine by the amino acid analog 1,2,4-triazole-3-alanine,[324] 2-methylhistidine,[326] azetidine-2-carboxylate,[328] and canavanine[327] results largely in the formation of altered subunits that are unable to dimerize and that can be detected in the periplasmic fluid by the use of antibodies to denatured monomers. Some mutants with defects in their structural genes for alkaline phosphatase produce anomolous subunits which behave in a similar fashion.[323] These studies emphasize the importance of several amino acids, especially histidine,[324,326] in the assembly process, and also illustrate that extrusion of these proteins through the membrane does not require such a high degree of specificity on the primary structure of the polypeptide chain as does dimerization.

Lazdunski and co-workers have also provided evidence that the apoenzyme undergoes profound conformational changes upon binding Zn^{2+} ions.[331] Apparently, the two dimer antigenic determinants that are exposed on the apoenzyme become masked and four additional ones become expressed. In addition, two other determinants, which were exposed on uncoiled monomers, also appear on the enzyme surface.[331] However, as might be expected, *activation* by Zn^{2+} is not blocked by antibodies to denatured monomers.[322] Anticooperative binding of phosphate to each subunit further results in the reappearance of the two antigenic determinants originally exposed on the apoenzyme, and loss of one of the two monomer determinants. Thus, the final active enzyme, containing bound Zn^{2+} and phosphate, appears to possess six determinants that can be recognized by homologous antiserum and an additional one which is recognized only by antiserum to the uncoiled monomer.[331] The presence of this latter determinant implies that the active alkaline phosphatase dimer is asymmetric.

Antibody directed towards the native enzyme inhibits catalytic activity by 30%.[320] In contrast, Fab fragments enhance Zn^{2+}-induced reactivation of a Zn^{2+}-deficient alkaline phosphatase produced by a mutant of *E. coli.* Furthermore, antibodies specific for the dimer form of the Zn^{2+}-deficient alkaline phosphatase activate the homologous enzyme but inhibit the wild type molecule. It seems probable that antibodies induce a confomational change in the mutant protein, producing a structure more akin to that of the wild type alkaline phosphatase.[332] Precedents for this type of effect certainly exist. It will be recalled that an inactive β-galactosidase produced by a *lac* z^- mutant of *E. coli* could be activated 900-fold by antibody raised to the wild type enzyme.[142,334]

C. Outer Membrane Proteins

1. Lipoprotein

Of the proteins in the outer membrane of *E. coli,* the most abundant, and certainly the best characterized, is the component commonly referred to as Braun's lipoprotein.[134] The lipoprotein molecule has a molecular weight of 7200 daltons and is composed of 58 amino acids whose primary sequence has been determined. The N-terminal amino acid is glycerylcysteine(1) to which are attached two ester-linked and one amide-linked fatty acids. One molecule in three is covalently linked to the peptidoglycan sacculus via bonds from the side chain of C-terminal lysine (58) to the L-center carboxyl group of mesodiaminopimelic acid. A lipoprotein molecule is probably attached to every tenth repeating unit in the cell wall. Aspects of the biosynthesis and assembly of lipoprotein are also well established and the molecule is thought to function, at least in part, in anchoring and maintaining the structural integrity of the outer membrane.[134,240,335]

Lipoprotein is an extremely interesting antigen. It is found in most members of the Enterobacteriaceae, including various species of *Escherichia, Edwardsiella, Citrobacter, Salmonella, Shigella, Klebsiella, Enterobacter, Serratia, Proteus,* and *Erwinia,*[108,336-340] and is extremely stable, retaining a highly ordered α-helical conformation over a wide pH range and in the presence of SDS. Extreme conditions, such as treatment with 8 *M* urea or 4% SDS, or heating to 90°C alter the native conformation, but renaturation is observed following removal of urea or detergent or upon cooling.[134,341,342] These physical properties are quite important from a practical viewpoint. Lipoprotein possesses only one aromatic amino acid per polypeptide chain, viz., tyrosine(56) and is not as amenable to quantitation by standard biochemical procedures as the average protein. Fortunately, the extreme stability of the molecule, together with its propensity to regain its native conformation, provide the bases for an alternative and popular assay involving the use of specific antibodies.[314,315,337,343-349]

Lipoprotein elicits a good antibody response in rabbits, especially if injected in the membrane-bound form.[108,128,339,347] Subtle differences in antibody response have been discerned when immunization is conducted with different molecular forms of purified lipoprotein, and these have been used to assess the important antigenic region of the polypeptide.[339] Thus, the major determinants present in lipoprotein molecules which have been released from the peptidogylcan by lysozyme (lipoprotein I) and by trypsin (lipoprotein II) appear to be the muropeptide attachment region and the C-terminal lysine(55), respectively. Very little cross-reaction is observed between lipoproteins I, II, and free lipoprotein (lipoprotein IV). However, antiserum raised to whole cells or to a form of the lipoprotein which terminates at threonine(54) (lipoprotein III) reacts strongly with all species of lipoprotein, suggesting the presence of a common determinant located towards the N terminus of the polypeptide.[339] The ester-linked fatty acids do not appear to contribute directly to this determinant region, but they may stabilize it, since irreversible denaturation can be achieved under deacylating conditions.[134]

Antibodies to the lipoprotein molecule have been detected in the sera of patients with severe enterobacterial infections[350] and also in rabbits with experimental pyelonephritis induced with an encapsulated strain of *E. coli.*[351] The latter observation emphasizes the point that proteins, such as lipoprotein, which are not exposed on the cell surface (see Section VI.A.3), can become expressed during infection and result in significant local synthesis of antibody. On the other hand, there is evidence to suggest that antibodies to lipoprotein are detected with less frequency than antibodies to the major surface antigen, lipopolysaccharide.[350] Moreover, the recurrent nature of many enterobacterial infections (e.g., infections of the urinary tract) suggests that antibody to lipoprotein is unlikely to have any protective effect. Rather it represents a marker for the presence of infection.[351]

Lipoprotein is a potent B-mitogen, inducing proliferation of mouse B-lymphocytes,[352] mouse and rat splenocytes,[353,354] and rabbit and bovine lymph node cells.[353] The effect on T-lymphocytes is less pronounced.[134,353] The N-terminal glycerylcysteine residue with attached fatty acids appears to be the biologically important part of the molecule, modifications to the C-terminal portion of lipoprotein having only marginal affect on activity.[353,354] This is reminiscent of the situation for lipopolysaccharide, in which the mitogenic activity has been shown to reside in the hydrophobic lipid-A region.[355] However, results from experimentally induced infections suggest that lipoprotein is not as powerful a mitogen as endotoxin.[351]

2. Other Outer Membrane Proteins

Many other proteins in the outer membrane of *E. coli* are also believed to be immunogenic. Solubilized outer membranes, for example, give a characteristic array of immunoprecipitates when analyzed by CIE against antienvelope immunoglobulins.[128] Furthermore, a full spectrum of outer membrane proteins is present in material precipitated from solubilized outer membranes of *E. coli* 026K60 by antiserum elicited in rabbits to different preparations of the bacterium.[356] Apart from lipoprotein, the predominant outer membrane proteins of the Enterobacteriaceae fall in the molecular weight range of 30,000 to 42,000 daltons.[356a] For convenience, they can be divided into two classes, viz., those which appear to be associated with the peptidoglycan layer, e.g., the matrix protein and other porins, and those which are not peptidoglycan associated. The best characterized member of the latter group is the 33,000-dalton, heat-modifiable protein of *E. coli* (the *omp A* gene product).[240] The actual molecular weight (and also number) of these outer membrane proteins can vary considerably from species to species and from strain to strain, resulting in a variety of different profiles when examined by SDS-polyacrylamide gel electrophoresis.[356a,356b] Among encapsulated strains of *E. coli,* the different patterns of major outer membrane proteins has recently been correlated with discrete O:K serotypes.[356c,356d] Evidence based upon the use of interfacial precipitin tests,[357] SDS-polyacrylamide gel immunoperoxidase studies,[357a,357b] and SDS-polyacrylamide gel-crossed immunoelectrophoresis[357c] appears to indicate that considerable cross reactivity exists between the peptidoglycan-associated proteins of members of the Enterobacteriaceae. Similar cross-reactivity has also been detected for the heat-modifiable proteins. It is interesting to note that this cross-reactivity is independent of molecular weight variation and has been detected for proteins which differ strikingly in their pattern of cyanogen bromide fragments[357d] or of tryptic peptides.[357e] However, it should be stressed that the principal techniques used to assess cross-reactivity (e.g., the SDS-polyacrylamide gel immunoperoxidase technique and SDS-polyacrylamide gel-crossed immunoelectrophoresis) involve the analysis of proteins which have been heated in SDS. Thus, the results may not reflect the true immunological relationship that exists in vivo. A related criticism can often be made of the specific antisera which have purportedly been generated to a number of *E. coli* outer membrane proteins, e.g., matrix proteins (the *omp C* and *omp F* proteins),[358-363] the *omp A* protein,[363,363a] phage-directed protein 2,[359,362] the λ-receptor (the *lam B* protein)[364-366] and protein e.[366a] In many instances, the specificities of the various immunoglobulin preparations have been poorly defined. It is also unclear whether many of the procedures adopted for purification of the corresponding immunogens allow the full native conformation of the polypeptides to be retained. In the author's hands, preparations of matrix protein, purified to near homogeneity by either the method of Rosenbusch,[358] or by procedures known to retain the instrinsic β-structure of the polypeptide,[367] fail to give immunoprecipitates against antienvelope immunoglobulins which cannot be attributed either to contaminating lipopolysaccharide or, more usually to contaminating lipoprotein.[108,128]

Few of the other outer membrane proteins of *E. coli* have been characterized in detail by immunological procedures. Interesting candidates for study might be protein K, a 40,000-dalton polypeptide which appears to be related to the presence of a capsule in *E. coli* strains causing purulent meningitis[356c] and the *tra T* protein, which is a 25,000-dalton polypeptide involved in surface exclusion and responsible for plasmid-specified serum resistance.[367a,367b]

The outer membrane proteins of pathogenic neisseria have received increasing attention over the past few years. In an important study in 1976, Johnston et al[253] demonstrated that *Neisseria gonorrhoeae* could be subdivided into 16 serologically distinct groups based upon the antigenic specificity of a membrane protein complex, which contained as its dominant component the major protein of the outer membrane. Each putative serotyping antigen had a unique molecular weight ranging from 32,000 to 39,000 daltons, and was not related to colony type.[253] Subsequent studies by Swanson[368] have revealed that proteins within this molecular weight range can be grouped into two main classes based on analysis of their ^{125}I-labeled tryptic peptides. It was speculated that this might reflect the existence of phase variations analogous to those observed in *Salmonella* flagellins.[369] The gonococcal serotyping antigen is expressed on the cell surface, as judged from results of fluorescent-[370] and ^{125}I-labeling[371] experiments, and can apparently produce strain-related immunity to gonococcal challenge in guinea pigs.[372] Furthermore, Hildebrandt et al.[373] have recently shown by transformation experiments that the ability of gonococci to resist complement-dependent serum killing, and thus the possible capacity to cause disseminated infections, is related to the presence of a distinct class of this protein of characteristic antigenicity and molecular weight (36,500 daltons).

Another gonococcal protein of similar molecular weight has recently been purified by Chen and co-workers.[374] This is the so-called L-antigen, antibodies to which have been demonstrated in 95% of infected females.[375] It is species specific, unlike the major protein of Johnston et al.,[253] and is unstable inasmuch as it does not appear to be expressed by isolates following subculture in vitro.[374] It is also heat labile and is sensitive to trypsin. Both its extraction characteristics and observed expression on the cell surface would tend to suggest that the L-antigen is an outer membrane component.[374] However, this has yet to be demonstrated conclusively.

An additional gonococcal surface antigen is thought to be involved in the adsorption of cells to human polymorphonuclear leukocytes.[376] Analysis has shown that this 28,000 to 29,000 dalton polypeptide is considerably reduced in isogenic strains which exhibit low levels of leukocyte association. Furthermore, antiserum raised to the corresponding polypeptide band observed on SDS-polyacrylamide gels reduces significantly the level of association of gonococci to leukocytes. Thus, far from restricting the ability of cells to become phagocytosed, this protein appears to actively enhance the bacterium-host cell interaction.[376] The molecule is clearly expressed on the cell surface and shows strain-dependent variation of molecular weight[376] which is similar to that observed for another outer membrane protein from *N. gonorrhoeae.* This latter component has been correlated with increased gonococcal aggregation and susceptibility to killing by proteolytic enzymes.[371,377-379] The data presently available do not allow for a definite correlation between these two functionally interesting antigens. However, it could be noted that they have closely similar molecular weights and that loss of each results in a gain of potential virulence characteristics.[374,378,380]

Although the above studies have greatly increased our understanding of the biological properties of gonococcal surface proteins, it should be recognized that the purified antigens have been investigated in only a few instances.[374] Indeed, some serotype preparations clearly represent a complex association of several antigenic species.[253] However, more definitive studies seem to be in the offing, since procedures have been developed recently which allow several of the major outer membrane proteins of *N. gonorrhoeae* to be isolated free of obvious contamination.[370,381]

Variations in the spectrum of outer membrane proteins observed following SDS-polyacrylamide gel electrophoresis have also been used to detect differences in group B meningococci[382,383] and in subspecies of *Bacteroides fragilis.*[384] In the former study 0.2 *M* LiCl or 0.2 *M* $CaCl_2$ extracts of whole cells were shown to contain several outer membrane proteins, the major one of which was thought to be responsible for serotype specificity.[382,385,386] Much of the anti-group B bacteriocidal antibody was also thought to be directed against this protein.[382,386] Doubts relating to the identity of the biologically active species have been largely dispelled in more recent studies. Frasch and Robbins[387] have purified the major serotype 2 protein antigen free of contaminating lipopolysaccharide and to apparent homogeneity as judged by SDS-polyacrylamide gel electrophoresis. (Serotype 2 is of special interest since it has been associated frequently with meningococcal disease.)[388-390] Purified preparations of this 41,000-dalton polypeptide are excellent immunogens and protect guinea pigs against infection by homologous serotype 2 strains of meningococcal groups B,C, and Y.[387] It should be noted that the protein isolated from organisms of serotype 2 is distinct from the corresponding 41,000-dalton polypeptide observed in the outer membrane of meningococci of serotype 11 as judged by comparative analysis of cyanogen bromide and chymotrytic peptides.[390a] However, there do seem to be some peptide sequences in common between the two proteins, and these are largely hydrophobic in nature. In contrast, most of the unique peptides are hydrophilic.[390a] Thus, it may be that the serotype specificity is determined largely by that hydrophilic part of the 41,000-dalton protein which is in direct contact with the aqueous environment. The results of the above experiments open up the very real possibility of using preparations of purified outer membrane proteins as protective vaccines against group B meningococcal disease. Furthermore, the positive identification of the 41,000-dalton serotyping antigen in CIE profiles of *N.*

meningitidis membranes[390b] should facilitate an analysis of the human immune response to this important envelope component.

Encouraging results have also been obtained in preliminary tests of the protective capacity of outer membrane proteins of *Salmonella typhimurium.*[390c,390d] Porin preparations containing the 34,000-, 35,000-, and 36,000-dalton proteins were shown to be good immunogens in both mice and rabbits and to give significant protection against experimental salmonellosis in mice.[390c] Interestingly, vaccination of mice with porins which had been covalently coupled to a nontoxic octasaccharide derived from the salmonella O-antigen afforded protection against an approximately tenfold higher challenge dose of *S. typhimurium* than was afforded to mice which had been vaccinated with porin or octasaccharide alone.[390d] Antibodies elicited by the conjugate possessed specificity for both oligosaccharide and porins and were active in passive immunization tests.[390d] The results of further trials with these novel and artificial O-antigen-porin vaccines will be awaited with interest.

Another unrelated protein antigen from the salmonella outer membrane is characterized by a high affinity for lipopolysaccharide.[390e] Recently, this protein has been purified to homogeneity from EDTA extracts of *Salmonella minnesota* and has been shown to be a basic polypeptide of molecular weight 15,000 daltons. The lipopolysaccharide-binding protein coprecipitates with purified lipopolysaccharide, a reaction which appears to be a consequence of ionic interactions between basic residues on the binding protein and negatively charged phosphate groups in the lipid A region of the endotoxin molecule. As might be anticipated from its affinity for lipopolysaccharide, this protein is expressed on the bacterial surface. Furthermore, antibodies to it can be adsorbed by strains of *Salmonella, Shigella, Escherichia,* and *Klebsiella,* suggesting that it may represent a common surface antigen of the Enterobacteriaceae. The biological function of this interesting protein is less clear, however. One speculation has been that it renders translocation of lipopolysaccharide from its site of synthesis in the inner membrane to its eventual cellular location in the outer membrane an essentially irreversible process[390e] (see also Section VI.B.1.c).

Proteins from the outer membrane of other bacteria have not received the same attention as those from species of *Escherichia, Neisseria,* and *Salmonella.* However, two proteins of the porin type and of molecular weights 36,000 and 39,000 daltons have been isolated from *Proteus mirabilis.* Both elicit strong and specific antibody responses in rabbits.[390f] In another study, three purified outer membrane proteins from *Pseudomonas aeruginosa* were tested for their ability to stimulate the proliferation of murine lymphocytes.[390g] One of these proteins was a 33,000-dalton heat-modifiable porin called protein F.[390h] The other two were lipoproteins termed proteins H and I. Protein I is an 8000-dalton lipopolypeptide analogous to Braun's lipoprotein,[390i] whereas protein H is a novel 21,000-dalton peptidoglycan-associated lipoprotein (analogs of which have also been detected in *E. coli).*[390j] In a manner reminiscent of Braun's lipoprotein, all three of the pseudomonad proteins were found to be B-mitogens.[390g] One might anticipate that the fatty acid moieties of proteins H and I would be essential for mitogenicity, since such has been found to be the case for Braun's lipoprotein (see previous section) and other bacterial cell surface amphiphiles.[390k] Of course, a similar situation is unlikely to apply to protein F, which apparently lacks acyl moieties.

D. Lipids

Several of the lipids commonly found in microbial membranes are immunologically active. These include the acidic phospholipids, such as cardiolipin, phosphatidylglycerol, and phosphatidylinositol,[391-400] and various glycolipids and phosphoglycolipids.[394,398,399,401-419] Antibodies to lipids can often be demonstrated in serum obtained following immunization with isolated membranes,[397,416,420,421] with cells,[397,409,412,422] or with lipids which have been reassociated with denatured membrane proteins.[339,410,418,419,423] However, purified lipids themselves do not elicit antibody responses in laboratory animals and should thus be regarded as haptens. A popular procedure for generating specific antilipid immunoglobulins involves immunizing with a mixture consisting of the relevant lipid and "auxiliary" lipids adsorbed to methylated bovine serum albumin.[392,396,397,419] The usual "auxiliary" lipids are phosphatidylcholine and cholesterol, neither of which is immunogenic. Both appear to assist in formation of immunologically active micellar structures.[395]

Gel precipitation techniques cannot be recommended for the analysis of lipids and their antibodies. Individual lipid molecules are univalent, and although they can act as multivalent antigens when aggregated following dispersal in aqueous solution, they often give poorly defined immunoprecipitates and artifactual reactions of identity in immunodiffusion tests against specific antibody. For these reasons, other techniques such as the microflocculation or complement fixation tests or the liposome permeability model are more commonly employed.[398]

One of the better characterized microbial lipids is cardiolipin.[391-393] This acidic phospholipid occurs widely in both Gram-positive and Gram-negative bacteria, and reacts with the Wassermann antibody found in syphilitic serum. The immunologically important regions of the molecule have been determined by analysis of chemically or enzymatically modified forms of natural cardiolipin or of synthetic cardiolipin analog.[391-393] Antibodies are directed largely against the polar head groups, a feature which appears to be common for all phospholipids. The two phosphate groups and the central free hydroxyl group are essential for serological activity. The distance between the two phosphate groups is also critical, but the hydrophobic fatty acid moieties seem to be less important since the degree of unsaturation does not influence the reaction with antibody.[391-393]

Antiphosphatidylglycerol antibodies have also been prepared and have been used to study the immunological relationship between phosphatidylglycerol and cardiolipin.[396,397] However, the results are conflicting. In an immunological study of *M. lysodeikticus* phospholipids, De Siervo[396] found that cardiolipin did not react with antiphosphatidylglycerol antiserum although phosphatidylglycerol could react with anticardiolipin immunoglobulins. Schiefer et al.[397] have since found entirely the opposite reactions for the corresponding lipids from *Mycoplasma hominis*! There are also conflicting reports as to whether cardiolipin cross-reacts with other phospholipids such as phosphatidylinositol and phosphatidic acid.[395,396,424] The reasons for these descrepancies are not immediately obvious. On the other hand, there does appear to be fairly universal agreement that the antigenic properties of amphiphilic phospholipid molecules are determined largely by the hydrophilic headgroups.

Another class of serologically active lipids are the glycolipids.[425] These occur as minor constitutents of most microbial membranes but are found in significant quantities in many species of mycoplasmas. Indeed, in some mycoplasmas, glycolipids appear to constitute one of the major antigenic classes. The role of glycolipids in the serology of *Mycoplasma pneumoniae,* for example, is well established. They react with specific antibody to fix complement and are intimately involved in the mechanism of immune lysis. Furthermore, they react with antibodies inhibiting growth and metabolism of the organism.[403,408-410,423] The serologically active glycolipids consist of digalactosyl diglyceride, trigalactosyl diglyceride, and diglycosyl and triglycosyl diglycerides containing both glucose and galactose.[406,408,417]

Acholoplasma laidlawii can also be killed by the combined action of antibodies and complement.[399,400,427] However, antibodies to membrane proteins and to phospholipids, as well as antibodies to glycolipids, appear to be involved in this instance.[399] The predominant complement-fixing activity of the glycolipids is associated with monoglucosyl diglyceride, diglucosyl diglyceride, and a phosphoglycolipid, glycerylphosphoryldiglucosyl diglyceride.[405,416,418] The phosphoglycolipid is also anticomplementary in the purified form and is thought to inhibit one of the complement components (C2) essential for lysis.[418] Serologically active glycolipids and phosphoglycolipids have been demonstrated in several other species of mycoplasmas,[398,412,415,419,421,422] and many have been shown to cross-react.[412,416,417] Glycolipids from mycoplasma have also been shown to possess determinants in common with glycolipids from Gram-positive bacteria and from plants.[405,408,429] This observation is not surprising, since glycolipids are widely distributed in nature and possess antigenic specificities which are dictated by the relatively simple hydrophilic sugar moieties.[398,416,419] A well-documented cross-reaction is that observed between *Treponema reiteri* and nervous tissue. In this instance, different galactose-containing glycolipids appear to be implicated.[413,430] *T. reiteri* possesses galactosyl diglycerides,[431] whereas nervous tissue contains galactosyl ceramides.

In contrast to *M. pneumonia* and *A laidlawii,* the thermophilic mycoplasma, *Thermoplasma acidophilum,* possesses glycolipids and phosphoglycolipids which fail to react in complement fixation tests.[416] Although the precise reasons for this phenomenon are unclear it may be due in part to the fact that the lipids of *T. acidophilum* are extremely hydrophobic and have a unique molecular structure (viz., they are symmetrically assembled diglycerol tetraethers which contain two sn-2,3-glycerol residues bridged through ether linkage by two C_{40}-isopranoid branched diols).[432] *T. acidophilum* also possesses a phenol-extractable compound which exhibits high activity in complement fixation tests.[416,433] Similar antigens have been observed in several other species of mycoplasma, and have been termed lipopolysaccharides because of certain similarities with the classical lipopolysaccharide of Gram-negative bacteria. However, it has been shown that mycoplasma lipopolysaccharides are really glycolipids with very long carbohydrate chains. For example, the lipopolysaccharide of *T. acidophilum* contains 24 mannose residues glycosidically linked to glucosylglycerol tetraether.[434] Unlike classical glycolipids, the mycoplasma lipopolysaccharides are immunogenic per se.[435]

Other serologically active microbial lipids include the glycolipids of streptococcal L-forms,[414] and the toxic dimycolates of trehalose (cord factors), sulfolipids, mycosides, and phosphatidylinositol mannosides found in mycobacteria and related organisms.[83,436] These have been described in detail elsewhere.[83,414,436]

VI. ANTIGENIC ARCHITECTURE OF MICROBIAL MEMBRANES

Over the past decade a great deal of effort has been devoted to an analysis of the molecular structure of bacterial membranes.[155] Various approaches have been employed and many of these are discussed elsewhere in this Series. One of the more fruitful approaches has been the use of antibodies directed against either membranes or isolated membrane components. Two basic immunochemical procedures may be distinguished in this respect. One involves immunoadsorption, the other is based upon the use of labeled antibodies, e.g., ferritin-antibody conjugates. There are distinct differences in type of information each experimental approach can offer.

A. Immunoadsorption

As the name implies, immunoadsorption involves the removal of antibodies by an adsorbing antigen. Comparative analysis of immunoglobulin populations present in adsorbed and in control antiserum allows conclusions to be drawn about the availability of antigens to homologus γ-globulins during the adsorption process. This type of experimental approach is well suited to determining the expression of antigens on membrane surfaces provided an analytical system, such as CIE, is available which can adequately resolve the multiplicity of membrane immunogens. The most usual way of conducting an immunoadsorption experiment is to make use of biological structures in which only one membrane surface is available to antibody at any one time. In the case of a Gram-positive organism, this would be equivalent to using a suspension of stable protoplasts or a homogenous population of closed plasma membrane vesicles possessing either a right-side-in or an inside-out orientation. However, it is not always possible to prepare membrane structures which are uniformly sealed and impermeable to antibody. Contrary to expectation, it is possible to derive much useful information from the study of partly sealed membrane structures, which, for the present, may be defined as a mixture of (1) topologically sealed membrane structures which are impermeable to antibody and (2) similar structures which have either become permeable to antibody or which possess a different topological orientation. It is important to distinguish this type of preparation from one in which most of the membrane structures are leaky or permeable to antibody. These latter preparations are of little use in the analysis of membrane architecture by immunoadsorption techniques.

1. Theoretical Considerations

a. Sealed Membrane Structures

A convenient method of performing immunoadsorption experiments is to incubate a fixed volume of antimembrane serum in a series of tubes with increasing volumes of a protoplast (or membrane) suspension. Following removal of agglutinated protoplasts or membranes by centrifugation, the various adsorbed antisera can be analyzed to assess the antibody titers to individual antigens. This can best be achieved by incorporating similar volumes of the various sera into antibody-containing gels of CIE test plates, which have been run in the first direction using a fixed amount of solubilized membranes. This process of progressive immunoadsorption is illustrated schematically in Figure 14. Immunoprecipitates corresponding to antigens which are available to antibody during adsorption behave in a manner which is clearly different from immunoprecipitates corresponding to immunogens which are inaccessible to homologous antibody. The differences are visually obvious and can be rationalized from a consideration of Equation 1 that may be legitimately rewritten as:

$$A_v = \frac{K_1 C}{B_l} \qquad (2)$$

where A_v is the peak area subtended by immunoprecipitate i following adsorption with v ml of membrane, B_l is the amount of anti-i immunoglobulin remaining after adsorption, and k_1 and C retain their previous meaning. If immunogen i is available to antibody during progressive adsorption experiments, B_l will decrease as the volume of adsorbing antigen is increased. Since C is constant by arrangement, it follows that adsorption of anti-i immunoglobulins will be reflected by an increase in peak area (A_v) across the adsorption series. Conversely, if antigen i is not available to antibody, B_l will be unaffected. In this situation, the peak area (A_v) of immunoprecipitate i will be unaffected by adsorption (Figure 14).

From the above arguments, it follows that antigens expressed on the outer surface of the plasma membrane should be detected by an increase in the peak areas of their respective immunoprecipitin arcs following adsorption of antimembrane serum with stable protoplasts, whereas antigens expressed solely on the inner face of the plasma

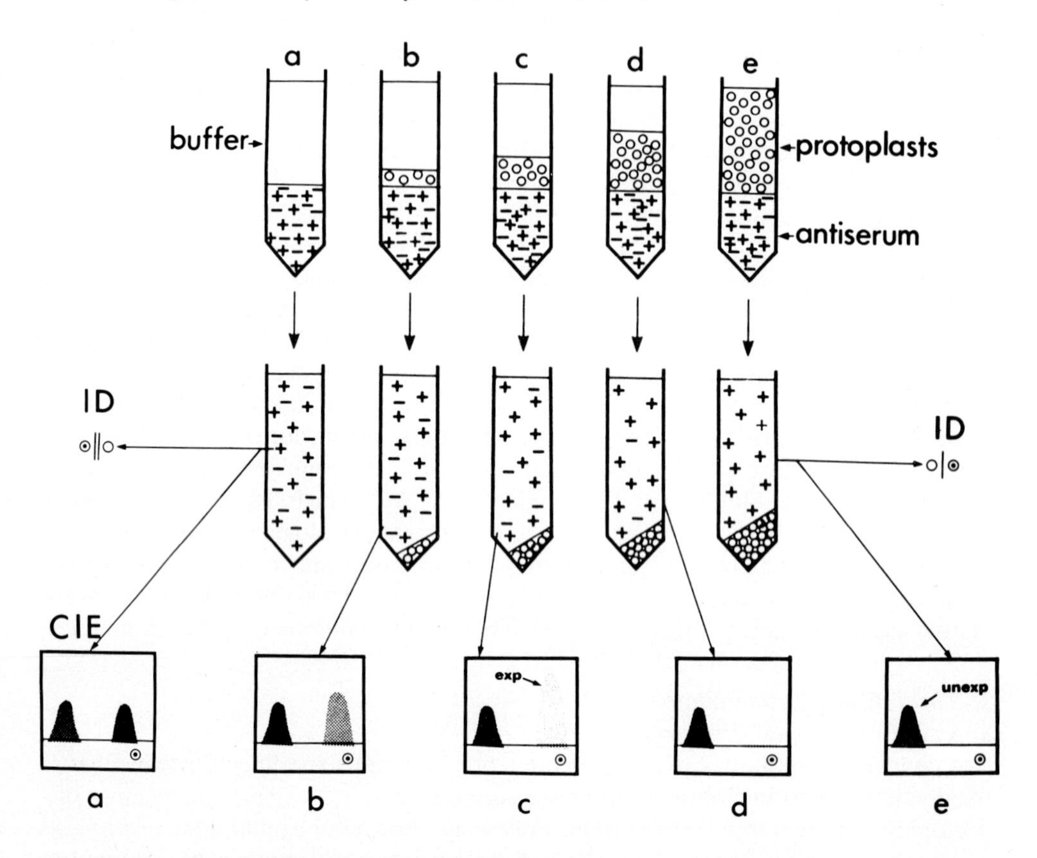

FIGURE 14. Schematic representation of a typical progressive immunoadsorption experiment. Increasing amounts of membrane (in this case protoplasts [○]) are incubated in a series of tubes (a to e) with a fixed amount of antimembrane serum (±) and sufficient buffer to equalize volumes. Antibodies designated thus (−) and directed to determinants expressed on the outer surface of the protoplasts are adsorbed during incubation and can be removed by centrifugation of the agglutinated protoplasts. This results in a series of supernatant fractions in which the concentration of (−)-immunoglobulins decreases progressively from a through e. In contrast, the concentration of antibodies designated (+), and directed to determinants which are not expressed on the protoplast surface, remains unaffected by adsorption. The adsorbed antisera can be analyzed by CIE by incorporating similar volumes into antibody gels of a series of CIE test plates. Providing a fixed amount of test antigen is used for analysis, the resultant immunoprecipitates behave in two distinctly different fashions. The peak area of immunoprecipitates corresponding to antigens *exp*osed on the protoplast surface increases in plates a through e because the concentration of homologous antibody (−) in the gels is decreasing. On the other hand, the peak areas of immunoprecipitates corresponding to antigens which are not exposed on the cell surface *(unexp)* stay constant because the concentration of homologous antibody (+) in gels a-e is the same. Corresponding patterns obtained by immunodiffusion (ID) are shown for serum samples a and e at the sides of the diagram. The anode is to the left and top of the CIE plates.

membrane should give immunoprecipitates whose peak areas remain unaffected by adsorption. Moreover, immunoprecipitates corresponding to antigens in the latter group should increase following adsorption with isolated membranes, since components on the inner face of the membrane are then available for interaction with antibody. Theoretically, any membrane immunogens which are not expressed on either membrane surface should be detected in adsorption experiments utilizing isolated membranes by the presence of immunoprecipitates of unaltered peak areas.

The situation for a transmembrane protein is somewhat more complex. If the antigen has a different set of determinants on each face of the membrane, then adsorption with protoplasts or with a homogeneous population of inside-out vesicles should result in the total removal of some, but not all, of the homologous antibody species. All else being equal, this might result in an immunoprecipitate which increases in peak area in the early stages of adsorption and then remains virtually unaffected by further adsorption. However, this assumes that the avidities of the different immunoglobulins for the transmembrane antigen are similar.

One further theoretical point is worthy of comment. Antigen molecules require two or more determinants in order to form insoluble precipitin arcs upon reaction with antibody. Thus, an immunogen which has only one of its many determinants in a cellular location that is inaccessible to antibody may not be detected following removal of immunoglobulins to the other exposed determinants.

b. Partly Sealed Membrane Structures

The fundamental difference between sealed and partly sealed membrane structures, as they have been defined, is the presence in the latter preparations of membrane fragments which are topologically distinct from the parent population. Consequently, antigens which are normally positioned on opposite surfaces of the membrane can interact with antibody to some extent during adsorption experiments. It follows that analysis by CIE of serum obtained following progressive adsorption experiments would probably result in a series of immunoplates in which most, if not all, of the immunoprecipitates were observed to increase in area. Obviously this does not give the clear visual assessment of surface expression that is provided by adsorption experiments conducted with preparations of completely sealed membranes or protoplasts. Fortunately, use can be made of the quantitative nature of CIE to determine precisely the extent to which each antigen is available to antibody.[159]

Owen and Kaback[159,160] have derived the following mathematical relationship and shown it to apply to the results of progressive immunoadsorption experiments conducted with partly sealed membrane structures (viz., membrane vesicles):

$$\frac{1}{A_v} = \frac{1}{A_o} - Kxv \qquad (3)$$

where A_o is the peak area subtended by immunogen *i* prior to adsorption, A_v is the peak area subtended by immunogen *i* following adsorption with *v* ml of membrane, *x* is the degree to which antigen *i* is expressed (i.e., its availability to antibody) and *K* is a constant which depends on *i*. This equation predicts a linear relationship between $1/A$ and *v* from which the value *Kx* can be computed. Comparison of *Kx* values obtained for antigen *i* following adsorption with control membranes and with disrupted membranes gives the term $x_{disrupted}/x_{control}$, where $x_{disrupted}$ and $x_{control}$ are the degrees to which antigen *i* is expressed in disrupted membranes and in control membranes, respectively. In other words, by careful measurement of peak areas it is possible to obtain a direct estimate of the extent by which disruption increases the availability of *i* to antibodies. In this manner, the distribution of each antigen on the various membrane surfaces can be calculated.[159,160]

Several aspects of this approach warrant further discussion. First, a primary assumption in the derivation of Equation 3 is that the amount of antibody adsorbed (B_v) is proportional to the amount of adsorbing antigen available, i.e., that

$$B_v = k_2xyv \qquad (4)$$

where k_2 is a proportionality constant which incorporates, among other things, the valency of *i*; *x* and *v* have their previous meaning; and *y* is the concentration of *i* in the membrane.[159] Whereas it is clear from a consideration of typical quantitative precipitin tests[437] that this assumption is valid to a first approximation for purified immunoglobulins under conditions of moderate antibody excess, it is equally obvious that it does not hold as equivalence is approached. In practice, the linear relationship between $1/A$ and v predicted by Equation 3 is found to hold provided less than 50% of the relevant antibody has been removed.[160] This corresponds to a situation where $A_v = 2\,A_o$, i.e., where the peak area has doubled.

Secondly, in order to obtain meaningful values of $x_{disrupted}/x_{control}$ it is obvious that the value of K must be identical in adsorption experiments conducted with disrupted membranes and with control membrane preparations. It should be noted that $K = k_2y/k_1C$ and is thus related to a potential variable, y (the concentration of antigen i in the membrane).[159] In order to eliminate error, it is imperative that the absolute concentrations of each antigen i be identical in both membrane preparations. The only practical way of doing this is to ensure that disrupted membranes are derived directly from the control membrane preparations and not from a different batch of similar membranes.

Finally, it has been assumed that the value $x_{disrupted}/x_{control}$ reflects only the degree to which antigen i becomes expressed following relief of the topological constraints imposed by the sealed nature of the membrane preparation. It could be argued that the methods used for disruption result in increased expression of antigenic determinants in open sheets of membrane. This is a difficult problem to resolve definitively. In some instances, it may be possible to compare the degree of antigen expression in untreated and disrupted membrane preparations which were initially devoid of any closed membrane structures. However, quality control is a problem in this approach and one usually ends up in a somewhat circular scientific argument. An alternative device is to disrupt the membrane preparations in a number of different ways and to compare the different values of $x_{disrupted}/x_{control}$. This approach was adopted by Owen and Kaback[159,160] during studies of the molecular architecture of membrane vesicles of *E. coli.* Values of $x_{disrupted}/x_{control}$ were found to be largely independent of the methods used to effect disruption (Figure 16), implying, but not proving, that antigen expression was a function of gross membrane morphology rather than considerations at the molecular level.

2. *Antigenic Structure of Micrococcus lysodeikticus Membranes*

By progressive adsorption experiments of the type outlined in Figure 14 and conducted with protoplasts[88,132] and isolated membranes[88] of *M. lysodeikticus,* it has been demonstrated that 12 membrane antigens of this organism have expression on the protoplast surface and that 14 others have determinants located exclusively on the cytoplasmic face of the membrane. The major antigens of the former group are the succinylated lipomannan and five glycoproteins. Antigens falling into the latter category include ATPase, succinate dehydrogenase, two NADH dehydrogenases and malate dehydrogenase (Figure 15).[88,132]

Thus, in a simple series of adsorption experiments it has been possible to establish that the plasma membrane of this organism is markedly asymmetric, and that it has an outer surface rich in antigens possessing carbohydrate residues, a feature, incidentally, shared with most mammalian cells. It has not been possible to determine whether any of the 12 antigens expressed on the protoplast surface possess a transmembrane orientation, i.e., have determinants on both surfaces. However, it can be stated with certainty that many of the important enzymes involved in respiration and energy transduction, viz., the identified dehydrogenases and ATPase, are located solely on the inner face of the membrane and do not possess major determinants on the external

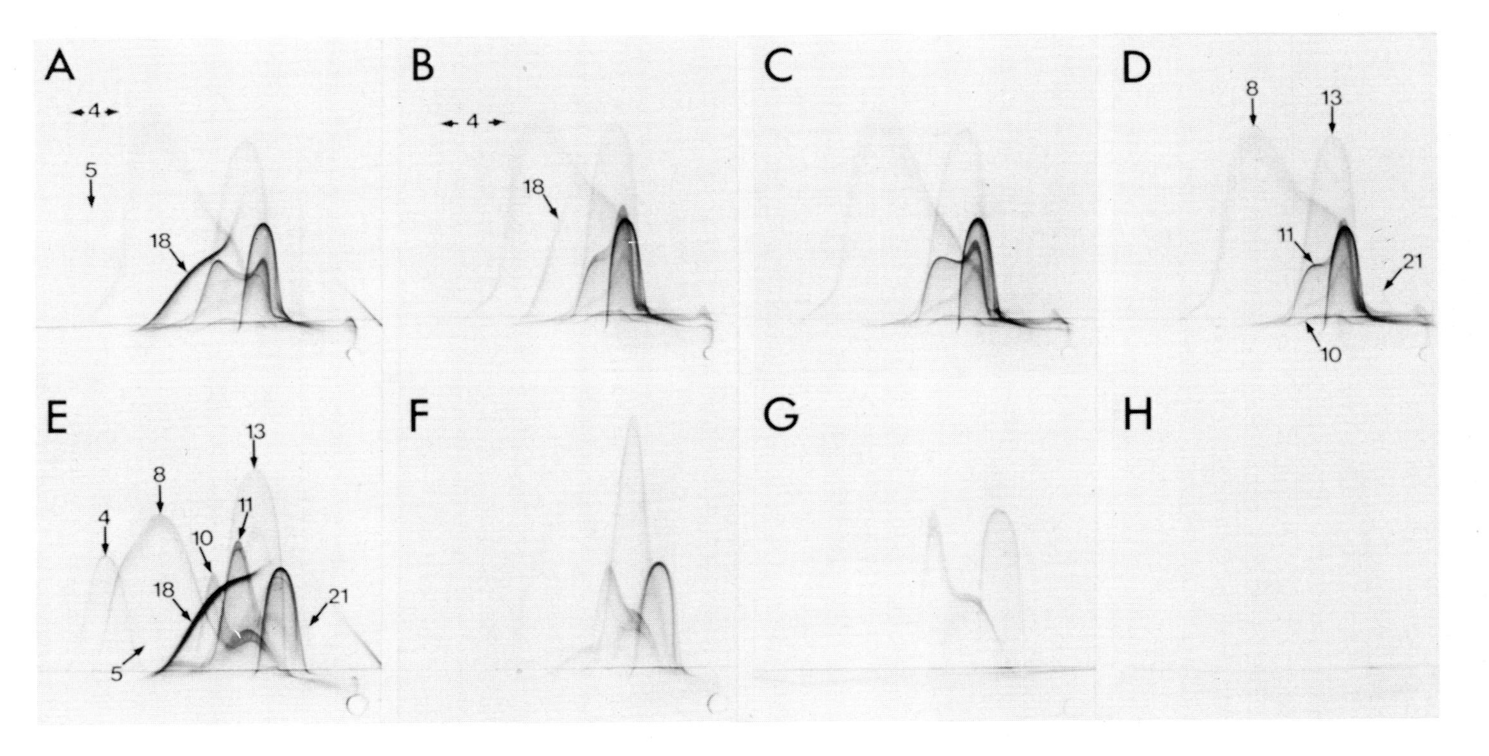

FIGURE 15. Effect of progressive adsorption of antimembrane serum with protoplasts and membranes of *M. lysodeikticus*. Antimembrane serum (1.0 m*l* of purified immunogloublin) was adsorbed with 0 m*l* (A) 0.5 m*l* (B), 1.0 m*l* (C) and 3.0 m*l* (D) of protoplast suspension, or with 0 mg (E), 1.1 mg (F), 4.5 mg (G) and 22.7 mg (H) of isolated plasma membranes. Similar volumes of immunoglobulin fractions were incorporated into agarose gels and were analyzed against Triton® X-100 extracts (57 μg of protein, A-D; 76 μg of protein, (E-H) of two different batches of *M. lysodeikticus* membranes. Note that immunoprecipitate numbers 4,5, and 18 (among others) increase in area upon adsorption with protoplasts (A-C) and are totally absent in D. Immunoprecipitate numbers 8,10,11,13, and 21 (among others) are unaffected by adsorption with protoplasts. The areas subtended by all immunoprecipitates increase following adsorption with isolated membranes (E-H). Thus, antigen numbers 4, 5, and 18 (among others) have expression on the protoplast surface, and antigen numbers 8, 10, 11, 13, and 21 (among others) are solely expressed on the cytoplasmic face of the membrane. No immunoprecipitates are detected in H suggesting the absence of immunogens buried totally in the membrane. Most of the numbered immunoprecipitates have been identified (see Table 2). Anode is to the left and top of all gels. (From Owen, P. and Salton, M. R. J.[88,132] Reprinted in part and with permission from *J. Bacteriol.*, 132, 974, 1977. Copyright by the American Society for Microbiology.)

membrane surface. Furthermore, there is no evidence from these[88,132] or from other studies[57] to suggest the existence of immunogens totally buried within the membrane (Figure 15).

Adsorption of antimembrane serum with whole cells of *M. lysodiekticus* has further established that immunogens on the protoplast surface are also expressed to some extent on the surface of the bacterium.[88] Thus, the succinylated lipomannan appears to be readily accessible to antibody in intact cells, an observation which is not at variance with an earlier report that this polymer is secreted into the external medium during the late logarithmic phase of growth.[38] Other carbohydrate containing antigens of the protoplast surface do not seem to share the same degree of exposure as the mannan and are probably located closer to the membrane surface.[88] Table 2 summarizes some of the data obtained from adsorption experiments conducted with *M. lysodeikticus.*

3. Antigenic Structure of Escherichia coli Membranes

The considerations detailed in Section VI.A.1.b have been applied to the analysis of membrane vesicles derived from *E. coli* ML308-225.[159,160] Before discussing the data, however, it is appropriate to place the study in some perspective. As the reader will remember from a previous Section, membrane vesicles derived from this particular organism have been widely used to study transport phenomena, and it has been consistently claimed by Kaback and his school[228,438] that the vesicles are topologically sealed, retaining the same orientation as the intact cell. Several workers disagree, and suggest that a significant proportion of the membrane may be in the form of open sheets of membrane or in the form of vesicles possessing the opposite polarity to that of the mother cell.[231,233,237] An alternative argument suggests that the membrane is a mosaic in the sense that a proportion of some membrane antigens become dislocated during the process of vesicle formation and reassociate in a nonrandom fashion to the other (usually outer) membrane surface.[222,232-236] As has been noted,[159] many of these arguments are based upon the results of experiments in which (1) the specific activities of marker enzymes are compared in intact vesicles and vesicles made permeable to substrate,[222,231,232] or (2) purportedly "monospecific" antibody is used either to selectively inhibit a certain enzyme activity or to remove by agglutination a portion of the vesicles possessing a certain topology.[233-237] These techniques suffer from serious disadvantages, not least of which is the tendency for the activities of many membrane-associated enzymes to be radically affected by reagents or techniques which perturb membrane structure.[16] In addition, there is the very real possibility that simple test tube agglutination procedures conducted on mixed populations of membrane vesicles may be biased by virtue of entrapment of unagglutinated vesicles within the loose three-dimensional immunoprecipitin lattice.[159] Progressive immunoadsorption procedures combined with CIE do not suffer from quite such serious problems. Moreover, the results obtained for a single antigen in any on progressive adsorption experiment are normally based upon the mean of several different adsorption tests (i.e., from the gradient of plots of $1/A$ against v), rather than from a single experimental point.

Owen and Kaback[159,160] have attempted to resolve the controversy surrounding vesicle structure by performing progressive adsorption experiments with (1) control membrane vesicle preparations and (2) identical vesicle preparations which had been physically disrupted in any one of a number of different ways. These included sonication, passage through a French pressure cell, dispersal in Triton® X-100 or solubilization in SDS under nondenaturing conditions.[130] Analysis by CIE of adsorbed antisera has allowed the expression of 14 different immunogens to be estimated from measurements of immunoprecipitin peak areas.[159,160] The identities of many of these antigens are also known and the reader is referred to Section III.A and Figures 2-4, 7, 8, 12 and 13 for details of characterization. A selection of the data pertaining to the adsorption

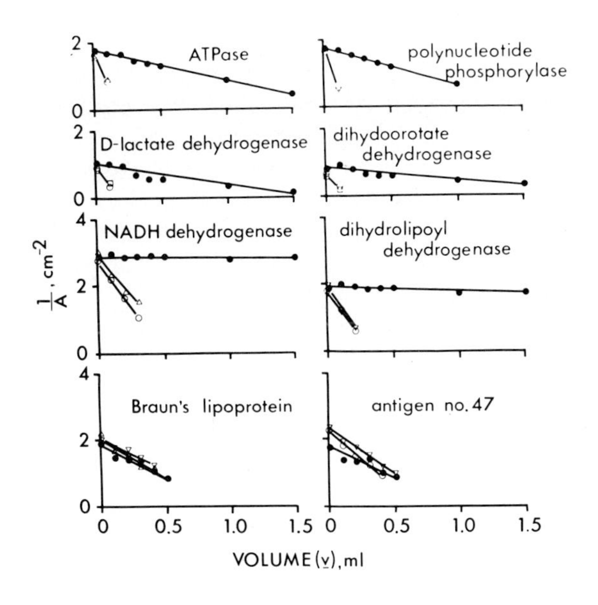

FIGURE 16. Effect of progressive immunoadsorption of antivesicle immunoglobulins with control and disrupted vesicles from *E. coli* ML308-225 on the peak areas of selected immunoprecipitates. Antivesicles immunoglobulin was adsorbed with increasing volumes (v) of an untreated control membrane vesicle suspension (●) or with identical aliquots of vesicles that had been disrupted by sonication (△), passage through a French press (▽), dispersal in SDS (○) or dispersal in Triton® X–100 (□). The adsorbed immunoglobulins were incorporated into agarose gels and analyzed against a Triton® X-100-EDTA extract of membrane vesicles (28 or 5.6 μg of protein) by CIE. The peak areas (A) were estimated from immunoplates that were stained either with Coomassie brilliant blue or by zymogram techniques. Peak areas obtained from gels run at the lower antigen loading (specifically those relating to D-lactate dehydrogenase and dihydroorotate dehydrogenase antigens) were normalized to correspond to an antigen concentration of 28 μg of protein. Note that $1/A$ is linearly related to v in most instances, and that, for any one antigen, the gradient ($-Kx$) for disrupted vesicles does not vary significantly with the method of disruption. For Braun's lipoprotein and antigen number 47, the slopes for disrupted and control vesicles are very similar, i.e., $x_{\text{disrupted}}/x_{\text{control}} \doteqdot 1$. For all other antigens shown, the gradients obtained for disrupted vesicles is about an order of magnitude greater than that for control vesicles. The percentage of antigen molecules accessible to antibody in intact vesicles can be computed from the gradients ($-Kx$), and these and other values are presented in Table 1. (From Owen, P. and Kaback, H.R.[159,160] Reprinted in part and with permission from *Biochemistry,* 18, 1422, 1979. Copyright by the American Chemical Society.)

experiments and plotted in graphical form ($1/A$ against v) is displayed in Figure 16, and a summary of these and other results can be found in Table 1. It should be noted that 10 immunogens including ATPase NADH dehydrogenase, dihydrolipoyl dehydrogenase, D-lactate dehydrogenase, dihydroorotate dehydrogenase, 6-phosphogluconate dehydrogenase, polynucleotide phosphorylase, and β-galactosidase exhibit minimal expression (11% or less) unless vesicles are disrupted. At least four other antigens, which include Braun's lipoprotein, are expressed to similar extents in untreated and disrupted vesicles. All the identified antigens in the former group would be expected to occupy a position on the inner surface of the plasma membrane or in the cytoplasm of the parent cell. Moreover, Braun's lipoprotein in the latter group of antigens is known to be located exclusively in the outer regions of the cell envelope.[134] From these data, it is apparent that the vesicles must be largely intact sealed structures which retain the topology of the intact cell and the antigenic asymmetry of its plasma membrane.

An interesting feature of the results displayed in Table 1 relates to the expression of the various enzyme antigens in intact vesicles. Expression varies from 2% for NADH dehydrogenase to about 11% for ATPase. This variability would seem to discredit the suggestion that expression results from the presence of inside-out vesicles and/or unsealed membrane fragments. A more likely explanation would appear to involve a process of dislocation in which some antigens normally located on the inner face of the plasma membrane become expressed to small, but differing extents on the outer surface of the membrane. Taken together, the results suggest that over 95% of the membrane in the vesicle preparations derived from *E. coli* ML308-225 are probably in the form of correctly oriented, asymmetric sealed sacculi which are impermeable to antibody. Furthermore, dislocation of antigens from the inner to the outer surface of the membrane during vesicle formation appears to be a relatively minor problem, and occurs to an extent not exceeding about 10%.[159,160]

Progressive adsorption experiments have also been conducted with intact cells and spheroplasts of *E. coli* ML308-225 in an attempt to determine which antigens are expressed on the outer surface of the outer membrane of this organism.[295] As might be expected, antibodies to lipopolysaccharide are readily adsorbed by whole cells. This is clearly illustrated in Figure 17, where the height of the immunoprecipitate corresponding to lipopolysaccharide increases in one-dimensional (rocket) immunoelectrophoresis experiments conducted with the purified antigen. In marked contrast, antibodies to another major component of the outer membrane, Braun's lipoprotein,[134] remain totally unaffected by adsorption, even with quantities of cells three times those required to completely remove antibodies to lipopolysaccharide (Figure 17). Similar experiments conducted with spheroplasts of *E. coli* ML308-225 or with the heptoseless mutant *E. coli* D21f2[439] give identical results, suggesting that lipoprotein does not have surface expression even in *E. coli* cells with an impaired outer membrane.[295] These results are at variance with an earlier immunochemical study of surface expression, in which it was implied that structural reorganization of the outer membrane allowed exposure of lipoprotein determinants in rough strains of the Enterobacteriaceae.[339] However, these latter conclusions should be treated with some caution since the bulk of the data were not derived from direct analysis of surface expression but from estimations of lipoprotein titers in rabbits immunized with various bacterial strains. It will be recalled from Section V.C.1 that antilipoprotein antibodies have been detected in rabbits immunized with heavily encapsulated strains of *E. coli.* This strongly suggests that antibody titers cannot be used as indicators of membrane structure. In the light of these observations it seems probable that lipoprotein is not expressed in the form of a determinant at or near the cell surface of *E. coli.* This feature should be accommodated in any model for the molecular organization of the outer membrane.[134,240,341,440]

Four other envelope antigens, apart from lipopolysaccharide, have also been detected on the cell surface of *E. coli* ML308-225 by immunoadsorption experiments.[295] Unfortunately, none of these have been identified to date (Table 1). Another major unidentified membrane immunogen (antigen number 47 in Figures 1 and 6) is not exposed on the cell surface but is expressed on the surface of membrane vesicles.[159]

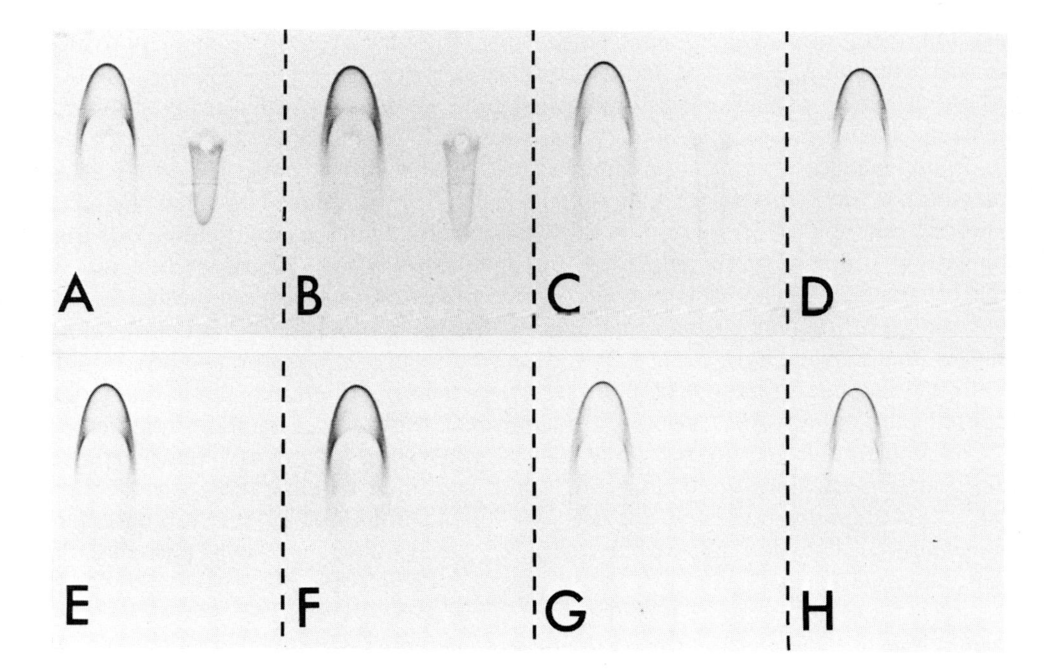

FIGURE 17. Effect of immunoadsorption with *E. coli* ML308-225 cells upon the concentration of antilipoprotein and antilipopolysaccharide immunoglobulins. Antivesicle serum (45 mg of protein) was incubated with 0 m*l* (A), 0.05 m*l* (B), 0.1 m*l* (C), 0.25 m*l* (D), 0.5 m*l* (E), 1.0 m*l* (F), 2.0 m*l* (G) and 3.0 m*l* (H) of a suspension of washed *E. coli* cells (A^{560nm}_{1cm} = 38). Adsorbed immunoglobulins were incorporated into agarose gels and analyzed in tests against Braun's lipoprotein (1 μg; left hand well in A-H) and lipopolysaccharide (0.8 μg; right hand well in each panel) by one dimensional "rocket" immunoelectrophoresis. Note that the immunoprecipitate corresponding to Braun's lipoprotein is totally unaffected by adsorption with quanitities of cells three times those required to fully remove antibodies to lipopolysaccharide. Unlike most plasma membrane antigens, lipopolysaccharide migrates cathodally under the test conditions. Anode is to the top of all gels.

Thus, it behaves in adsorption experiments in a manner similar to Braun's lipoprotein, and is probably located at a membrane surface exposed to the periplasm.

4. *Antigenic Structure of Mycoplasma Membranes*

In their study of *A. laidlawii* membranes, Johannson and Hjertén[57] showed that only one immunoprecipitate in the CIE reference profile increased following adsorption of antimembrane serum with intact cells. These data imply that only one of the 15 to 20 Tween® 20-soluble membrane antigens detected for this organism (viz., T_2 in Table 2) has expression on the outer surface of the membrane. (It will be remembered that this organism lacks a cell wall.[433]) All the other antigens are apparently expressed on the inner face of the membrane since corresponding antibodies can be adsorbed by lysed cells.[57] The antigenic simplicity of the cell surface, as implied from these studies, is surprising, and a reappraisal of the situation may be necessary, especially in the light of more recent reports of one and possibly two glycoprotein antigens in the membrane of this organism (Table 3).[137] In an unrelated study, Alexander and Kenny[114] have shown that a major unidentified antigen of *Mycoplasma arginini* membranes is expressed on the cell surface.

The distribution of phospholipids in the membranes of some species of mycoplasmas has also been probed by immunoadsorption procedures. Thus, antiphosphatidylglycerol antibodies have been employed to show that most of the phosphatidylglycerol molecules in *M. hominis* are masked, but become expressed following treatment of whole cells with pronase.[397] Assuming that pronase does not induce any structural rearrangement of membrane lipids, these results suggest that at least some of the phosphatidylglycerol molecules are located in the outer leaflet of the plasma membrane. Similar types of experiments have been employed to show that the phosphoglycolipid of *M. mycoides* is also located largely in the outer leaflet, and that phosphoglycolipid determinants contribute significantly to the antigenic architecture of the cell surface.[419] The fact that glycolipids have been strongly implicated in the mechanism of complement-mediated killing of some mycoplasmas by antibody would suggest that they too have surface expression (see Section V.D). It should be stressed, however, that interpretations of data pertaining to the distribution of lipids, and based upon the use of antibodies, are rarely unequivocal. Moreover, limitations imposed by the haptenic nature of lipids usually require that expression be assessed by simple agglutination reactions or similar tests. These and other considerations (see Section VI.C) have greatly restricted the use of antilipid antibodies in structural studies. Fortunately, other more informative techniques are available for probing phospholipid distribution.[441,442]

5. *Antigenic Structure of* Rhodopseudomonas sphaeroides *Membranes and Chromatophores*

Konings and colleagues[174] have attempted to compare the orientation of plasma membrane vesicles and chromatophores from *R. sphaeroides* by progressive immunoadsorption experiments of the type outlined in Section VI.A.1.b. The fraction of ATPase, succinate dehydrogenase, and NADH dehydrogenase which was accessible to antibody on the outer surface of the membrane was calculated from a comparison of antigen expression in untreated and detergent-solubilized membranes. It was observed that about 29%, 20%, and 30%, respectively, of the ATPase, succinate dehydrogenase and NADH dehydrogenase antigens were available to antibody in untreated plasma membrane vesicles, compared with values of 94%, 115%, and 95% in control chromatophore preparations. These results appear to confirm a variety of other data suggesting that the plasma membrane vesicles from this organism have predominantly a right-side in orientation and that chromatophore membranes have a polarity opposite to that of the intact cell.[174]

B. Labeled-Antibody Techniques

An alternative immunochemical approach to the study of membrane ultrastructure involves the use of immunoglobulin molecules which are coupled, usually covalently, to other compounds or reporter groups whose intrinsic properties allow for convenient detection of the resultant conjugate. The location of a particular antigen is determined from the position of the reporter group following incubation of the test sample with labeled antibody. Invariably, such procedures require that labeled specimens be examined by (electron) microscopic techniques. One of the first labeled-antibody methods to be developed was the immunofluorescence technique.[443] Here, specific antibodies are coupled either directly or indirectly to a fluorescent dye, such as fluorescein or rhodamine. The position of the relevant antigen is deduced from the pattern of fluorescence observed following incubation of the test sample with the fluorescently-labeled antibody. Although the immunofluorescence technique has yielded a great deal of valuable information about the distribution and mobility of surface antigens in mammalian cells,[444] its application to microbiology has been largely restricted to the analysis of structures such as cell walls,[445,446] whose gross morphology can be recognized under the light microscope. The technique is therefore of limited use in the study of bacterial membranes and their structure, and alternative procedures have to be adopted for meaningful resolution. One interesting approach is to couple antibody to a relatively stable enzyme, such as horseradish peroxidase, which can be detected by a convenient histochemical reaction. The detection system for horseradish peroxidase, for example, results in the formation of amorphous electron dense deposits which can be visualized readily following electron microscopic examination.[447-449] The cellular location of a number of bacterial antigens have been examined in this way.[450-454] These include the lipopolysaccharides of *Bacteroides oralis* and *Fusobacterium nucleatum*,[454] and amino acid transport proteins of *E. coli*.[450] In the former study, it has been clearly demonstrated that antigenic determinants of the endotoxin molecule are expressed on the surface of the outer membrane of both organisms.[454] These observations are in complete agreement with results obtained by different procedures for other organisms.[455-462] However, the resolution achieved in the latter study does not allow the amino acid transport proteins to be located with precision within the bacterial envelope structure.[450] As a general rule, antigens cannot be located by the peroxidase-labeled antibody procedure with the same definition and degree of accuracy as another, more popular method, namely the ferritin-labeled antibody technique.[42-44]

1. Ferritin-Labeled Antibody

Singer[42] originally introduced the ferritin-labeled antibody technique in 1959 in what has since been recognized as one of the more important technical advances in the study of cell-surface immunology. The procedure involves the covalent coupling of specific antibody to the electron-dense ferroprotein, ferritin.[42,43] The addition of the ferritin molecule adds only a relatively small degree of uncertainty to the precise localization of antigenic components,[44] and allows even singly bound ferritin-antibody conjugates to be visualized following electron microscopic examination of labeled preparations. As a result, the ferritin-labeled antibody technique has become an extremely popular method for establishing the cellular location of discrete antigens. It is also one of the few techniques which enables the position of membrane immunogens to be pinpointed in the lateral plane of the lipid bilayer. Further details of uses and modifications of this extremely powerful technique can be found in several recent review articles.[463-465] Of these, the text by Wagner[464] offers valuable and comprehensive coverage of related methodology.

Whereas the ferritin-labeled antibody method has been used extensively to establish the cellular location of a variety of bacterial surface antigens (*e.g.*, wall material, pili, and capsules,[464,466-472]), its application to the study of microbial membranes and their components is more restricted. This is a reflection, in part, of the difficulties involved in preparing monospecific antisera to many membrane components.

a. Plasma Membrane Components

Oppenheim and Salton[274] have used the ferritin-labeled antibody procedure in a careful study of the localization of ATPase in *M. lysodeikticus*. Monospecific antiserum was raised to highly purified F_1. ATPase, and the ferritin-antibody conjugate shown to react specifically with the 10 nm particles previously associated with ATPase activity.[473] An examination of thin sections of isolated membranes and protoplasts revealed that ferritin-labeled antiATPase was capable of reacting only with the inner and not with the outer surface of the plasma membrane. It is interesting to note that a large portion of the isolated membranes in these preparations appeared to be in the form of vesicles. Moreover, the polarity of the vesicles was opposite to that of protoplasts, in as much as the ATPase particles, as evidenced from the distribution of ferritin, were exclusively

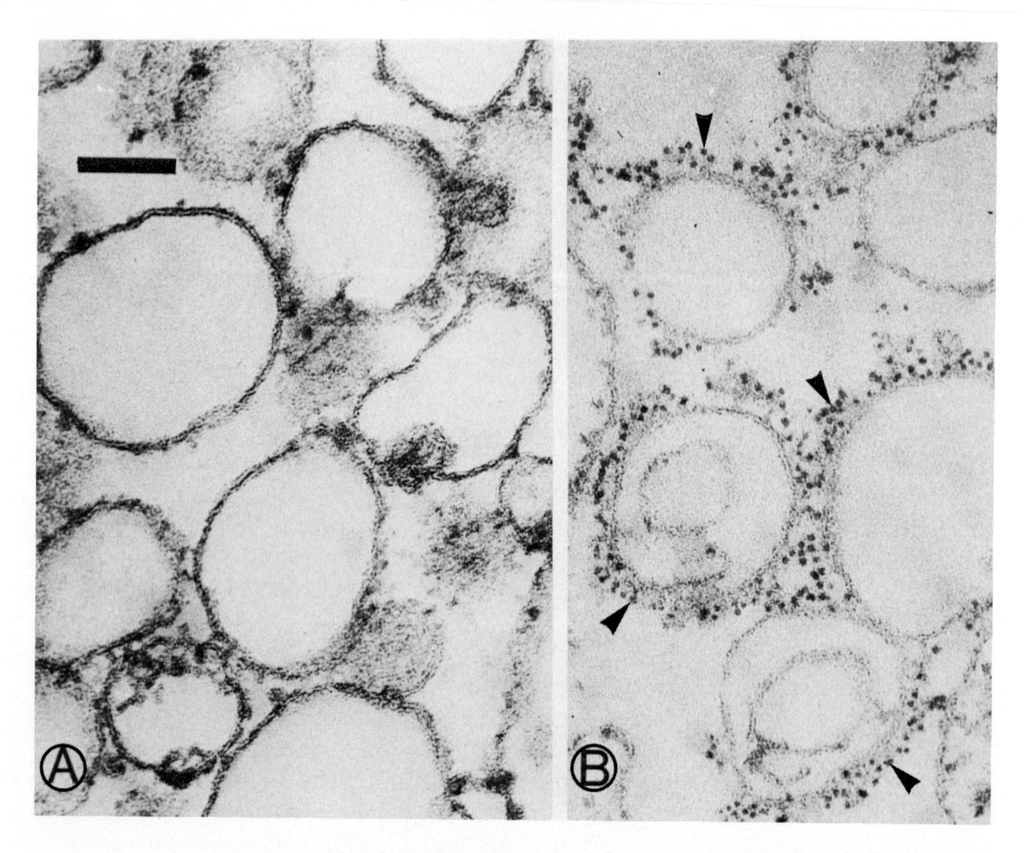

FIGURE 18. Ferritin-labeling studies of bacterial membranes. Panels A and B depict electron micrographs of sectioned plasma membranes of *M. lysodeikticus.* Preparation A is unlabeled, whereas B has been labeled with ferritin-anti-ATPase conjugate. Ferritin particles (arrowed) are present only on one side of the membrane in the labeled preparation. This can be shown to represent the inner surface of the plasma membrane of the intact cell. Bar represents 0.1 μm. (C) depicts an electron micrograph of a thin section of a *L. fermentii* protoplast which has been treated with ferritin-antilipoteichoic acid conjugate. Ferritin particles (arrowed) can be visualized over the entire outer surface of the plasma membrane. Bar represents 0.5 μm. (From Oppenheim, and Salton [274] and Van Driel, *et al.*[474] Figures 18A and 18B reprinted with permission of the authors and the journal from *Biochim. Biophys. Acta,* 298, 297, 1973; and Figure 18C from *J. Ultrastruct. Res.*, 43, 483, 1973. Copyright by Elsevier Scientific Publishing Co. and Academic Press, Inc., respectively.)

located on the outermost surface (see Figure 18A).[274] Three independent experimental approaches, namely ferritin-labeled antibody technique,[274] CIE,[88,132] and ^{125}I-labeling experiments,[285] have now independently established that the membrane associated F_1. ATPase from this organism is expressed solely on the inner face of the plasma membrane. Moreover, the observation by Oppenheim and Salton[274] that isolated "mesosomal" membrane vesicles from *M. lysodeikticus* do not label with ferritin-conjugated anti-ATPase is fully compatible with the results of enzyme[245] and quantitative immunochemical analysis[155] suggesting that these membranous structures have dramatically reduced complements of ATPase.

In contrast to ATPase, the lipoteichoic acids of Gram-positive bacteria are known to be located on the outer surface of the cytoplasmic membrane. This was firmly established by van Driel and her colleagues[474] in an elegant electron microscopic study of *Lactobacillus fermentii* and *L. casei.* Protoplasts were shown to bind ferritin-anti-lipoteichoic acid conjugate uniformly over their outer membrane surface (Figure 18). Isolated membrane preparations also bound ferritin-labeled antibody. However, in this particular instance it is difficult to establish from the micrographs whether one or both

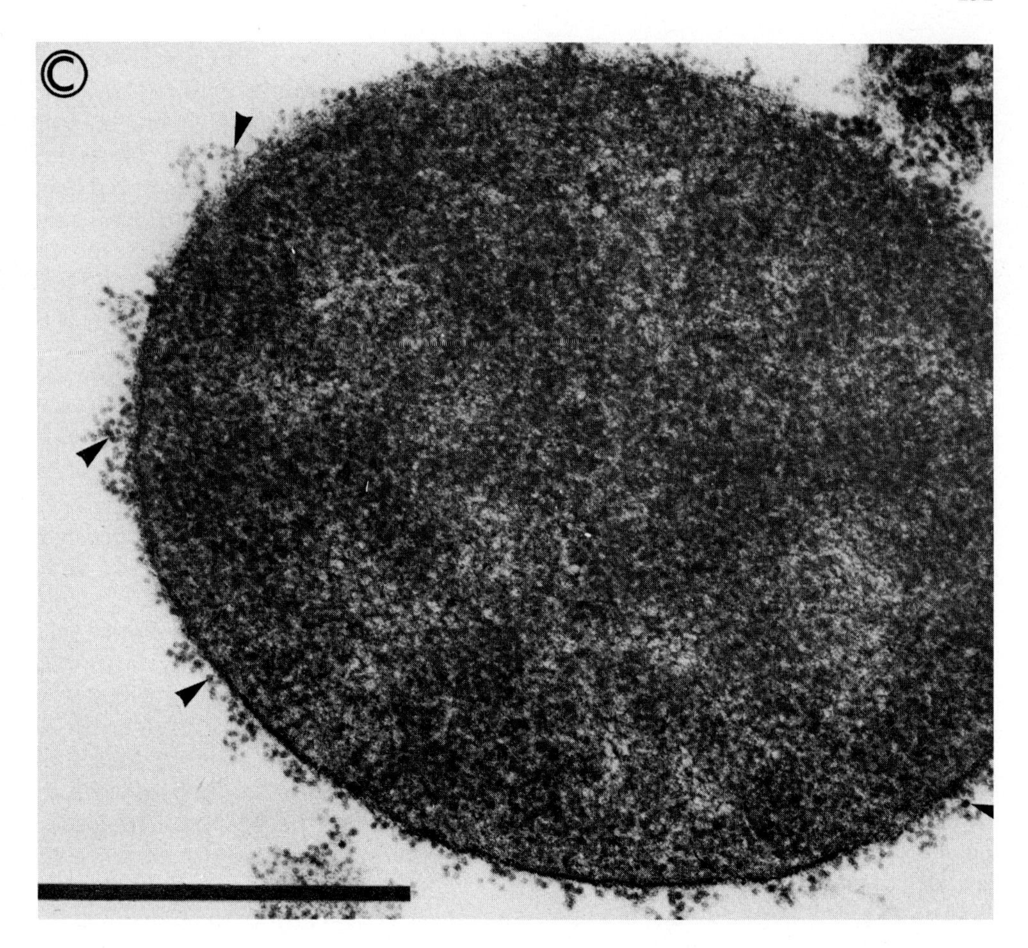

FIGURE 18C

membrane surfaces are capable of reacting with conjugate.[474] It was additionally established that *L. fermentii* possessed a much more even distribution of lipoteichoic acid determinants on its cell surface than *L. casei.* This was interpreted in terms of a model in which lipoteichoic acid chains penetrated the wall of *L. fermentii* to a greater extent than the wall of *L. casei.*[474] Certain aspects of these results were confirmed and extended in subsequent immunoferritin studies in which thin sections of *L. plantarum* were prepared in such a way as to retain the *in situ* antigenic reactivity of lipoteichoic acids.[74] Antigen chains were shown to extend from the outer surface of the membrane through the cell wall and, in some regions, into the external medium.[74]

b. Exported Proteins

Immunoferritin studies[475-478] have provided direct morphological evidence for the developing concensus for vectorial synthesis of exported proteins from membrane-associated ribosomes (see also Section V.B). One of the first pieces of experimental evidence seems to have been the report by Coulter and Mukherjee[475] that disrupted cells of *Staphylococcus aureus* contained α-toxin in close association with the plasma membrane. The authors interpreted the results of their ferritin-labeling experiments to suggest that α-toxin was modifying or controlling the permeability of the staphylococcal cell membrane.[475] In retrospect, it seems equally possible that the ferritin-antibody conjugate was detecting α-toxin molecules in the process of synthesis and secretion.

More recent and convincing evidence comes from the immunological studies of

MacAlistair et al.[476,477] These authors monitored the distribution of both β-galactosidase and alkaline phosphatase in frozen thin sections of whole cells of *E. coli.* β-Galactosidase, as determined by the distribution of the corresponding ferritin-antibody conjugate, was evenly dispersed throughout the cell. In contrast, the ferritin markers for both monomer and dimer forms of alkaline phosphatase were located close to or in the envelope structure.[477] These observations are not incompatible with the notion that exported proteins such as alkaline phosphatase are synthesized on membrane bound polysomes and traverse the membrane during synthesis, and that cytoplasmic proteins, such as β-galactosidase, are synthesized on free ribosomes. The same authors also showed in a related ferritin-labeling study that inactive subunits of alkaline phosphatase, as well as the active dimeric form of the enzyme, were expressed on the cell surface of a deep rough strain of *E. coli.* A similar phenomenon was not observed for the parent strain which possessed a smooth lipopolysaccharide.[476] It was suggested that structured regions of the cell envelope might provide a template for dimerization of the inactive subunits into active enzyme.[476]

A recent study of penicillinase secretion in *Bacillus licheniformis* 749/C has shed light on a slightly different aspect of the export process.[478] By treating mildly fixed homogenized cells, isolated membrane fractions and frozen thin sections of cells with ferritin-antipenicillinase IgG conjugate, Ghosh and Ghosh[478] showed that the inner face of the plasma membrane labeled more evenly and extensively than the outer membrane surface. Furthermore, the antigenic sites on the outer surface were distributed in discrete patches which extended through the cell wall. This particular result brings to mind an early immunoferritin study in which several secreted products of Gram-positive bacteria were observed to emanate from localized zones of lysis or "channels" in the cell wall.[479] These and other considerations[480,481] open up the interesting possibility that not all proteins which are destined for export may be successfully secreted or positioned correctly in the envelope. The frequency of errors might depend, for example, on the availability and/or location of the putative hydrophilic channels, or on the fluidity of the membrane.

c. Outer Membrane Components

The first bacterial membrane component to be located by the ferritin-labeling technique was lipopolysaccharide.[455,456] In 1966, Shands[456] showed that this antigen was apparently located on both faces of the outer membrane of penicillin-prepared spheroplasts of *Salmonella typhimurium.* More recent studies have confirmed that lipopolysaccharide is indeed a major surface antigen of Gram-negative microorganisms.[457-462] However, the observation by Shands[456] that endotoxin molecules are, in addition, located on the inner face of the outer membrane is misleading. Mühlradt and Golecki[461] have monitored the distribution of lipopolysaccharide in isolated cell walls and outer membranes of *S. typhimurium,* and have shown that removal of the peptidoglycan layer by either lysozyme or trypsin at physiological temperatures leads to an artifactual reorganization of the lipopolysaccharide molecules. Apparently, an intact murein layer normally maintains and stabilizes an asymmetric distribution of lipopolysaccharide, with the endotoxin molecules localized preferentially on the outer surface of the outer membrane. Removal of the peptidogylcan structure relaxes these constraints and, providing the membrane is sufficiently fluid, allows randomization of the lipopolysaccharide molecules.[461] Although this interpretation seems reasonable, it does not appear to explain the recent observation that lipopolysaccharide can be detected on both sides of the outer membrane (and also on the inner membrane) of intact cells of *S. typhimurium* which have been processed by a direct post-embedding staining method.[481a] However, these results are in themselves controversial[481b] and obviously at variance with a variety of evidence suggesting an asymmetric distribution of endotoxin molecules on the outer membrane.[461,481b]

Other immunoferritin studies have been directed towards an understanding of the mechanism of translocation of lipopolysaccharide from its site of synthesis in the plasma membrane to the outer membrane.[458-460,462] This has been achieved by immunochemical analyses in which the antigenic properties of the lipopolysaccharide molecule are manipulated in a fashion analogous to the use of radioisotopes in pulse-labeling experiments. Newly synthesized lipopolysaccharide has been distinguished from existing lipopolysaccharide in two different ways. First, use has been made of bacteriophage ϵ15, which converts the galactosidic linkage between O-antigen repeats of *S. anatum* from an α-to a β-configuration.[459] Newly synthesized lipopolysaccharide could thus be monitored after bacteriophage infection and glycosidic conversion by employing ferritin conjugated to immunoglobulins specific for the β-galactosidically linked antigen.[459] In a second study, a uridine 5′-diphosphogalactose-4-epimerase negative strain of *S. typhimurium* has been employed.[458,462] This mutant cannot synthesize UDP-galactose in the absence of added galactose and forms a lipopolysaccharide which totally lacks the outer core and O-antigen repeating units. If galactose is added to the growth medium, UDP-galactose is synthesized via galactose-1-phosphate and a complete, smooth lipopolysaccharide is produced. It was then possible by immunoadsorption procedures to generate immunoglobulins specific for the complete lipopolysaccharide. This could be used in ferritin-labeling experiments to monitor the appearance of complete lipopolysaccharide in cells shifted to galactose-supplemented medium.[458] The results obtained from both approaches are very similar. Analysis of thin sections[458-460] and of freeze-fractured cells[458,460] indicate that new lipopolysaccharide is inserted into the outer membrane very shortly after phage infection[459] or addition of galactose to the medium.[458] Moreover, insertion occurs at a limited number of sites which are distributed evenly over the cell surface. These were estimated to number about 30/cell for *S. anatum* and about 220 per cell for *S. typhimurium.*[458] Significantly, over 70% of these insertion sites occur directly over localized zones of adhesion between the inner and outer membranes known as Bayer junctions.[458,459] Very similar observations have been made concerning the mode of insertion into the outer envelope of capsular polysaccharide[483] and certain outer membrane proteins of the porin variety.[484] Time course studies have further indicated that it takes about 3 min for newly synthesized lipopolysaccharide to cover the entire surface of *S. typhimurium* at 37°C. This implies a lateral diffusion coefficient for lipopolysaccharide (3×10^{-13} cm^2/sec) which is some 10^4 times lower than that of phospholipids in biological membranes.[482] At 0°C the endotoxin molecules remain immobilized at the translocation sites.[460,462]

Thus, the ferritin-labeled antibody procedure has revealed that several different components, *viz.*, lipopolysaccharide,[458-460,462] capsular polysaccharide,[483] and proteins[484] become incorporated into the outer envelope by a common pathway which involves translocation through a relatively small number of adhesion sites distributed randomly over the cell surface. Morever, these observations, together with a consideration of apparent lateral mobilities of membrane components,[460,480,484] suggest that the outer membrane of Gram-negative bacteria grows predominantly by a process of diffuse intercalation occurring all over the cell surface, rather than by a zonal growth mechanism.[485]

d. Lipids

An interesting and unique approach to the study of cholesterol distribution in the membranes of both prokaryotic and eukaryotic cells has been developed by Pendleton and co-workers.[486] Utilizing the facts that the bacterial toxin, cereolysin (from *Bacillus cereus*) is antigenically related to tetanolysin (from *Clostridium tetani*) and that both toxins interact specifically with cholesterol, they treated membranes sequentially with cereolysin and ferritin-labeled antitetanolysin, and were able to assess from the density of the ferritin marker the actual distribution of the cholesterol molecules. As predicted, membranes which contain relatively high concentrations of cholesterol, such as those of *Mycoplasma gallisepticum,* were intensely labeled, whereas *M. lysodeikticus* and *Sarcina lutea,* both of which lack sterols, were unlabeled. The cholesterol-containing membranes of *Acholeplasma laidlawii* behaved anomolously, however, and did not appear to bind the cereolysin-ferritin-antibody conjugate.[486] It was suggested that cholesterol might be located in an environment which is inaccessible to the 52,000-dalton toxin molecule.[487] Certainly, there are indications that lipids of some species of mycoplasma, including *A. laidlawii,* might be partially masked by protein.[397,488] As pointed out by the authors[486] and by Bernheimer,[487] the method for detecting cholesterol has several limitations, not least of which is the possibility that hemolytic toxins induce a reorganization of the sterols in the membrane.

2. Lectins

Although a full discussion of lectins is obviously out of the scope of this review, it is relevant to note that these sugar-specific macromolecules often offer attractive alternatives to antibody for the analysis and detection of certain glycoproteins and polysaccharides.[140,243,489] Concanavalin A, which displays primary sugar specificities for α-D-mannopyranosyl and α-D-glucopyranosyl residues, is one of the most popular lectins. It has been used directly to visualize the organization of teichoic acid in the cell wall of *B. subtilis*[490] and has also been employed in the form of the ferritin-lectin[491] and peroxidaselectin[492] conjugates to study the distribution of surface carbohydrate in various species of mycoplasma.

C. Comparison of Methods

At this stage it is instructive to compare the relative merits of the ferritin-labeled antibody technique and immunoadsorption procedures. In practical terms, the former method is certainly more demanding. Ferritin has to be extensively purified and the serum must be specific for the antigen to be studied. Difficulties are often encountered in raising monospecific antiserum to membrane immunogens, and this has restricted the use of the immunoferritin technique in the study of bacterial membrane architecture. Furthermore, preparations that have been labeled with ferritin-antibody conjugate need to be examined ultimately by electron microscopy. This involves sophisticated equipment and additional expertise. It can also raise some problems in interpretation, since samples have to be subjected to (cryo)fixation and thin sectioning, or negative staining or freeze-etching prior to examination in the electron microscope.

In contrast, immunoadsorption experiments may be conducted with either monospecific or polyvalent antisera. The latter is often more convenient and desirable since the amount of information that can be gleaned from such experiments increases with the complexity of the antibody preparation. Adsorbed sera can be analyzed readily using fairly basic equipment. For example, CIE, which is probably one of the best of the available methods, requires little more than a water-cooled immunoelectrophoresis cell, a power pack, and a steady hand. Additional advantages of immunoadsorption-CIE relate to the quantitative nature of the approach and the rapidity and ease with which results can be obtained. As is evident from earlier discussions, identification of individual immunoprecipitates in terms of component antigens is not a major problem and can be achieved in a number of ways which do not rely necessarily on possession of either the purified antigen or its homologous antibody (see Section II.C). However, analysis is restricted generally to those antigens which can be solubilized under nondenaturing conditions and which form insoluble immune complexes in the presence of detergents. The possibility that some antigen-antibody reactions are affected by nonionic detergents can not be ruled out at present.

The types of information that can be gleaned from the use of immunoferritin and immunoadsorption techniques can also be distinguished. Immunoadsorption is an averaging technique in the sense that results relating to the expression, partial expression, or nonexpression of antigens pertain to the total surface area exposed to antibody. Thus, no conclusions can be made about the distribution of immunogens in the lateral plane of the membrane. Of course, it is possible to determine the location of immunogens in the vertical plane of the membrane, provided structures of suitable orientation and integrity are available (see Section VI. A). Indeed, the number of membrane components whose surface expression may be assessed in a single series of experiments is limited only by the quality of the antimembrane serum and approximates the number of discrete immunoprecipitates observed in the test system adopted. On the other hand, the ferritin-labeled antibody method is not an averaging technique and can give information regarding the distribution of antigens in both the lateral and vertical planes of the membrane (Figure 18). However, it is not possible to analyze unequivocally the location of more than one antigen at a time by this method.

It should also be remembered that neither technique is capable of detecting components which fail to elicit reasonable immune responses in the host animal (e.g., some lipids and low molecular weight peptides). Nor can they detect as surface components immunogens that are either exposed on membrane surfaces in forms other than determinants, or which are masked and therefore inaccessible to homologous antibody for steric reasons. It may be possible to detect antigens which fall into the latter category by CIE analysis of membrane preparations which have been radiolabeled with small chemical probes.[54,493] However, it should be noted that agar gel precipitation techniques are normally capable of resolving only antigens which possess two or more determinants. Molecules which possess only a single determinant generally form soluble immune complexes upon interaction with antibody (see also Section VI.A.1.a).

It is evident from a consideration of Section VI.A.3 that progressive immunoadsorption experiments can be used to quantitate the orientation and integrity of membranous structures, provided the precise cellular locations of some of the membrane antigens are known or can be justifiably assumed. Immunoferritin procedures can also be used towards the same end, although this approach is not so convenient. One important limitation should be borne in mind, however, when using antibodies as probes for structural integrity. Immunoglobulins are large molecules: IgG has a molecular weight of 150,000 daltons and a molecular diameter of 13 to 14 nm.[494,495] In this context, it is instructive to consider a hypothetical collection of vesicles which have D-lactate dehydrogenase, succinate dehydrogenase, and NADH dehydrogenase, for example, on the inner face of the membrane. It is apparent that such vesicles would probably be impervious to antibody raised against either of the three enzymes even though the membrane vesicles contained holes 13 nm or less through the bilayer. Superficially, the vesicles might appear to be sealed. A very different picture might emerge if integrity were assessed by enzyme assay. From a consideration of space-filling models, it is apparent that the enzyme substrates D-lactate, succinate, and NADH would experience difficulty only in diffusing through hydrophilic pores or holes which were less than about 0.8, 1.0, and 2.5 nm in diameter, respectively. Thus, many substrate molecules are probably capable of diffusing simultaneously into a membrane vesicle through a single hole whose diameter excludes an antibody molecule.

The possible existence of cross-reacting antigens should always be considered when using antibodies as probes to establish molecular structure. This comment applies most forcibly to immunological experiments designed to assess either the location of lipids or the distribution of antigens which possess carbohydrate determinants. For example, a significant percentage of the antilipoteichoic acid antibodies are directed against the polyglycerolphosphate backbone of the molecule.[79] These cross react, not surprisingly, with cardiolipin,[496] which also possesses a glycerolphosphate repeat. Thus, attempts to ascertain the distribution of cardiolipin in the membranes of most Gram-positive bacteria might very well be hampered by the presence of cross-reacting lipoteichoic acids. The converse situation (i.e., that cardiolipin interferes with the assessment of lipoteichoic acid), is less likely to apply, since the phospholipid headgroups probably do not share the same degree of surface exposure as the lengthy hydrophilic chains of the lipoteichoic acids. The scope of antibodies as specific probes for phospholipid distribution[397,419,497-499] is also very severely limited by the degree of cross-reactions observed between various phospholipids such as phosphatidylglycerol, cardiolipin, and phophatidylinositol (see Section V. D). Moreover, antibodies to cardiolipin have been reported to react strongly with DNA,[500] a reaction which could cripple attempts to assess the topology of the phospholipid in all but the most stable of protoplasts. The potential for glycolipids and phosphoglycolipids to cross-react with antibodies to other carbohydrate-containing antigens of the cell surface should also be borne in mind.[398]

VII. FUTURE PROSPECTS

Undoubtedly, a great deal of future interest will focus on the serological properties of outer membrane proteins. These components have received scant attention, largely as a result of problems of solubilization (see Section V.C). Fortunately, these have been overcome to a large extent and a number of different techniques capable of resolving antigens of the outer membrane have been developed recently. CIE, for example, allows a certain degree of resolution of outer membrane antigens.[128] Unfortunately, the identities of many of the outer membrane components resolved by this method remain uncertain. However, this situation could be remedied readily by analysis on SDS-polyacrylamide gels of individual immunoprecipitates excised from wet CIE immunoplates. This particular approach (i.e., CIE) is not limited to the analysis of antigens which are soluble in Triton®X-100. Chaotropic agents[129,156] or SDS,[130] for example, may be employed where necessary. Of course, one has to tolerate the attendant problem of optimizing conditions for extraction and at the same time retaining full immunological specificity. Nevertheless, this approach does appear to have several advantages over methods which involve the analysis of products of interfacial precipitin tests performed with solubilized membranes and antisera.[356,357]

Other possibilities include the detection of outer membrane antigens following their resolution by SDS-polyacrylamide gel electrophoresis. Such methods are based upon the observation that many proteins retain (or can be induced to regain) sufficient native antigenic character following electrophoresis in the presence of SDS to allow their detection by antibody. For example, the outer membrane proteins of *E. coli* 026K60 have been analyzed recently by SDS-polyacrylamide gel electrophoresis-crossed immunoelectrophoresis.[501] This combination technique, in which the first direction electrophoresis is conducted in SDS-gels, has obvious potential. However, certain questions relating to quantitation and sensitivity remain unresolved. Distortion effects are also apparent in all but a few instances.[101-103,502-504] Another variation, which may prove to be of exceptional value in the analysis of (outer) membrane proteins, involves the direct detection of antigens in SDS-polyacrylamide gels. Thus, samples can be analyzed by SDS-polyacrylamide gel electrophoresis in the normal manner, the gels fixed, and antigens detected by densitometry following incubation with relevant antibody.[505] The sensitivity of the method may be increased by the use of enzyme-linked anti-IgG conjugates,[505,506] ^{125}I-labeled protein A[507] or ^{125}I-labeled antibodies. Alternatively, it may prove more convenient to analyze membrane antigens in unfixed SDS-polyacrylamide gels by direct transfer onto either IgG-coated polyvinyl sheets[508] or diazobenzyloxymethyl paper.[509] In these instances, detection could be effected by the use of ^{125}I-labeled specific antibodies. SDS-polyacrylamide gel electrophoresis may also become increasingly popular as a method for isolating potential immunogens, since it has been established that proteins eluted from discrete regions of dry, stained SDS-gels often generate an antibody population which will recognize the native polypetide.[376,504]

One can also foresee an increased use for CIE in the analysis of membrane-bound receptors. The feasibility of such an approach was demonstrated clearly by Blomberg and his colleagues[149] in studies of the adrenaline receptors of rat liver plasma membranes. Putative receptor proteins were identified by their ability to bind [^{14}C]-epinephrine, even after precipitation with antibodies.[149] Technical problems associated with fluorography of radiolabeled components which are embedded in an agarose matrix have been overcome to a large extent[510] and it should now be possible to detect receptors or binding proteins, e.g., the penicillin binding proteins,[304,511] even when they are present in low copy numbers. These comments apply with equal force to the study of any membrane antigen which can be specifically labeled in some manner. For example, in instances where the antireceptor titers are likely to be very low, it may be feasible to analyze receptor-ligand complexes using antibodies directed against the ligand.[512,513] Carrier proteins may be amenable to analysis following covalent coupling to solutes or solute analogs containing reporter groups.[438] In addition, the disposition of certain proteins in membranes seems to be affected by the electrochemical gradient[514] and it may thus be possible to detect carrier molecules immunologically by virtue of the changes in degree of expression or in conformation incurred during transport. Precedents for this exist. The ADP,ATP translocator protein of beef-heart mitochondria[515] has been detected in two distinct conformational states, which appear to reflect different stages in the transport of adenine nucleotides across the membrane.[516] In one conformation, the carrier faces the cytosol and binds the ligand carboxyatractylate with high affinity. In the other, it faces the mitochondrial matrix and binds the bongkrekate ligand. Antibody populations specific for each conformation have been generated, although in this instance neither type of antibody was observed to inhibit ADP.ATP transport.[516] Antibodies to another transport protein, the Na^+, K^+-ATPase, have also been reported to be conformation-sensitive.[517,518]

Progressive immunoadsorption experiments performed in conjunction with CIE provide a powerful method for studying antigen topology and membrane structure, and the application of this approach to the analysis of other membrane systems, both biological and artificial, seems likely. Analysis of the membrane components contained within discrete regions of CIE immunoprecipitates may also provide a valuable insight into the nature and complexity of any functional mosiacs that might exist in the bacterial membrane. It should be emphasized that monospecific antiserum is not a prerequisite for such a study; analysis can be conducted on immunoplates developed against polyvalent serum, providing that regions which contain unrelated and intersecting immunoprecipitates are avoided.

Antibodies have been used on a number of occasions to study the three-dimensional structure of folded proteins. The antigenic structures of water-soluble proteins such as lysozyme,[519] bovine serum albumin,[520] and β-galactosidase[521] have been analyzed in this manner, either by studying the reaction of defined fragments with antiserum against the native protein, or by testing the intact protein against antisera raised to individual fragments. To the author's knowledge this type of experimental approach has not been employed in the study of a bacterial membrane protein, except to a limited extent in the case of Braun's lipoprotein.[339] Interesting candidates for such a study include the so-called integral membrane proteins, which normally have a substantial part of their mass in contact with the hydrophobic fatty acyl chains of the bilayer, e.g., bacteriorhodopsin, enzymes involved in phospholipid metabolism, certain carrier proteins, etc. An understanding of the important structural features of protective membrane antigens such as the 41,000-dalton polypeptide of serotype 2 meningococci would also be of great value (see Section V.C.2). Progress that has been made recently in the purification of amphiphilic proteins[219,370,381,387,522-529] may facilitate studies of this type.

The results of several recent studies[508,530,531] have clearly demonstrated that radiolabeled antibodies can be used as specific probes to detect translation products in isolated bacterial colonies. Of special interest to the membranologist is the report by Henning et al.[531] that colonies can be screened rapidly for the presence of a specific membrane protein (in this particular instance, outer membrane protein II* of *E. coli*) by a relatively simple radioimmunoassay that involves initial transfer of colonies onto filter paper. Colonies containing specific gene products are detected by autoradiography following treatment with organic solvents and radioiodinated IgG. The method is extremely sensitive; colonies containing less than 500 copies of specific protein per cell can be detected. This immunological screening technique could be used to identify directly clones containing specific DNA segments for foreign membrane proteins, and may be extremely useful when direct selection procedures are not available.[531]

VIII. SUMMARY AND CONCLUSIONS

Crossed immunoelectrophoresis has added a new dimension to the analysis of the bacterial membrane and has clearly revealed the antigenic complexity of this multifunctional cell structure. When resolved by CIE, membrane antigens usually retain and express sufficient intrinsic biological activity to facilitate identification in a manner which does not rely on possession of either the purified antigen or its homologous specific antibody. This may be achieved by zymograms if the antigen in question is an enzyme or it may be accomplished by utilizing the affinity of the antigen for specific ligands, by analysis for characteristic prosthetic groups or co-factors (e.g., ^{59}Fe), from a consideration of the interaction of the antigen with lectins or other affinity adsorbents, or by analysis of the behavior of the antigen in different gel matrices or in the presence of different ionic detergents. The membranes of *M. lysodeikticus, E. coli, N. gonorrhoeae, R. sphaeroides,* and recently *Laidlawii faecalis* [531a, 531b] have been studied in detail by these methods and comprehensive reference immunoprecipitate profiles have been established in a number of instances. Of these, the reference profile for membrane vesicles of *E. coli* ML308-225 is probably the most detailed; 14 of the 52 discrete antigens detected by CIE have been identified and several others have been partially characterized. The area subtended by individual immunoprecipitates in the CIE system is proportional to the antigen/antibody ratio and this empirical relationship has been exploited in progressive immunoadsorption experiments designed to assess and quantitate the disposition of antigens in the plasma membrane of *A. laidlawii* and *M. lysodeikticus,* and in membrane vesicles derived from *E. coli* and *R. sphaeroides.* Although this approach has provided valuable and detailed information about the expression and (asymmetric) distribution of antigens across the bilayer, the method does not readily yield data relating to the (relative) positioning of immunogens in the lateral plane of the membrane (except, perhaps, if used in conjunction with bifunctional cross-linking reagents). Fortunately, the ferritin-labeled technique is well-suited for this purpose and has been used to investigate aspects of the synthesis and/or precise cellular location of F_1·ATPase, lipoteichoic acids, alkaline phosphatase, penicillinase, lipopolysaccharide, and certain proteins of the porin type. Important biochemical properties of a number of bacterial membrane proteins, e.g., ATPase, NADH dehydrogenase, and D-lactate dehydrogenase have been studied in detail with the aid of antibodies, as have the antigenic structures of Braun's lipoprotein and alkaline phosphatase. Specific antibodies have also played an unique role in the selective resolution of precursors of membrane-associated proteins.

It has become fashionable recently to study the protein antigens of the outer membrane of Gram-negative bacteria and some progress has been made in resolution of the important surface antigens of pathogenic neisseria. The 41,000-dalton polypeptide of the outer membrane of *N. meningitidis* serotype 2 has been purified to homogeneity and appears to have good potential as a protective vaccine against infection by group B meningococci. Certain phospholipids and glycolipids are also immunologically active, the latter being intimately involved in the mechanism of complement-mediated immune lysis of *Mycoplasma pneumoniae*. However, problems exist in the immunological analysis of most bacterial lipids. These relate to the haptenic nature of the molecules and to the high degree of cross-reactivity with other components of the microbial cell.

Future prospects include a detailed understanding of the antigenic structure of amphiphilic membrane proteins and of other important surface antigens. It should also be possible to derive much useful information about the nature of discrete functional regions within the membrane from an analysis of antigen complexes which have been selectively precipitated from detergent-solubilized membranes by specific antisera. One can also foresee an increased use for antibodies as specific and sensitive probes to detect or screen for gene products, such as cell surface receptors, which are present in extremely low copy numbers.

IX. APPENDIX

Below is a listing of 18 enzyme stains which have proved of value in the author's laboratory during immunochemical analysis of bacterial membranes and related cell fractions. Methodology pertaining to other enzyme stains can be found in References 59, 60, 88, 108, 146, and 147. Many of the staining reactions described here are not original and represent simple adaptions or modifications of zymograms described by other workers.[59,60,532-534]

Zymogram stains are performed normally on CIE immunoplates which have been washed and pressed[87] (but not dried). Petri dishes provide ideal reaction vessels if 5- × 5-cm glass immunoplates are used. Following immersion of the immunoplate in the histochemical stain and full development of the zymogram, gels should be pressed and washed several times in 100 m*M* NaC1 to remove residual reaction ingredients. It is often possible to repeat the above process using a different histochemical stain each time. Gels should be rinsed finally in distilled water and dried with the aid of a hair drier or in a hot oven in order to provide a permanent record. Zymograms may be counterstained with Coomassie brilliant blue to confirm the identity of the enzyme-active immunoprecipitates in the reference immunoprecipitate profile.

Most of the dehydrogenase enzymes can be monitored conveniently by reduction of the artificial electron acceptor 2,2′5,5′-tetra-*p*-nitrophenyl-3,3′-(3,3′-dimethoxy-4,4′-diphenylene)-ditetrazolium chloride, commonly referred to as tetranitrobluetetrazolium (TNBT). This tetrazolium salt is sparingly soluble in aqueous buffers, and solutions should be filtered prior to use. A convenient approach is to add 10 mg of TNBT to 100 m*l* of 50 m*M* *tris* (hydroxymethyl) methylamine·HCl buffer at the appropriate pH and to filter the solution after 10 min of vigorous mixing. The filtrate, which will be referred to as TNBT·Tris for convenience, should be used the same day. Care should be taken to readjust the pH of this solution after the addition of substrate. It should also be noted that the molarities indicated below for the various nucleotide and napthol derivatives are probably overestimates since they are computed from values for the anhydrous molecular weight. Commercial samples regularly contain varying amounts of water and/or acetone. This is of marginal importance, however, since in practice the concentrations of many of the reaction ingredients, especially substrates, can be decreased by up to an order of magnitude without grossly affecting development of the zymogram.

(i) D-Lactate dehydrogenase (EC 1.1.1.28)—Reaction mixture contains 18.0 m*l* of TNBT·Tris (pH 7.5), 2.0 m*l* of 1.0 *M* lithium-D-lactate (pH 7.5), 5.0 mg of β-nicotinamide adenine dinucleotide (NAD), and about 0.5 mg of phenazine methosulfate. The final concentration of substrate and NAD are 100 m*M* and 375 μ*M*, respectively. Incubate in the dark at 30°C.

Comments: Lithium-DL-lactate may be substituted for the expensive D-lactate enantiomorph in some instances, e.g., in the analysis of extracts of *Escherichia coli* in which the inducible L-lactate dehydrogenase is absent. Examples of relevant zymograms performed on membranes of *E. coli* can be found in Figure 13 of this text and also in References 108 and 128. Similar zymograms performed on envelopes from *Neisseria gonorrhoeae* are depicted in References 106 and 107. In these instances, development was usually complete within 30 min.

(ii) Malate dehydrogenase (EC 1.1.1.37)—Reaction mixture contains 18.0 m*l* of TNBT·Tris (pH 7.5), 2.0 m*l* of 1.0 *M* sodium-DL-malate (pH 7.5), 10 mg of NAD, and about 0.5 mg of phenazine methosulfate. The final concentrations of substrate and NAD are 100 m*M* and 750 μ*M*, respectively. Incubate in the dark at 30°C.

Comments: Examples of relevant zymograms performed on plasma membranes of *Micrococcus lysodeikticus, E. coli,* and *Rhodopseudomonas sphaeroides* can be found in References 132, 128, and 174, respectively. In most instances, development was complete within 2hr.

(iii) Isocitrate dehydrogenase (EC 1.1.1.42)—Reaction mixture contains 18.0 m*l* of TNBT·Tris (pH 7.0), 2.0 m*l* of 100 m*M* trisodium-DL-isocitrate (pH 7.0), 4.0 mg of β-nicotinamide adenine dinucleotide phosphate (monosodium salt; NADP), 4.0 mg of $MnCl_2 \cdot 4H_20$, and about 0.5 mg of phenazine methosulfate. The final concentrations of substrate, NADP, and $MnCl_2$ are 10 m*M*, 260 μ*M* and 1 m*M*, respectively. Incubate in the dark at 30°C.

Comments: An example of a relevant zymogram performed on a soluble cytoplasmic fraction from *M. lysodeikticus* can be found in Reference 88. In this particular instance, development was complete within 10 min.

(iv) 6-Phospho-D-gluconate dehydrogenase (EC 1.1.1.43)—Reaction mixture contains 20.0 m*l* of TNBT·Tris (pH 7.0), 40 mg of tri-monocyclohexylammonium 6-phospho-D-gluconate, 10 mg of NADP, and about 0.5 mg of phenazine methosulfate. The final concentrations of substrate and NADP are 3.5 m*M* and 650 μ*M*, respectively. Incubate in the dark at 30°C.

Comments: An example of a relevant zymogram performed on plasma membranes from *E. coli* can be found in Reference 128. In this particular instance development was usually complete within 15 min.

(v) D-Glucose-6-phosphate dehydrogenase (EC 1.1.1.49)—Reaction mixture contains 20.0 m*l* of TNBT·Tris (pH 7.0), 40 mg of monosodium D-glucose-6-phosphate, 10 mg of NADP, and about 0.5 mg of phenazine methosulfate. The final concentrations of substrate and NADP are 7.1 m*M* and 650 μ*M*, respectively. Incubate in the dark at 30°C.

Comments: This is normally considered to be a soluble cytoplasmic enzyme. An example of a relevant zymogram performed on the cytoplasm of *N. gonorrhoeae* can be found in Reference 107. In this particular instance, development was usually complete within 30 min.

(vi) Glycerol-3-phosphate dehydrogenase (EC 1.1.99.5)—Reaction mixture contains 18.0 m*l* of TNBT·Tris (pH 7.0), 2.0 m*l* of 1.0 *M* disodium DL-α-glycerophosphate (pH 7.0) and about 0.5 mg of phenazine methosulfate. The final concentration of substrate is 100 m*M*. Incubate in the dark at 30°C.

Comments: This is normally an inducible enzyme in *E. coli.* An example of a relevant zymogram performed on plasma membranes from this organism can be found in Reference 128. In this particular instance, development was complete within 1 to 2 hr.

(vii) Dihydro-L-orotate dehydrogenase (EC 1.3.3.1)—Reaction mixture contains 18.0 ml of TNBT·Tris (pH 7.2), 2.0 m*l* of 1.0 *M* dihydro-L-orotic acid (pH 7.2), 10 mg of NAD, and about 0.5 mg of phenazine methosulfate. The final concentrations of substrate and NAD are 100 m*M* and 750 μ*M,* respectively. Incubate in the dark at 30°C.

Comments: Examples of relevant zymograms performed on membranes of *E. coli* can be found in References 108 and 128. In these instances, development was usually complete within 1 hr.

(viii) Succinate dehydrogenase (EC 1.3.99.1)—Reaction mixture contains 18.0 m*l* of TNBT·Tris (pH 7.2), 2.0 m*l* of 1.0 *M* disodium succinate (pH 7.2), and about 0.5 mg of phenazine methosulfate. The final concentration of substrate is 100 m*M*. Incubate in the dark at 30°C.

Comments: Examples of relevant zymograms may be found in texts describing the immunochemical properties of membranes derived from *M. lysodeikticus,*[132] *Bacillus subtilis,*[115] *E. coli*[128] and *R. sphaeroides.*[174] Development is usually complete within 2 to 3 hr.

(ix) Glutamate dehydrogenase (EC 1.4.1.4)—Reaction mixture contains 18.0 m*l* of TNBT·Tris (pH 7.0), 2.0 m*l* of 1.0 *M* monosodium L-glutamate (pH 7.0), 10 mg of NADP, and about 0.5 mg of phenazine methosulfate. The final concentrations of substrate and NADP are 100 m*M* and 650 μ*M,* respectively. Incubate in the dark at 30°C.

Comments: Examples of relevant zymograms performed on membranes of *E. coli* can be found in References 108, 128, and 153. Similar zymograms performed on a cytoplasmic fraction from *N. gonorrhoeae* may be found in Reference 107. Two enzymes possessing this apparent activity have been detected in the plasma membranes of *E. coli* K12.[108] Development is usually complete within 1 hr.

(x) NADPH dehydrogenase (EC 1.6.99.1)—Reaction mixture contains 19.0 m*l* of TNBT·Tris (pH 7.0) and 1.0 m*l* of 5 m*M* NADPH (pH 7.0). The final concentration of substrate is 250 μ*M*. Incubate at room temperature.

Comments: An example of a relevant zymogram performed on membranes from *E. coli* may be found in Figure 7 of this text and in Reference 153. Development is usually complete within 15 min. Enzymes which display "NADPH diaphorase" activity will also be detected by this method.

(xi) NADH dehydrogenase (EC 1.6.99.3)—Reaction mixture contains 19.0 m*l* of TNBT·Tris (pH 7.0) and 1.0 m*l* of 5 m*M* NADH (pH 7.0). The final concentration of substrate is 250 μ*M*. Incubate at room temperature.

Comments: A pH in excess of 7 should be avoided as chemical reduction of TNBT by NADH becomes pronounced in alkaline solution. Development is extremely rapid and is usually complete within 15 min. Examples of relevant zymograms may be found in Figures 4, 11, and 12 of this text and also in reports dealing with the immunochemical properties of fractions isolated from cells of *M. lysodeikticus,*[88,132] *E. coli,*[108,128,153] *N. gonorrhoeae*[106,107] and *R. sphaeroides.*[174] In most instances two or more enzyme-active immunoprecipitates have been detected for the plasma membrane.[106-108,132,153,174] However, it should be remembered that enzymes such as dihydrolipoyl dehydrogenase, which display "NADH diaphorase" activity, will also be detected by this method (see Figure 12). The NADH dehydrogenase antigen in the membranes of *E. coli* K12, unlike corresponding enzyme in *E. coli* ML308-225 (see Figure 11 and 12), gives a poorly defined immunoprecipitate following CIE.[128] There is some evidence that it associates with unrelated membrane components following solubilization.

(xii) Catalase (EC 1.11.1.6)—Reaction mixture contains 17.6 m*l* of Tris·HCl (50 m*M*, pH 7.0), 1.4 m*l* of 0.6 M $Na_2S_2O_3$ and 1.0 m*l* of 3% H_2O_2. The final concentrations of $Na_2S_2O_3$ and H_2O_2 are 42 m*M* and 0.15%, respectively. Incubate at room temperature for about 15 min or until bubbles are visible, and then remove gel and immerse in 20 m*l* of 45 m*M* KI.

Comments: This procedure results in a negatively stained image on a blue background. The reaction products are diffusable and gels should be photographed immediately or dried rapidly with a stream of hot air if a permanent record is required. Catalase is normally considered to be a soluble cytoplasmic enzyme, and a relevant zymogram performed on a cytoplasm fraction from *M. lysodeikticus* can be found in Reference 88.

(xiii) Polynucleotide phosphorylase (EC 2.7.7.8)—Reaction mixture contains 21.0 m*l* of Tris·HCl (50 m*M*, pH 7.5), 125 mg of adenosine 5′-diphosphate (disodium salt, ADP), 2.5 m*l* of 100 m*M* $MgSO_4$, and 1.5 m*l* of 2% Pb $(No_3)_2$. The final concentrations of ADP, $MgSO_4$ and $Pb(NO_3)_2$ are 10.6 m*M*, 10 m*M*, and 0.12%, respectively. Incubate at 37°C for 2 hr, and then press and wash the gel a number of times in 100 m*M* NaCl. Final development is achieved by immersion of the immunoplate in 0.1% Na_2S for about 30 min.

Comments: Immunoprecipitates possessing polynucleotide phosphorylase (or ADPase) activity stain brown. Examples of relevant zymograms performed on the cytoplasm of *M. lysodeikticus* and membrane vesicles from *E. coli* can be found in References 88 and 108, respectively.

(xiv) Alkaline phosphatase (EC 3.1.3.1)—Dissolve 5.0 mg of napthol AS-MX phosphoric acid (disodium salt) in 0.2 m*l* of dimethyl formamide and add 19.8 m*l* of Tris·HCl (50 m*M*, pH 8.7). The final concentration of substrate is 600 μ*M*. Filter the reaction mixture if necessary and then incubate at 30°C for 2 to 3 hr. Remove immunoplate and immerse for 30 min in 20 m*l* of a filtered solution (1 mg/m*l*) of diazotized *o*-aminoazotoluene (Fast Garnet GBC salt) in 100 m*M* acetate buffer (pH 4.8).

Comments: Enzyme-active immunoprecipitates stain blue-violet.[59]

(xv) Acid phosphatase (EC 3.1.3.2)—Reaction mixture contains 5.0 mg of napthol AS-BI phosphoric acid (sodium salt) in 20 m*l* of acetate buffer (100 m*M*, pH 4.8). The final concentration of substrate is 500 μ*M* Incubate at 30°C for 2 to 3 hr. Remove immunoplate and immerse for 30 min in 20 m*l* of a filtered solution (1 mg/m*l*) of diazotized *o*-aminoazotoluene (Fast Garnet GBC salt) in 100 m*M* acetate buffer (pH 4.8).

Comments: Enzyme-active immunoprecipitates stain blue violet.[59]

(xvi) Chymotrypsin-like proteases (EC 3.4.21.1)—Dissolve 5.0 mg of N-acetyl-DL-phenylalanine-β-napthyl ester in 2.0 m*l* of dimethyl formamide and add 18.0 m*l* of Tris·HCl (50 m*M*, pH 7.5) containing 10 mg of 4-chlorotoluene-1,5′-diazonium napthalene disulfonate (Fast Red TR salt). Filter if necessary and incubate at 37°C. The final concentration of substrate is 750 μ*M*.

Comments: Enzyme-active immunoprecipitates have been detected following CIE analysis of both inner and outer membranes of *E. coli.*[128] Development in these instances was very slow and was only observed after several hours of incubation. Enzyme-active precipitates stain red.

(xvii) Trypsin-like proteases (EC 3.4.21.4)—Dissolve 20 mg of α-*N*-benzoyl-DL-arginine-β-napthylamide·HCl (BANA) in 2.0 m*l* of dimethyl formamide and add 18.0 m*l* of Tris·HCl (50 m*M*, pH 7.5) containing 20 mg of Fast Red TR salt. Filter if necessary and incubate at 37°C. The final concentration of substrate is 2.3 m*M*.

Comments: Enzyme-active immunoprecipitates stain red. Examples of relevant zymograms performed on extracts of decapod Crustaceae can be found in References 535 and 536.

(xviii) ATPase (EC 3.6.1.3)—Reaction mixture contains 21.0 m*l* of Tris·HCl (50 m*M*, pH 7.5), 125 mg of ATP (disodium salt), 2.5 m*l* of 100 m*M* $MgSO_4$, and 1.5 m*l* of 2% $Pb(NO_3)_2$. The final concentrations of ATP, $MgSO_4$ and $Pb(NO_3)_2$ are 9.1 m*M*, 10 m*M*, and 0.12%, respectively. Incubate at 37°C for 2 hr and then press and wash gels a number of times in 100 m*M* NaCl. Final development is achieved by immersion of the immunoplates of 0.1% Na_2S.

Comments: Immunoprecipitates possessing ATPase activity stain brown. Examples of relevant zymograms performed on plasma membranes of *M. lysodeikticus* can be found in Figure 10 of this text and in Reference 124. Similar zymograms performed on membranes of *E. coli* and *R. sphaeroides* may be found in References 108 and 132, and in Reference 174, respectively.

ACKNOWLEDGMENTS

The scientific work presented in this review could not have been accomplished without the help and advice of many colleagues and friends. In particular, the author would like to express his gratitude to John Arbuthnott, Helen Doherty, Ron Kaback, Martin Nachbar, Joel Oppenheim, Milton Salton, Julie Siegel, and Cyril Smyth. The author would also like to thank Milton Salton and Tony Wicken for kindly supplying the electron micrographs for Figure 18; Karl-Erik Johansson, Wil Konings, Lars Rutberg, and Tony Wicken for sending preprints of manuscripts; Cyril Smyth for helpful discussions; Pauline Christie for photographic assistance; Corinne Harrison and Lesley O'Riordan for careful secretarial service; and Helen Doherty and Gill Owen for help in proofreading the manuscript. Work in the author's laboratory is supported in part by a grant from the Medical Research Council of Ireland.

REFERENCES

1. **Heidelberger, M. and Kendall, F.G.,** A quantitative study of precipitin reaction between Type III pneumococcus polysaccharide and purified homologous antibody, *J. Exp. Med.,* 50, 809, 1929.
2. **Kabat, E. A.,** Michael Heidelberger as a carbohydrate chemist, *Carbohydr. Res.,* 40, 1, 1975.
3. **Heidelberger, M.,** Quantitative absolute methods in the study of antigen-antibody reactions, *Bacteriol. Rev.,* 3, 49, 1939.
4. **Landsteiner, K.,** *The Specificity of Serological Reactions* (rev. ed.), Harvard University Press, Cambridge, Mass., 1947.
5. **Kabat, E. A.,** The nature of an antigenic determinant, *J. Immunol.,* 97, 1, 1966.
6. **Heidelberger, M.,** Structure and immunological specificity of polysaccharides, *Fortschr. Chem. Org. Naturstoffe,* 18, 503, 1960.
7. **Heidelberger, M.,** Some contributions of immunochemistry to biochemistry and biology, *Annu. Rev. Biochem.,* 36, 1, 1967.
8. **Lüderitz, O., Staub, A. M., and Westphal, O.,** Immunochemistry of O and R antigens of *Salmonella* and related *Enterobacteriaceae, Bacteriol. Rev.,* 30, 192, 1966.
9. **Lüderitz, O., Westphal, O., Staub, A. M., and Nikaido, H.,** Isolation and chemical and immunological characterization of bacterial lipopolysaccharides, in *Microbial Toxins,* Vol. 4, **Weinbaum, G., Kadis, S., and Ajl, S. J., Eds.,** Academic Press, New York, 1971, 145.
10. **Westphal, O.,** Bacterial Endotoxins, *Int. Arch. Allergy Appl. Immunol.,* 49, 1, 1975.
11. **Jann, K., and Westphal, O.,** Microbial polysaccharides, in *The Antigens,* Vol. 3, **Sela, M., Ed.,** Academic Press, New York, 1975, Chap. 1.
12. **Weibull, C.,** The isolation of protoplasts from *Bacillus megaterium* by controlled treatment with lysozyme, *J. Bacteriol.,* 66, 688, 1953.
13. **Weibull, C.,** Characterization of protoplasmic constituents of *Bacillus megaterium, J. Bacteriol.,* 66, 696, 1953.
14. **Salton, M. R. J.,** Structure and function of bacterial cell membranes, *Annu. Rev. Microbiol.,* 21, 417, 1967.
15. **Salton, M. R. J.,** Isolation and characterization of bacterial membranes, *Trans. N. Y. Acad. Sci., Ser. 2,* 29, 764, 1967.
16. **Salton, M. R. J.,** Bacterial membranes, *C.R.C. Crit. Rev. Microbiol.* 1, 161, 1971.
17. **Schnaitman, C. A.,** Protein composition of the cell wall and cytoplasmic membrane of *Escherichia coli, J. Bacteriol.,* 104, 890, 1970.
18. **Osborn, M. J. and Munson, R.,** Separation of the inner (cytoplasmic) and outer membranes of gram-negative bacteria, in *Methods in Enzymology,* Vol. 31A, **Fleischer, S. and Packer L., Eds.,** Academic Press, New York, 1974, 642.
19. **Yamato, I., Anraku, Y., and Hirosawa, K.,** Cytoplasmic membrane vesicles of *Escherichia coli.* I. A simple method for preparing the cytoplasmic and outer membranes, *J. Biochem. (Tokyo),* 77, 705, 1975.
20. **Vennes, J. W. and Gerhardt, P.,** Antigenic analysis of cell structures isolated from *Bacillus megaterium, J. Bacteriol.,* 77, 581, 1959.
21. **Ouchterlony, O.,** Diffusion in gel methods for immunological analysis. I, in *Progress in Allergy,* Vol. 5, S. Karger, Basel, 1958, 1.
22. **Ouchterlony, O.,** Diffusion in gel methods for immunological analysis. II, in *Progress in Allergy,* Vol. 6, S. Karger, Basel, 1962, 30.
23. **Williams, C. A. and Chase, M. W., Eds.,** *Methods in Immunology and Immunochemistry,* Vol. 3, Academic Press, New York, 1971, chap. 14.
24. **Laurell, C. -B.,** Antigen – antibody crossed electrophoresis, *Anal. Biochem.,* 10, 358, 1965.
25. **Kahane, I., and Razin, S.,** Immunological analysis of *Mycoplasma* membranes, *J. Bacteriol.,* 100, 187, 1969.
26. **Argaman, M., and Razin, S.,** Antigenic properties of *Mycoplasma* organisms and membranes, *J. Gen. Microbiol.,* 55, 45, 1969.
27. **Fukui, Y., Nachbar, M. S., and Salton, M. R. J.,** Immunological properties of *Micrococcus lysodeikticus* membranes, *J. Bacteriol.,* 105, 86, 1971.
28. **Salton, M. R. J.,** On the basis of behaviour in single cells: cell surface components in relation to bacterial behaviour, in *Future of the Brain Sciences,* Plenum Press, New York, 1969, 1.
29. **Salton, M. R. J.,** The bacterial membrane, in *Biomembranes,* Vol. 1, **Manson, L. A., Ed.,** Plenum Press, New York, 1971, 1.
30. **Salton, M. R. J.,** Membrane associated enzymes in bacteria, in *Advances in Microbial Physiology,* Vol. 11, **Rose, A. H. and Tempest, D. W., Eds.,** Academic Press, London, 1973, 213.
31. **Machtiger, N. A., and Fox, C. F.,** Biochemistry of bacterial membranes, *Annu. Rev. Biochem.,* 42, 575, 1973.

32. **Braun, V. and Hantke, K.,** Biochemistry of bacterial cell envelopes, *Annu. Rev. Biochem.,* 43, 89, 1974.
33. **Salton, M. R. J.,** Molecular bacteriology, in *Molecular Microbiology,* **Kwapinski, J. B. G., Ed.,** John Wiley & Sons, New York, 1974, 387.
34. **Braun, V.,** Structure-function relationships of the gram-negative bacterial cell envelope, in *Relations between Structure and Function in the Prokaryotic Cell,* **Stanier, R. Y., Rogers, H. J., and Ward, J. B., Eds.,** Cambridge University Press, Cambridge, England, 1978, 111.
35. **Salton, M. R. J.,** Structure and function of bacterial plasma membranes, in *Relations between Structure and Function in the Prokaryotic Cell,* **Stanier, R. Y., Rogers, H. J., and Ward, J. B., Eds.,** Cambridge University Press, Cambridge, England, 1978, 201.
36. **Rottem, S. and Razin, S.,** Electrophoretic patterns of membrane proteins of *Mycoplasma, J. Bacteriol.,* 94, 359, 1967.
37. **Salton, M. R. J., Schmitt, M. D., and Trefts, P. E.,** Fractionation of isolated bacterial membranes, *Biochem. Biophys. Res. Commun.,* 29, 728, 1967.
38. **Owen, P. and Salton, M. R. J.,** Isolation and characterisation of a mannan from mesosomal membrane vesicles of *Micrococcus lysodeikticus, Biochim. Biophys. Acta,* 406, 214, 1975.
39. **Fukui, Y., Nachbar, M. S., and Salton, M. R. J.,** Immunochemistry and peptide mapping of *Micrococcus lysodeikticus* membrane proteins, *Biochim. Biophys. Acta,* 241, 30, 1971.
40. **Fukui, Y. and Salton, M. R. J.,** Common peptides in *Micrococcus lysodeikticus* membrane proteins, *Biochim. Biophys. Acta,* 288, 65, 1972.
41. **Whiteside, T. L. and Salton, M. R. J.,** Antibody to adenosine triphosphatase from membranes of *Micrococcus lysodeikticus, Biochemistry,* 9, 3034, 1970.
42. **Singer, S. J.,** Preparation of an electron-dense antibody conjugate, *Nature (London),* 183, 1523, 1959.
43. **Singer, S. J., and Schick, A. F.,** The properties of specific stains for electron microscopy prepared by the conjugation of antibody molecules with ferritin, *J. Biophys. Biochem. Cytol.,* 9, 519, 1961.
44. **McLean, J. D. and Singer, S. J.,** A technique for the specific staining of macromolecules and viruses with ferritin-antibody conjugates, *J. Molec. Biol.* 56, 633, 1971.
45. **Clarke, H. G. M. and Freeman, T.,** A quantitative immuno-electrophoresis method (Laurell electrophoresis), in *Protides of the Biological Fluids,* Vol. 14, Peeters, H., Ed., Elsevier, Amsterdam, 1967, 503.
46. **Laurell, C.-B.,** Electrophoretic microheterogeneity of serum α_1-antitrypsin, *Scand. J. Clin. Lab. Invest.,* 17, 271, 1965.
47. **Laurell, C.-B.,** Quantitative estimation of proteins by electrophoresis in agarose gels containing antibodies, *Anal. Biochem.,* 15,45, 1966.
48. **Laurell, C.-B.,** Quantitative estimation of proteins by electrophoresis in antibody-containing agarose gel, in *Protides of the Biological Fluids,* Vol. 14., **Peeters, H., Ed.,** Elsevier, Amsterdam, 1967, 499.
49. **Laurell, C.-B.,** Electroimmuno assay, *Scand. J. Clin. Lab. Invest.,* 29 (Suppl. 124), 21, 1972.
50. **Laurell, C.-B., Ed.,** Electrophoretic and Electro-immuno-chemical analysis of proteins, *Scand. J. Clin. Lab. Invest.,* 29 (Suppl. 124), 1972.
51. **Axelsen, N. H., Krøll, J., and Weeke, B., Eds.,** *A Manual of Quantitative Immunoelectrophoresis,* Universitetsforlaget, Oslo, 1973.
52. **Axelsen, N. H., Ed.,** *Quantitative Immunoelectrophoresis: New Developments and Applications,* Universitetsforlaget, Oslo, 1975.
53. **Axelsen, N. H.,** A survey of quantitative immunoelectrophoretic methods (electrophoresis in antibody (or antigen)-containing agarose gel) with examples from *Candida* immunochemistry, in *Automation in Microbiology and Immunology,* **Hedén, C.-G. and Illeni, T., Eds.,** John Wiley & Sons, New York, 1975, 357.
54. **Bjerrum, O. J., and Bøg-Hansen, T. C.,** Immunochemical gel precipitation techniques in membrane studies, in *Biochemical Analysis of Membranes,* **Maddy, A. H., Ed.,** Chapman and Hall, London, 1976, 378.
55. **Verbruggen, R.,** Quantitative immunoelectrophoretic methods: a literature survey, *Clin. Chem. (Winston-Salem, N.C.),* 21, 5, 1975.
56. **Bjerrum, O. J. and Lundahl, P.,** Detergent-containing gels for immunological studies of solubilized erythrocyte membrane components, *Scand. J. Immunol.,* 2 (Suppl. 1), 139, 1973.
57. **Johansson, K. E. and Hjertén, S.,** Localization of the Tween 20-soluble membrane proteins of *Acholeplasma laidlawii* by crossed immunoelectrophoresis, *J. Mol. Biol.,* 86, 341, 1974.
58. **Bjerrum, O. J. and Bøg-Hansen, T. C.,** The immunochemical approach to the characterization of membrane proteins: human erythrocyte membrane proteins analysed as a model system, *Biochim. Biophys. Acta,* 455, 66, 1976.
59. **Uriel, J.,** Characterization of precipitates in gels, in *Methods in Immunology and Immunochemistry,* Vol. 3, **Williams, C. A., and Chase, M. W., Eds.,** Academic Press, New York, 1971, 294.

60. **Brogren, C.-H. and Bøg-Hansen, T. C.,** Enzyme characterization in quantitative immunoelectrophoresis: an enzymological study of human serum esterases, *Scand. J. Immunol.*, 4, Suppl. 2, 37, 1975.
61. **Owen, P. and Smyth, C. J.,** Enzyme analysis by quantitative immunoelectrophoresis, in *Immunochemistry of Enzymes and their Antibodies,* **Salton, M. R. J., Ed.,** John Wiley & Sons, New York, 1977, 147.
62. **Weber, K. and Osborn, M.,** The reliability of molecular weight determinations by dodecylsulphate-polyacrylamide gel electrophoresis, *J. Biol. Chem.*, 244, 4406, 1969.
63. **O'Farrell, P. H.,** High resolution two-dimensional electrophoresis of proteins, *J. Biol. Chem.*, 250, 4007, 1975.
64. **Helenius, A. and Simons, K.,** Solubilization of membranes by detergents, *Biochim, Biophys. Acta,* 415, 29, 1975.
65. **Weeke, B.,** Crossed immunoelectrophoresis, *Scand. J. Immunol.*, 2 (Suppl. 1), 47, 1973.
66. **Simmons, D. A. R.,** Immunochemistry of *Shigella flexneri* O-antigens: a study of structural and genetic aspects of the biosynthesis of cell-surface antigens, *Bacteriol. Rev.*, 35, 117, 1971.
67. Bacterial lipopolysaccharide; chemistry, immunology, biological responses, models of enterotoxic activity and clinical aspects, *J. Infect. Dis.*, 128 (Suppl. S9-S303), 1973.
68. **Nikaido, H.,** Biosynthesis and assembly of lipopolysaccharide and the outer membrane layer of gram-negative cell wall, in *Bacterial Membranes and Walls,* Vol. 1, **Leive, L., Ed.,** Marcel Dekker, New York, 1973, 131.
69. **Ørskov, I., Ørskov, F., Jann, B., and Jann, K.,** Serology, chemistry, and genetics of O and K antigens of *Escherichia coli, Bacteriol. Rev.*, 41, 667, 1977.
70. **Mäkelä, P. H. and Mayer, H.,** Enterobacterial common antigen, *Bacteriol. Rev.*, 40, 591, 1976.
71. **Männel, D. and Mayer, H.,** Isolation and chemical characterisation of the enterobacterial common antigen, *Eur. J. Biochem.*, 86, 361, 1978.
72. **Männel, D. and Mayer, H.,** Serological and immunological properties of isolated enterobacterial common antigen, *Eur. J. Biochem.*, 86, 371, 1978.
73. **Hofstad, T.,** Serological responses to antigens of *Bacteroidaceae, Microbiol. Rev.*, 43, 103, 1979.
74. **Wicken, A. J. and Knox, K. W.,** Lipoteichoic acids: a new class of bacterial antigen, *Science,* 187, 1161, 1975.
75. **Lambert, P. A., Hancock, I. C., and Baddiley, J.,** Occurrence and function of membrane teichoic acids. *Biochim. Biophys. Acta,* 472,1, 1977.
76. **Smith, H.,** Microbial surfaces in relation to pathogenicity, *Bacteriol. Rev.*, 41, 475, 1977.
77. **Heath, E. C.,** Complex polysaccharides, *Annu. Rev. Biochem.*, 40, 29, 1971.
78. **Heidelberger, M.,** Immunochemistry of bacterial polysaccharides, in *Research in Immunochemistry and Immunobiology,* Vol. 3, **Kwapinski, J. B. G., Ed.,** University Park Press, Baltimore, 1973, 1.
79. **Knox, K. W. and Wicken, A. J.,** Immunological properties of teichoic acids, *Bacteriol. Rev.*, 37, 215, 1973.
80. **Fox, E. N.,** M proteins of group A streptococci, *Bacteriol. Rev.*, 38, 57, 1974.
81. **Akatov, A. K.,** Antigennaia struktura *Staphylococcus aureus, Zh. Mikrobiol. Epidemiol. Immunobiol.*, 10, 55, 1973.
82. **Lederer, E., Adam, A., Ciorbaru, R., Petit, J.-F., and Wietzerbin, J.,** Cell walls of *Mycobacteria* and related organisms; chemistry and immunostimulant properties, *Molec. Cell. Biochem.*, 7, 87, 1975.
83. **Barksdale, L. and Kim, K.-S.,** *Mycobacterium, Bacteriol. Rev.*, 41, 217, 1977.
84. **Daniel, T. M. and Janicki, B. W.,** Mycobacterial antigens: a review of their isolation, chemistry, and immunological properties, *Microbiol. Rev.*, 42, 84, 1978.
85. **Petit, J.-F. and Lederer, E.,** Structure and immunostimulant properties of mycobacterial cell wall, in *Relations between Structure and Function in the Prokaryotic Cell,* **Stanier, R. Y., Rogers, H. J., and Ward, J. B., Eds.,** Cambridge University Press, Cambridge, England, 1978, 177.
86. **Barber C. and Eylan, E.,** Immunochimie de *Salmonella typhimurium:* variations de la fraction protéique de l'antigène somatique, *Rev. Immunol.*, 36, 1, 1972.
87. **Weeke, B.,** General remarks on principles, equipment, reagents and procedures, *Scand. J. Immunol.*, 2 (Suppl. 1), 15, 1973.
88. **Owen, P. and Salton, M. R. J.,** Membrane asymmetry and expression of cell surface antigens of *Micrococcus lysodeikticus* established by crossed immunoelectrophoresis, *J. Bacteriol.*, 132, 974, 1977.
89. **Axelsen, N. H. and Bock, E.,** Identification and quantitation of antigens and antibodies by means of quantitative immunoelectrophoresis. A survey of methods, *J. Immunol. Methods,* 1, 109, 1972.
90. **Bock, E. and Axelsen, N. H.,** The reaction of partial identity in quantitative immunoelectrophoretic patterns, *J. Immunol. Methods,* 2, 75, 1972.
91. **Grubb, A.,** Analysis of the immunochemical relationship between antigens by electrophoresis in agarose gels containing antibodies, *Scand. J. Clin. Lab. Invest.*, 29 (Suppl. 124), 59, 1972.

92. **Axelsen, N. H. and Bock, E.,** Further studies on the reaction of partial identity in quantitative immunoelectrophoretic patterns, *J. Immunol. Methods,* 2, 393, 1973.
93. **Axelsen, N. H.,** Intermediate gel in crossed and in fused rocket immunoelectrophoresis, *Scand. J. Immunol.,* 2 (Suppl. 1), 71, 1973.
94. **Axelsen, N. H., Bock, E., and Krøll, J.,** Comparison of antigens: the reaction of "identity", *Scand J. Immunol.,* 2 (Suppl. 1), 91, 1973.
95. **Bock, E. and Axelsen, N. H.,** Comparison of antigens: the reaction of partial identity, *Scand. J. Immunol.,* 2 (Suppl. 1), 95, 1973.
96. **Axelsen, N. H., Bock, E., and Krøll, J.,** Comparsion of antisera, *Scand. J. Immunol.,* 2 (Suppl. 1), 101, 1973.
97. **Closs, O., Harboe, M., and Wassum, A. M.,** Cross-reactions between Mycobacteria. 1. Crossed immunoelectrophoresis of soluble antigens of *Mycobacterium lepraemurium* and comparison with BCG, *Scand. J. Immunol.,* 4 (Suppl. 2), 173, 1975.
98. **Negassi, K., Closs, O., and Harboe, M.,** False spurs in quantitative immunoelectrophoresis, *Scand. J. Immunol.,* 8, 279, 1978.
99. **Cann, J. R.,** A phenomenological theory of rocket immunoelectrophoresis, *Biophys. Chem.,* 3, 206, 1975.
100. **Cann, J. R.,** A phenomenological theory of rocket and crossed immunoelectrophoresis, *Immunochemistry,* 12, 473, 1975.
101. **Lundahl, P. and Liljas, L.,** Crossed immunoelectrophoresis. Polyacrylamide gel electrophoresis followed by electrophoresis into an agarose gel containing antibody, *Anal. Biochem.,* 65, 50, 1975.
102. **Converse, C. A. and Papermaster, D. S.,** Membrane protein analysis by two-dimensional immunoelectrophoresis, *Science,* 189, 469, 1975.
103. **Kirkpatrick, F. H. and Rose, D. J.,** Crossed immunoelectrophoresis from sodium dodecyl sulfate-polyacrylamide gels into antibody-containing agarose: an improved method for evaluation of the number of immunochemical determinants in polypeptides, with reference to spectrin, *Anal. Biochem.,* 89, 130, 1978.
104. **Høiby, N.,** Cross-reactions between *Pseudomonas aeruginosa* and thirty-six other bacterial species, *Scand. J. Immunol.,* 4 (Suppl. 2), 187, 1975.
105. **Høiby, N.,** Normally occurring precipitating antibodies against *Pseudomonas aeruginosa:* prevalence, specificities and titres, *Scand J. Immunol.,* 4 (Suppl. 2), 197, 1975.
106. **Smyth, C. J., Friedman-Kien, A. E., and Salton, M. R. J.,** Antigenic analysis of *Neisseria gonorrhoeae* by crossed immunoelectrophoresis, *Infect. Immun.,* 13, 1273 1976.
107. **Smyth, C. J. and Salton, M. R. J.,** Crossed immunoelectrophoresis: a new approach to high resolution analysis of gonococcal antigens and antibodies, in *The Gonococcus,* Roberts, R. B., Ed., John Wiley & Sons, New York, 1977, chap. 14.
108. **Owen, P. and Kaback, H. R.,** Immunochemical analysis of membrane vesicles from *Escherichia coli, Biochemistry,* 18, 1413, 1979.
109. **Owen, P.,** unpublished data, 1978.
110. **Pillion, D. J. and Czech, M. P.,** Antibodies against intrinsic adipocyte plasma membrane proteins activate D-glucose transport independent of interaction with insulin binding sites, *J. Biol. Chem.,* 253, 3761, 1978.
111. **Owen, P. and Kaback, H. R.,** unpublished data, 1978.
112. **Crowle, A. J., Revis, G. J., and Jarrett, K.,** Preparatory electroimmunodiffusion for making precipitins to selected native antigens, *Immunol. Commun.,* 1, 325, 1972.
113. **Krøll, J. and Andersen, M. M.,** Specific antisera produced by immunization with precipitin lines, *J. Immunol. Methods,* 13, 125, 1976.
114. **Alexander, A. G. and Kenny, G. E.,** Characterization of membrane and cytoplasmic antigens of *Mycoplasma arginini* by two-dimensional (crossed) immunoelectrophoresis, *Infect. Immun.,* 15, 313, 1977.
115. **Rutberg, B., Hederstedt, L., Holmgren, E., and Rutberg, L.,** Characterization of succinic dehydrogenase mutants of *Bacillus subtilis* by crossed immunoelectrophoresis, *J. Bacteriol.,* 136, 304, 1978.
116. **Wróblewski, H.,** Dissolution sélective de protéines de la membrane de *Spiroplasma citri* par le désoxycholate de sodium, *Biochimie,* 57, 1095, 1975.
117. **Wróblewski, H. and Ratanasavanh, D.,** Étude par immunoélectrophorèse bidimensionnelle de la composition antigénique de la membrane de quelques souches de mycoplasmes, *Can. J. Microbiol.,* 22, 1048, 1976.
118. **Wróblewski, H., Johansson, K.-E., and Hjérten, S.,** Purification and characterization of spiralin, the main protein of the *Spiroplasma citri* membrane, *Biochim. Biophys. Acta,* 465, 275, 1977.
119. **Wróblewski, H., Johansson, K.-E., and Burlot, R.,** Crossed immunoelectrophoresis of membrane proteins from *Acholeplasma laidlawii* and *Spiroplasma citri, Int. J. Syst. Bacteriol.,* 27, 97, 1977.

120. **Johansson, K.-E. and Wróblewski, H.,** Crossed immunoelectrophoresis, in the presence of Tween 20 or sodium deoxycholate, of purified membrane proteins from *Acholeplasma laidlawii, J. Bacteriol,* 136, 324, 1978.
121. **Alexander, A. G. and Kenny, G. E.,** Application of charge shift electrophoresis to antigenic analysis of mycoplasmic membranes by two-dimensional (crossed) immunoelectrophoresis, *Infect. Immun.,* 20, 861, 1978.
122. **Helenius, A. and Simons, K.,** Charge shift electrophoresis: simple method for distinguishing between amphiphilic and hydrophilic proteins in detergent solution, *Proc. Natl. Acad. Sci. U.S.A.,* 74, 529, 1977.
123. **Bhakdi, S., Bhakdi-Lehnen, B., and Bjerrum, O. J.,** Detection of amphiphilic proteins and peptides in complex mixtures: charge-shift crossed immunoelectrophoesis and two-dimensional charge-shift electrophoresis, *Biochim. Biophys. Acta,* 470, 35, 1977.
124. **Schnaitman, C. A.,** Solubilization of the cytoplasmic membrane of *Escherichia coli* by Triton X-100, *J. Bacteriol.,* 108, 545, 1971.
125. **Schnaitman, C. A.,** Effect of ethylenediaminetetraacetic acid, Triton X-100, and lysozyme on the morphology and chemical composition of isolated cell walls of *Escherichia coli, J. Bacteriol.,* 108, 553, 1971.
126. **Filip, C., Fletcher, G., Wulff, J. L., and Earhart, C. F.,** Solubilization of the cytoplasmic membrane of *Escherichia coli* by the ionic detergent sodium-lauryl sarcosinate, *J. Bacteriol.,* 115, 717, 1973.
127. **Hindennach, I. and Henning, U.,** The outer membrane proteins of the *Escherichia coli* outer cell envelope membrane: preparative isolation of all major membrane proteins, *Eur. J. Biochem.,* 59, 207, 1975.
128. **Smyth, C. J., Siegel, J., Salton, M. R. J., and Owen, P.,** Immunochemical analysis of inner and outer membranes of *Escherichia coli* by crossed immunoelectrophoresis, *J. Bacteriol.,* 133, 306, 1978.
129. **Salton, M. R. J. and Urban, C.,** Selective extraction of gonococcal envelope antigens with thiocyanates, *FEMS Microbiol. Lett.* 4, 303, 1978.
130. **Owen, P. and Doherty, H.,** Immunochemical analysis of Triton X-100-insoluble residues from membranes of *Micrococcus lysodeikticus, J. Bacteriol.* 140, 881, 1979.
131. **Norrild, B., Bjerrum, O. J., and Vestergaard, B. F.,** Polypeptide analysis of individual immunoprecipitates from crossed immunoelectrophoresis, *Anal. Biochem.,* 81, 432, 1977.
132. **Owen, P. and Salton, M. R. J.,** Antigenic and enzymatic architecture of *Micrococcus lysodeikticus* membranes established by crossed immunoelectrophoresis, *Proc. Natl. Acad. Sci. U.S.A.,* 72, 3711, 1975.
133. **Thirkhill, C. E. and Kenny, G. E.,** Antigenic analysis of three strains of *Mycoplasma arginini* by two-dimensional immunoelectrophoresis, *J. Immunol.* 114, 1107, 1975.
134. **Braun, V.,** Covalent lipoprotein from the outer membrane of *Escherichia coli, Biochim. Biophys. Acta,* 415, 335, 1975.
135. **Krøll, J.,** Tandem-crossed immunoelectrophoresis, *Scand. J. Immunol.,* 2 (Suppl. 1), 57, 1973.
136. **Bøg-Hansen, T. C.,** Crossed immuno-affinoelectrophoresis: an analytical method to predict the result of affinity chromatography, *Anal. Biochem.,* 56, 480, 1973.
137. **Johansson, K.-E., Wróblewski, H., Blomqvist, I. I., and Hjertén, S.,** The cell membrane of *Acholeplasma laidlawii:* purification and characterization of the protein D_{12} and some observations on the NADH dehydrogenase, submitted for publication.
138. **Owen, P. and Salton, M. R. J.,** Submicrogram quantitation of an acidic polysaccharide by rocket immunoelectrophoresis and rocket affinoelectrophoresis, *Anal. Biochem.,* 73, 20, 1976.
139. **Oppenheim, J. D., Owen, P., Nachbar, M. S., Colledge, K., and Kaplan, H. S.,** Identification and quantitation of solubilized I blood group substance by wheat germ agglutinin using quantitative immunoelectrophoresis, *Immunol. Commun.* 6, 167, 1977.
140. **Owen, P., Oppenheim, J. D., Nachbar, M. S., and Kessler, R. E.,** The use of lectins in the quantitation and analysis of macromolecules by affinoelectrophoresis, *Anal. Biochem.,* 80, 446, 1977.
141. **Bøg-Hansen, T. C., Bjerrum, O. J., and Brogren, C.-H.,** Identification and quantitation of glycoproteins by affinity electrophoresis, *Anal. Biochem.,* 81, 78, 1977.
142. **Cinader, B.,** Enzyme-antibody interactions, in *Methods in Immunology and Immunochemistry,* Vol. 4, **Williams, C. A. and Chase, M. W., Eds.,** Academic Press, New York, 1977, 313.
143. **Harboe, N. and Ingild, A.,** Immunization, isolation of immunoglobulins, estimation of antibody titre, *Scand. J. Immunol.,* 2 (Suppl. 1), 161, 1973.
144. **Uriel, J.,** Characterization of enzymes in specific immuneprecipitates, *Ann. N. Y. Acad. Sci.,* 103, 956, 1963.
145. **Uriel, J.,** Immunoelectrophoretic analysis of enzymes, in *Antibodies to Biologically Active Molecules,* Cinader, B., Ed., Pergamon Press, Oxford, 1967, 181.
146. **Harris, H. and Hopkinson, D. A.** *Handbook of Enzyme Electrophoresis in Human Genetics,* North Holland Publishing, Amsterdam, 1976.

147. **Harris, H. and Hopkinson, D. A.,** *Handbook of Enzyme Electrophoresis in Human Genetics. Supplement,* North Holland Publishing, Amsterdam, 1977.
148. **Owen, P. and Kaback, H. R.,** unpublished data, 1979.
149. **Blomberg, F. and Berzins, K.,** Epinephrine-binding plasma-membrane antigens in rat liver, *Eur. J. Biochem.,* 56, 319, 1975.
150. **Orme-Johnson, W. H.,** Iron-sulphur proteins: structure and function, *Annu. Rev. Biochem.,* 42, 159, 1973.
151. **Hatefi, Y. and Stiggall, D. L.,** Metal-containing flavoprotein dehydrogenases, in *The Enzymes,* Vol. 13C, Boyer, P. D., Ed., Academic Press, New York, 1976, chap. 4.
152. **Haddock, B. A. and Jones, C. W.,** Bacterial respiration, *Bacteriol. Rev.,* 41, 47, 1977.
153. **Owen, P., Kaczorowski, G., and Kaback, H. R.,** Resolution and identification of iron-containing antigens in membrane vesicles from *Escherichia coli, Biochemistry,* 19, 596, 1980.
154. **Crowe, B. A. and Owen, P.,** Immunochemical analysis of iron-containing antigens in the plasma membrane of *Micrococcus lysodeikticus,* manuscript in preparation, 1979.
155. **Salton, M. R. J. and Owen, P.,** Bacterial membrane structure, *Annu. Rev. Microbiol.,* 30, 451, 1976.
156. **Collins, M. L. P. and Salton, M. R. J.,** Solubility characteristics of *Micrococcus lysodeikticus* membrane components in detergents and chaotropic salts analyzed by immunoelectrophoresis, *Biochim. Biophys. Acta,* 553, 40, 1979.
157. **Wadström, T.,** Biological properties of extracellular proteins from *Staphylococcus, Ann. N. Y. Acad. Sci.,* 236, 343, 1974.
158. **Hederstedt, L., Holmgren, E., and Rutberg, L.,** Characterization of a succinate dehydrogenase complex solubilized from the cytoplasmic membrane of *Bacillus subitlis* with the nonionic detergent Triton X-100, *J. Bacteriol.,* 138, 370, 1979.
159. **Owen, P. and Kaback, H. R.,** Molecular structure of membrane vesicles from *Escherichia coli, Proc. Natl. Acad. Sci. U.S.A.,* 75, 3148, 1978.
160. **Owen, P. and Kaback, H. R.,** Antigenic architecture of membrane vesicles from *Escherichia coli, Biochemistry,* 18, 1422, 1979.
161. **Høiby, N.,** The serology of *Pseudomonas aeruginosa* analysed by means of quantitative immunoelectrophoretic methods. I. Comparison of thirteen O groups of *Ps. aeruginosa,* with a polyvalent *Ps. aeruginosa* antigen-antibody reference system, *Acta Pathol. Microbiol. Scand., Sect.* B, 83, 321, 1975.
162. **Høiby, N.,** The serology of *Pseudomonas aeruginosa* analysed by means of quantitative immunoelectrophoresis methods. II. Comparison of the antibody response in man against thirteen O groups of *Ps. aeruginosa, Acta Pathol. Microbiol. Scand. Sect. B,* 83, 328, 1975.
163. **Høiby, N.,** The serology of *Pseudomonas aeruginosa* analysed by means of quantitative immunoelectrophoretic methods. III. Reproducibility of polyvalent *Ps. aeruginosa* reference standard-antigen, *Acta. Pathol. Microbiol. Scand., Sect. B,* 83, 433, 1975.
164. **Høiby, N.,** The serology of *Pseudomonas aeruginosa* analysed by means of quantitative immunoelectrophoretic methods, IV. Production of polyvalent pools of antiserum against *Ps. aeruginosa* (reference standard-antibody), *Acta Pathol. Microbiol. Scand., Sect. C.,* 84, 372, 1976.
165. **Høiby, N.,** The serology of *Pseudomonas aeruginosa* analysed by means of quantitative immunoelectrophoretic methods. V. Thermostability, resistance to degradation by plasmin activity and storage conditions of a polyvalent *Ps. aeruginosa* reference standard-antigen, *Acta Pathol. Microbiol. Scand., Sect. C,* 84, 383, 1976.
166. **Larsson, P., Hanson, L. A., and Kaijser, B.,** Immunodiffusion studies on some *Proteus* strains, *Acta Pathol. Microbiol. Scand., Sect. B,* 81, 641, 1973.
167. **Lyerly, D. and Kreger, A.,** Purification and characterization of a *Serratia marcescens* metalloprotease, *Infect. Immun.,* 24, 411, 1979.
168. **Hertz, J. B., Høiby, N., Andersen, V., and Baekgaard, P.,** Crossed immunoelectrophoretic analysis of *Bordetella pertussis* antigens and of corresponding antibodies in human sera, *Acta Pathol. Microbiol. Scand., Sect. B,* 84, 386, 1976.
169. **Høiby, N., Hertz, J. B., and Andersen, V.,** Cross-reactions between *Bordetella pertussis* and twenty-eight other bacterial species, *Acta Pathol. Microbiol. Scand., Sect. B,* 84, 395, 1976.
170. **Hoff, G. E., and Høiby, N.,** Cross-reactions between *Neisseria meningitidis* and twenty-seven other bacterial species, *Acta Pathol. Microbiol. Scand., Sect. B,* 86, 87, 1978.
171. **Penn, C. W., Parsons, N. J., Veale, D. R., and Smith, H.,** Correlation with different immunotypes of gonococcal antigens associated with growth *in vivo, J. Gen. Microbiol.,* 105, 153, 1978.
172. **Parsons, N. J., Penn, C. W., Veale, D. R., and Smith, H.,** More than one antigen contributes to the immunogenicity of *Neisseria gonorrhoeae* in the guinea pig chamber model, *J. Gen. Microbiol.,* 113, 97, 1979.

173. **Buckmire, F. L.,** Identification and quantitation of capsular antigen in capsulated and noncapsulated strains of *Haemophilus influenzae* Type b by crossed-immunoelectrophoresis, *Infect. Immun.,* 13, 1733, 1976.
174. **Elferink, M. G. L., Hellingwerf, K. J., Michels, P. A. M., Seyen, H. G., and Konings, W. N.,** Immunochemical analysis of membrane vesicles and chromatophores of *Rhodopseudomonas sphaeroides* by crossed immunoelectrophoresis, *FEBS Lett.,* 107, 300, 1979.
174a. **Collins, M. L. P., Mallon, D. E., and Niederman, R. A.,** Crossed immunoelectrophoretic analysis of chromatophore membranes from *Rhodopseudomonas sphaeroides, J. Bacteriol.,* 139, 1089, 1979.
175. **Wayne, L. G. and Sramek, H. A.,** Antigenic differences between extracts of actively replicating and synchronized resting cells of *Mycobacterium tuberculosis, Infect. Immun.,* 24, 363, 1979.
176. **Roberts, D. B., Wright, G. L., Jr., Affronti, L. F., and Reich, M.,** Characterization and comparison of mycobacterial antigens by two-dimensional immunoelectrophoresis, *Infect. Immun.,* 6, 564, 1972.
177. **Wright, G. L. and Roberts, D. B.,** Two-dimensional immunoelectrophoresis of mycobacterial antigens, *Am. Rev. Resp. Dis.,* 109, 306, 1974.
178. **Janicki, B. W., Wright, G. L., Jr., Good, R. C., and Chaparas, S. D.,** Comparison of antigens in sonic and pressure cell extracts of *Mycobacterium tuberculosis, Infect. Immun.,* 13, 425, 1976.
179. **Thorel, M. F.,** Utilisation d'une méthode d'immunoélectrophorèse bidimensionnelle dans l'étude des antigènes de *Mycobacterium simiae* et *M. habana, Ann. Inst. Pasteur,* 127B, 41, 1976.
180. **Kronvall, G., Closs, O., and Bjune, G.,** Common antigen of *Mycobacterium leprae, M. lepraemurium, M. avium,* and *M. fortuitum* in comparative studies using two different types of antisera, *Infect. Immun.,* 16, 542, 1977.
181. **Harboe, M., Closs, O., Svindahl, K., and Deverill, J.,** Production and assay of antibodies against one antigenic component of *Mycobacterium bovis* BCG, *Infect. Immun.,* 16, 662, 1977.
182. **Caldwell, H. D., Kuo, C.-C., and Kenny, G. E.,** Antigenic analysis of *Chlamydiae* by two-dimensional immunoelectrophoresis. I. Antigenic heterogeneity between *C. trachomatis* and *C. psittaci, J. Immunol.,* 115, 963, 1975.
183. **Caldwell, H. D., Kuo, C.-C., and Kenny, G. E.,** Antigenic analysis of *Chlamydiae* by two-dimensional immunoelectrophoresis. II. A trachoma-LGV-specific antigen, *J. Immunol.,* 115, 969, 1975.
184. **Thirkill, C. E. and Kenny, G. E.,** Serological comparison of five arginine-utilizing *Mycoplasma* species by two dimensional immunoelectrophoresis, *Infect. Immun.,* 10, 624, 1974.
185. **Johansson, K.-E., Blomquist, I., and Hjertén, S.,** Purification of membrane proteins from *Acholeplasma laidlawii* by agarose suspension electrophoresis in Tween 20 and polyacrylamide and dextran gel electrophoresis in detergent-free media, *J. Biol. Chem.,* 250, 2463, 1975.
186. **Fenske, J. D. and Kenny, G. E.,** Role of arginine deiminase in growth of *Mycoplasma hominis, J. Bacteriol.,* 126, 501, 1976.
187. **Axelsen, N. H.,** Antigen-antibody crossed electrophoresis (Laurell) applied to the study of the antigenic structure of *Candida albicans, Infect. Immun.,* 4, 525, 1971.
188. **Axelsen, N. H. and Svendsen, P. J.,** *Candida* precipitins characterised by a modified antigen-antibody crossed electrophoresis, in *Protides of the Biological Fluids,* Vol. 19, Peeters, H., Ed., Pergamon Press, Elmsford, N. Y., 1972, 561.
189. **Axelsen, N. H.,** Quantitative immunoelectrophoretic methods as tools for a polyvalent approach to standardization in the immunochemistry of *Candida albicans, Infect. Immun.,* 7, 949, 1973.
190. **Syverson, R. E., Buckley, H. R., and Campbell, C. C.,** Cytoplasmic antigens unique to the mycelial or yeast phase of *Candida albicans, Infect. Immun.,* 12, 1184, 1975.
191. **Syverson, R. E. and Buckley, H. R.,** Cell wall antigens in soluble cytoplasmic extracts of *Candida albicans* as demonstrated by crossed immuno-affinoelectrophoresis with concanavalin A, *J. Immunol. Methods,* 18, 149, 1977.
192. **Holmberg, K., Nord, C.-E., and Wadström, T.,** Serological studies of *Actinomyces israelii* by crossed immunoelectrophoresis: standard antigen-antibody system for *A. israelii, Infect. Immun.,* 12, 387, 1975.
193. **Holmberg, K., Nord, C.-E., and Wadström, T.,** Serological studies of *Actinomyces israelii* by crossed immunoelectrophoresis: taxonomic and diagnostic applications, *Infect. Immun.,* 12, 398, 1975.
194. **Bout, D., Fruit, J., and Capron, A.,** Application de la chromatographie d'affinité à l'isolement des fractions antigéniques d'Aspergillus fumigatus supportant une activité chymotrypsique, *C. R. Seances Acad. Sci., Ser. D,* 276, 2341, 1973.
195. **Sweet, G. H., Wilson, D. E., and Gerber, J. D.,** Application of electroimmunodiffusion and crossed electroimmunodiffusion to the comparative serology of a microorganism *(Histoplasma capsulatum), J. Immunol.,* 111, 554, 1973.
196. **Walbaum, S., Biquet, J., Vaucelle, T., and Tran van Ky, P.,** Observations preliminaires sur l'emploi des techniques d'immunoélectrophorèse bidimensionelle et d'électroimmunodiffusion dans l'étude et la standardisation des extraits antigéniques de *Micropolyspora faeni, Bull. Soc. Fr. Mycol. Med.,* 2, 3, 1973.

197. **Walbaum, S., Vaucelle, T., and Biquet, J.,** Analyse de l'extrait de *Micropolyspora faeni* par immunoélectrophorèse en double dimension. Localisation des activités chymotrypsiques, *Pathol. Biol.,* 21, 355, 1973.
198. **Hornock, L. and Szécsi, Á.,** Demonstration of antifungal antibodies in hyperimmune mouse ascites fluid by means of crossed immunoelectrophoresis, *J. Gen. Microbiol.,* 100, 213, 1977.
199. **Huppert, M., Spratt, N. S., Vukovich, K. R., Sun, S. H., and Rice, E. H.,** Antigenic analysis of coccidioidin and spherulin determined by two-dimensional immunoelectrophoresis, *Infect. Immun.,* 20, 541, 1978.
200. **Huppert, M., Alder, J. P., Rice, E. H., and Sun, S. H.,** Common antigens among systemic disease fungi analyzed by two-dimensional immunoelectrophoresis, *Infect. Immun.,* 23, 479, 1979.
201. **Axelsen, N. H.,** Human precipitins against a microorganism *(Candida albicans)* demonstrated by means of quantitative immunoelectrophoresis, *Clin. Exp. Immunol.,* 9, 749, 1971.
202. **Svendsen, P. J. and Axelsen, N. H.,** A modified antigen-antibody crossed electrophoresis characterizing the specificity and titre of human precipitins against *Candida albicans, J. Immunol. Methods,* 1, 169, 1972.
203. **Axelsen, N. H. and Kirkpatrick, C. H.,** Simultaneous characterization of free *Candida* antigens and *Candida* precipitins in a patients serum by means of crossed immunoelectrophoresis with intermediate gel, *J. Immunol. Methods,* 2, 245, 1973.
204. **Axelsen, N. H., Kirkpatrick, C. H., and Buckley, R. H.,** Precipitins to *Candida albicans* in chronic mucocutaneous candidiasis studied by crossed immunoelectrophoresis with intermediate gel: correlation with clinical and immunological findings, *Clin. Exp. Immunol.,* 17, 385, 1974.
205. **Axelsen, N. H., Buckley, H. R. Drouhet, E., Budtz-Jørgensen, E., Hattel, T., and Anderson, P. L.,** Crossed immunoelectrophoretic analysis of precipitins to *Candida albicans* in deep *Candida* infection: possibilities for standardization in diagnostic *Candida* serology, *Scand. J. Immunol.,* 4 (Suppl. 2), 217, 1975.
206. **Axelsen, N. H.,** Analysis of human candida precipitins by quantitative immunoelectrophoresis: a model for analysis of complex microbial antigen-antibody systems, *Scand. J. Immunol.,* 5, 177, 1976.
207. **Høiby, N. and Axelsen, N. H.,** Identification and quantitation of precipitins against *Pseudomonas aeruginosa* in patients with cystic fibrosis by means of crossed immunoelectrophoresis with intermediate gel, *Acta Pathol. Microbiol. Scand., Sect. B,* 81, 298, 1973.
208. **Høiby, N., Jacobsen, L., Jørgensen, B. A., Lykkegaard, E., and Weeke, B.,** *Pseudomonas aeruginosa* infection in cystic fibrosis. Occurrence of precipitating antibodies against *Pseudomonas aeruginosa* in relation to the concentration of sixteen serum proteins and the clinical and radiographical status of the lungs, *Acta Paediatr. Scand.,* 63, 843, 1974.
209. **Høiby, N.,** *Pseudomonas aeruginosa* infection in cystic fibrosis. Relationship between mucoid strains of *Pseudomonas aeruginosa* and the humoral immune response, *Acta Pathol. Microbiol. Scand., Sect. B.,* 82, 551, 1974.
210. **Høiby, N. and Mathiesen, L.,** *Pseudomonas aeruginosa* infection in cystic fibrosis: distribution of B and T lymphocytes in relation to the humoral immune response, *Acta Pathol. Microbiol. Scand., Sect. B,* 82, 559, 1974.
211. **Høiby, N. and Wiik, A.,** Antibacterial precipitins and autoantibodies in serum of patients with cystic fibrosis, *Scand. J. Respir. Dis.,* 56, 38, 1975.
212. **Schøitz, P. O. and Høiby, N.,** Precipitating antibodies against *Pseudomonas aeruginosa* in sputum from patients with cystic fibrosis: specificities and titres determined by means of crossed immunoelectrophoresis with intermediate gel, *Acta Pathol. Microbiol. Scand., Sect. C,* 83, 469, 1975.
213. **Høiby, N. and Olling, S.,** *Pseudomonas aeruginosa* infection in cystic fibrosis: bacteriocidal effect of serum from normal individuals and patients with cystic fibrosis on *Ps. aeruginosa* strains from patients with cystic fibrosis or other diseases, *Acta. Pathol. Microbiol. Scand., Sect C,* 85, 107, 1977.
214. **Hoff, G. E. and Høiby, N.,** Crossed immunoelectrophoretic analysis of *Neisseria meningitidis* antigens and corresponding antibodies in patients with meningococcal disease, *Acta Pathol. Microbiol. Scand., Sect C,* 86, 1, 1978.
215. **Salton, M. R. J., Friedman-Kien, A. E., and Urban, C.,** Detection of human antibodies to *Neisseria gonorrhoeae* envelope antigens by crossed immunoelectrophoresis, *FEMS Microbiol. Lett.,* 4, 307, 1978.
216. **Axelsen, N. H., Harboe, M., Closs, O., and Godal, T.,** BCG antibody profiles in tuberculoid and lepromatous leprosy, *Infect. Immun.,* 9, 952, 1974.
217. **Kronvall, G., Bjune, G., Stanford, J., Menzel, S., and Samuel, D.,** Mycobacterial antigens in antibody responses of leprosy patients, *Int. J. Lepr.,* 43, 306, 1975.
218. **Closs, O. and Kronvall, G.,** Experimental murine leprosy, IX. Antibodies against *Mycobacterium lepraemurium* in C3H and C57BL mice with murine leprosy and in patients with lepromatous leprosy, *Scand. J. Immunol.,* 4, 735, 1975.

219. **Kaczorowski, G., Kohn, L. D., and Kaback, H. R.,** Purification and properties of D-lactate dehydrogenase from *Escherichia coli* ML308-225, in *Methods in Enzymology,* Vol. 52D, **Fleischer, S. and Packer, L., Eds.,** Academic Press, New York, 1978, 519.
220. **Sakamoto, N., Kotre, A. M., and Savageau, M. A.,** Glutamate dehydrogenase from *Escherichia coli:* purification and properties, *J. Bacteriol.,* 124, 775, 1975.
221. **Karibian, D.,** Dihydro-orotate dehydrogenase of *Escherichia coli* K12: effects of Triton X-100 and phospholipids, *Biochim. Biophys. Acta,* 302, 205, 1973.
222. **Weiner, J. H.,** The localization of glycerol-3-phosphate dehyrogenase in *Escherichia coli, J. Membr. Biol.,* 15, 1, 1974.
223. **Blobel, G. and Dobberstein, B.** Transfer of proteins across membranes. I. Presence of proteolytically processed and unprocessed nascent immunoglobulin light chains on membrane-bound ribosomes of murine myeloma, *J. Cell Biol.,* 67, 835, 1975.
224. **Blobel, G. and Dobberstein, B.,** Transfer of proteins across membranes. II. Reconstitution of functional rough microsomes from heterologous components, *J. Cell Biol.,* 67, 852, 1975.
225. **Kaback, H. R.,** Transport across isolated bacterial cytoplasmic membranes, *Biochim. Biophys. Acta,* 265, 367, 1972.
226. **Kaback, H. R.,** Transport studies in bacterial membrane vesicles, *Science,* 186, 882, 1974.
227. **Kaback, H. R.,** Molecular biology and energetics of membrane transport, *J. Cell Physiol.* 89, 575, 1976.
228. **Stroobant, P. and Kaback, H. R.,** Ubiquinone-mediated coupling of NADH dehydrogenase to active transport in membrane vesicles from *Escherichia coli, Proc. Natl. Acad. Sci. U.S.A.,* 72, 3970, 1975.
229. **Harold, F. M.,** Conservation and transformation of energy by bacterial membranes, *Bacteriol. Rev.,* 36, 172, 1972.
230. **Mitchell, P.,** Performance and conservation of osmotic work by proton-coupled solute porter systems, *J. Bioenerg.* 4, 63, 1973.
231. **van Thienen, G. and Postma, P. W.,** Coupling between energy conservation and active transport of serine in *Escherichia coli, Biochim. Biophys. Acta,* 323, 429, 1973.
232. **Futai, M.,** Orientation of membrane vesicles from *Escherichia coli* prepared by different procedures, *J. Membr. Biol.* 15, 15, 1974.
233. **Hare, J. F., Olden, K., and Kennedy, E. P.,** Heterogeneity of membrane vesicles from *Escherichia coli* and their subfractionation with antibody to ATPase, *Proc. Natl. Acad. Sci. U.S.A.,* 71, 4843, 1974.
234. **Futai, M. and Tanaka, Y.,** Localization of D-lactate dehydrogenase in membrane vesicles prepared by using a French press or ethylene-diaminetetraacetate-lysozyme from *Escherichia coli, J. Bacteriol.,* 124, 470, 1975.
235. **Wickner, W.,** Fractionation of membrane vesicles from coliphage M13-infected *Escherichia coli, J. Bacteriol.,* 127, 162, 1976.
236. **Adler, L. W. and Rosen, B. P.,** Functional mosaicism of membrane proteins in vesicles of *Escherichia coli, J. Bacteriol.,* 129, 959, 1977.
237. **Yamato, I., Futai, M., Anraku, Y., and Nonomura, Y.,** Cytoplasmic membrane vesicles of *Escherichia coli.* II. Orientation of the vesicles studied by localization of enzymes, *J. Biochem. (Tokyo),* 83, 117, 1978.
238. **Siegel, L. M., Murphy, M. J., and Kamin, H.,** Reduced nicotinamide adenine dinucleotide phosphate-sulfite reductase of enterobacteria. I. The *Escherichia coli* hemoflavoprotein: molecular parameters and prosthetic groups, *J. Biol. Chem.,* 248, 251, 1973.
239. **Siegel, L. M., Davis, P. S., and Kamin, H.,** Reduced nicotinamide adenine dinucleotide phosphate-sulfite reductase of Enterobacteria. III. The *Escherichia coli* hemoflavoprotein: catalytic parameters and the sequence of electron flow, *J. Biol. Chem.,* 249, 1572, 1974.
240. **DiRienzo, J. M., Nakamura, K., and Inouye, M.,** The outer membrane proteins of gram-negative bacteria: biosynthesis, assembly, and functions, *Annu. Rev. Biochem.,* 47, 481, 1978.
241. **Hollifield, W. C., Jr. and Neilands, J. B.,** Ferric enterobactin transport system in *Escherichia coli* K-12. Extraction, assay, and specificity of the outer membrane receptor, *Biochemistry,* 17, 1922, 1978.
242. **Owen, P. and Salton, M. R. J.,** A succinylated mannan in the membrane system of *Micrococcus lysodeikticus, Biochem. Biophys. Res. Commun.,* 63, 875, 1975.
243. **Sharon, N. and Lis, H.,** Lectins: cell-agglutinating and sugar-specific proteins, *Science,* 177, 949, 1972.
244. **Nicolson, G. L. and Blaustein, J.,** The interaction of *Ricinus communis* agglutinin with normal and tumor cell surfaces, *Biochim. Biophys. Acta,* 266, 543, 1972.
245. **Owen, P. and Freer, J. H.,** Isolation and properties of mesosomal membrane fractions from *Micrococcus lysodeikticus, Biochem. J.,* 129, 907, 1972.
246. **Holmgren, E., Hederstedt, L., and Rutberg, L.,** Role of heme in synthesis and membrane-binding of succinic dehydrogenase in *Bacillus subtilis, J. Bacteriol.,* 138, 377, 1979.

247. **Buchanan, T. M., Swanson, J., Holmes, K. K., Kraus, S. J., and Gotschlich, E. C.,** Quantitative determination of antibody to gonococcal pili. Changes in antibody levels with gonococcal infection, *J. Clin. Invest.,* 52, 2896, 1973.
248. **Jephcott, A. E.,** Gonorrhoea. Recent advances in routine laboratory procedures, in *Recent Advances in Sexually Transmitted Diseases,* **Morton, R. S. and Harris, J. R. W., Eds.,** Churchill Livingstone, Edinburgh, 1975, 44.
249. **Roberts, R. B., Ed.,** *The Gonococcus,* John Wiley & Sons, New York, 1977.
250. **Danielsson, D., Olcén, P., and Sandström, E.,** Serological methods of diagnosis, in FEMS Symp. 2, *Gonorrhoea: Epidemiology and Pathogenesis,* **Skinner, F. A., Walker, P. D. and Smith, H., Eds.,** Academic Press, London, 1977, 27.
251. **Sandström, E.,** Studies on the Serology of *Neisseria gonorrhoeae,* Ph.D. thesis, Karolinska Institute, Stockholm, 1979.
252. **Danielsson, D., Thyressan, N., Falk, V., and Barr, J.,** Serologic investigation of the immune response in various types of gonococcal infection, *Acta Dermatol. Venereol.,* 52, 467, 1972.
253. **Johnston, K. H., Holmes, K. K., and Gotschlich, E. C.,** The serological classification of *Neisseria gonorrhoeae.* I. Isolation of the outer membrane complex responsible for serotypic specificity, *J. Exp. Med.,* 143, 741, 1976.
254. **Pollack, J. D., Somerson, N. L., and Senterfit, L. B.,** Isolation, characterization and immunogenicity of *Mycoplasma pneumoniae* membranes, *Infect. Immun.,* 2, 326, 1970.
255. **Hollingdale, M. R. and Lemcke, R. M.,** Membrane antigens of *Mycoplasma hominis, J. Hyg.,* 70, 85, 1972.
256. **Johansson, K.-E.,** Fractionation of membrane proteins from *Acholeplasma laidlawii* by preparative agarose-suspension electrophoresis, in *Protides of the Biological Fluids,* Vol. 21, Peeters, H., Ed., Pergamon Press, Oxford, 1974, 151.
257. **Hjertén, S., and Johansson, K.-E.,** Selective solubilization with Tween 20 of membrane proteins from *Acholeplasma laidlawii, Biochim, Biophys. Acta,* 288, 312, 1972.
258. **Fries, E.,** Determination of Triton X-100 binding to membranc proteins by polyacrylamide gel electroporesis, *Biochim. Biophys. Acta,* 455, 928, 1976.
259. **Blomberg, F. and Raftell, M.,** Enzyme polymorphism in rat-liver microsomes and plasma membranes. 1. An immunochemical study of multienzyme complexes and other enzyme-active antigens, *Eur. J. Biochem.,* 49, 21, 1974.
260. **Raftell, M. and Blomberg, F.,** Enzyme polymorphism in rat-liver microsomes and plasma membranes. 2. An immunochemical comparison of enzyme-active antigens solubilized by detergents, papain or phospholipases, *Eur. J. Biochem.,* 49, 31, 1974.
261. **Berzins, K., Blomberg, F., and Perlmann, P.,** Soluble and membrane-bound enzyme-active antigens of rat-liver lysosomes, *Eur. J. Biochem.,* 51, 181, 1975.
262. **Berzins, K.,** Membrane Antigens from Rat Liver. Immunological Resolution and Functional Characterization, Ph.D. thesis, University of Stockholm, Stockholm, 1976.
263. **Ruíz-Herrera, J. and DeMoss, J. A.,** Nitrate reductase complex of *Escherichia coli* K-12: participation of specific formate dehydrogenase and cytochrome b_1 components in nitrate reduction, *J. Bacteriol.,* 99, 720, 1969.
264. **MacGregor, C. H.,** Solubilization of *Escherichia coli* nitrate reductase by a membrane-bound protease, *J. Bacteriol.,* 121, 1102, 1975.
265. **MacGregor, C. H., Schnaitman, C. A., Normansell, D. E., and Hodgins, M. G.,** Purification and properties of nitrate reductase from *Escherichia coli* K12, *J. Biol. Chem.,* 249, 5321, 1974.
266. **MacGregor, C. H.,** Anaerobic cytochrome b_1 in *Escherichia coli:* association with and regulation of nitrate reductase, *J. Bacteriol.,* 121, 1111, 1975.
267. **Enoch, H. G. and Lester, R. L.,** The purification and properties of formate dehydrogenase and nitrate reductase from *Escherichia coli, J. Biol. Chem.,* 250, 6693, 1975.
268. **MacGregor, C. H.,** Biosynthesis of membrane-bound nitrate reductase in *Escherichia coli;* evidence for a soluble precursor, *J. Bacteriol.,* 126, 122, 1976.
269. **Schmitt, M., Rittinghaus, K., Scheurich, P., Schwalera, U., and Dose, K.,** Immunological properties of membrane-bound adenosine triphosphatase: immunological identification of rutamycin-sensitive $F_0 \cdot F_1$ ATPase from *Micrococcus luteus* ATCC 4698 established by crossed immunoelectrophoresis, *Biochim. Biophys. Acta,* 509, 410, 1978.
270. **Oppenheim, J. D. and Nachbar, M. S.,** Immunochemistry of bacterial ATPase, in *Immunochemistry of Enzymes and their Antibodies,* **Salton, M. R. J., Ed.,** John Wiley & Sons, New York, 1977, 89.
271. **Sternweis, P. C.,** The ϵ subunit of *Escherichia coli* coupling factor 1 is required for its binding to the cytoplasmic membrane, *J. Biol. Chem.,* 253, 3123, 1978.
272. **Kagawa, Y., Sone, N., Hirata, H. and Yoshida, M.,** Reconstitution of ATPase and carriers from thermophilic biomembranes, *Trends Biochem. Sci.,* 4, 31, 1979.

273. **Whiteside, T. L., DeSiervo, A. J., and Salton, M. R. J.,** Use of antibody to membrane adenosine triphosphatase in the study of bacterial relationships, *J. Bacteriol.*, 105, 957, 1971.
274. **Oppenheim, J. D. and Salton, M. R. J.,** Localization and distribution of *Micrococcus lysodeikticus* membrane ATPase determined by ferritin labeling, *Biochim. Biophys. Acta,* 298, 297, 1973.
275. **Hanson, R. L. and Kennedy, E. P.,** Energy-transducing adenosine triphosphatase from *Escherichia coli:* purification, properties, and inhibition by antibody, *J. Bacteriol.,* 114, 772, 1973.
276. **Nelson, N., Kanner, B. I., and Gutnick, D. L.,** Purification and properties of Mg^{2+} Ca^{2+} adenosine triphosphatase from *Escherichia coli, Proc. Natl. Acad. Sci. U.S.A.*, 71, 2720, 1974.
277. **Kanner, B. I., Nelson, N., and Gutnick, D. L.,** Differentiation between mutants of *Escherichia coli* K12 defective in oxidative phosphorylation, *Biochim. Biophys. Acta,* 396, 347, 1975.
278. **Houghton, R. L., Fisher, R. J., and Sanadi, D. R.,** Energy-linked and energy-independent transhydrogenase activities in *Escherichia coli* vesicles, *Biochim. Biophys. Acta,* 396, 17, 1975.
279. **Vogel, G. and Steinhart, R.,** ATPase of *Escherichia coli:* purification, dissociation, and reconstitution of the active complex from isolated subunits, *Biochemistry,* 15, 208, 1976.
280. **Ralt, D., Nelson, N., and Gutnick, D.,** Specific immunoprecipitation of ATPase from *Escherichia coli, FEBS Lett.*, 91, 85, 1978.
281. **Müller, H. W., Schmitt, M., Schneider, E., and Dose, K.,** Immunological and reconstitution studies on the adenosine triphosphatase complex from *Rhodospirillum rubrum, Biochim. Biophys. Acta,* 545, 77, 1979.
282. **Monteil, H. and Schreiber, B.,** Action inhibitrice des immunsérums sur les formes L stables de Proteus P18, *Ann. Microbiol. (Paris),* 124A, 193, 1973.
283. **Monteil, H. and Roussel, G.,** Immunogenic properties of membrane-bound ATPase from stable *Proteus* P18 L-forms: a kinetic study of inhibition by specific antibodies, *Biochem. Biophys. Res. Commun.*, 63, 313, 1975.
283a. **Mollinedo, F., Larraga, V., Coll, F. J., and Muñoz, E.,** Role of the subunits of the energy-transducing adenosine triphosphatase from *Micrococcus lysodeikticus* membranes studied by proteolytic digestion and immunological approaches, *Biochem. J.*, 186, 713, 1980.
283b. **Yoshida, M., Sone, N. Hirata, H., Kagawa, Y., and Ui, N.,** Subunit structure of adenosine triphosphatase. Comparison of the structure in thermophilic bacterium PS3 with those in mitochondria, chloroplasts, and *Escherichia coli, J. Biol. Chem.*, 254, 9525, 1979.
284. **Okamoto, H., Sone, N., Hirata, H., Yoshida, M., and Kagawa, Y.,** Purified proton conductor in proton translocating adenosine triphosphatase of a thermophilic bacterium, *J. Biol. Chem.*, 252, 6125, 1977.
285. **Salton, M. R. J., Schor, M. T., and Ng, M. H.,** Internal localization of *Micrococcus lysodeikticus* membrane ATPase by iodination with ^{125}I, *Biochim. Biophys. Acta,* 290, 408, 1972.
286. **Bragg, P. D.,** Purification and properties of menadione reductase of *Escherichia coli, Biochim. Biophys. Acta,* 96, 263, 1965.
287. **Bragg, P. D. and Hou, C.,** Reduced nicotinamide adenine dinucleotide oxidation in *Escherichia coli* particles. II. NADH dehydrogenase, *Arch. Biochem. Biophys.*, 119, 202, 1967.
288. **Gutman, M., Schejter, A. and Avi-Dor, Y.,** The preparation and properties of the membranal DPNH dehydrogenase from *Escherichia coli, Biochim. Biophys. Acta,* 162, 506, 1968.
289. **Hendler, R. W.,** Respiration and protein synthesis in *Escherichia coli* membrane-envelope fragments. V. On the reduction of nonheme iron and the cytochromes by nicotinamide adenine dinucleotide and succinate, *J. Cell. Biol.,* 51, 664, 1971.
290. **Hendler, R. W. and Burgess, A. H.,** Fractionation of the electron-transport chain of *Escherichia coli, Biochim. Biophys. Acta,* 357, 215, 1974.
291. **Dancey, G. F., Levine, A. E., and Shapiro, B. M.,** The NADH dehydrogenase of the respiratory chain of *Escherichia coli.* I. Properties of the membrane-bound enzyme, its solubilization, and purification to near homogeneity, *J. Biol. Chem.*, 251, 5911, 1976.
292. **Dancey, G. F. and Shapiro, B. M.,** The NADH dehydrogenase of the respiratory chain of *Escherichia coli.* II. Kinetics of the purified enzyme and the effects of antibodies elicited against it on membrane-bound and free enzyme, *J. Biol. Chem.*, 251, 5921, 1976.
293. **Dancey, G. F. and Shapiro, B. M.,** Specific phospholipid requirement for activity of the purified respiratory chain NADH dehydrogenase of *Escherichia coli, Biochim. Biophys. Acta,* 487, 368, 1977.
294. **Owen, P. and Smyth, C. J.,** unpublished data, 1976.
295. **Owen, P. and Kaback, H. R.,** Identification of antigen no. 19/27 as dihydrolipoyl dehydrogenase, and its involvement in ubiquinonemediated, NADH-dependent transport phenomena in membrane vesicles from *Escherichia coli* ML308-225, *FEMS Microbiol. Lett.*, 7, 347, 1980.
296. **Patel, L., Schuldiner, S., and Kaback, H. R.,** Reversible effects of chaotropic agents on the proton permeability of *Escherichia coli* membrane vesicles, *Proc. Natl. Acad. Sci. U.S.A.*, 72, 3387, 1975.
297. **Massey, V.,** Lipoyl dehydrogenase, in *The Enzymes,* Vol. 3, **Boyer, P. D., Lardy, H., and Myrbäck, K. Eds.,** Academic Press, New York, 1963, 275.

298. **Williams, C. H., Jr.,** Flavin-containing dehydrogenases, in *The Enzymes,* Vol. 13C, **Boyer, P. D., Ed.,** Academic Press, New York, 1976, 89.
299. **Guest, J. R. and Creaghan, I. T.,** Further studies with lipoamide dehydrogenase mutants of *Escherichia coli* K12, *J. Gen. Microbiol.* 81, 237, 1974.
300. **Kohn, L. D. and Kaback, H. R.,** Mechanism of active transport in isolated bacterial membrane vesicles. XV. Purification and properties of the membrane bound D-lactate dehyrogenase from *Escherichia coli, J. Biol. Chem.,* 248, 7012, 1973.
301. **Short, S. A., Kaback, H. R. Hawkins, T., and Kohn, L. D.,** Immunochemical properties of the membrane-bound D-lactate dehydrogenase from *Escherichia coli, J. Biol. Chem.,* 250, 4285, 1975.
302. **Short, S. A., Kaback, H. R. and Kohn, L. D.,** Localization of D-lactate dehydrogease in native and reconstituted *Escherichia coli* membrane vesicles, *J. Biol. Chem.,* 250, 4291, 1975.
303. **Kaback, H. R.,** Bacterial membranes, in *Methods in Enzymology,* Vol. 22, **Colowick, S. P., Ed.,** Academic Press, New York, 1971, 99.
304. **Blumberg, P. M. and Strominger, J. L.,** Interaction of penicillin with the bacterial cell: penicillin-binding proteins and penicillin-sensitive enzymes, *Bacteriol. Rev.,* 38, 291, 1974.
305. **Ghuysen, J.-M.,** The concept of the penicillin target from 1965 until today, *J. Gen. Microbiol.,* 101, 13, 1977.
306. **Nguyen-Distèche, M., Frère, J.-M., Dusart, J., Lehy-Bouille, M., Ghuysen, J.-M., Pollock, J. J. and Iacono, V. J.,** The peptidoglycan crosslinking enzyme system in *Streptomyces* R61, K15 and *rimosus:* immunological studies, *Eur. J. Biochem.,* 81, 29, 1977.
307. **Buchanan, C. E., Hsia, J., and Strominger, J. L.,** Antibody to the D-alanine carboxypeptidase of *Bacillus subitlis* does not cross-react with other penicillin-binding proteins, *J. Bacteriol.,* 131, 1008, 1977.
308. **Richmond, M. H.** Enzyme/antiserum interactions of β-lactamases, in *Immunochemistry of Enzymes and their Antibodies,* **Salton, M. R. J., Ed.,** John Wiley & Sons, New York, 1977, 29.
309. **Ogawara, H.,** Penicillin-binding proteins of *Escherichia coli:* comparison of a strain carrying an R-factor and the parent strain, *Biochim. Biophys. Acta,* 491, 223, 1977.
310. **Cancedda, R. and Schlesinger, M. J.,** Localization of polyribosomes containing alkaline phosphatase nascent polypeptides on membranes of *Escherichia coli, J. Bacteriol.,* 117, 290, 1974.
311. **Randall, L. L. and Hardy, S. J. S.,** Synthesis of exported proteins by membrane-bound polysomes from *Escherichia coli, Eur. J. Biochem.* 75, 43, 1977.
312. **Randall, L. L., Hardy, S. J. S., and Josefsson, L.-G.,** Precursors of three exported proteins in *Escherichia coli, Proc. Natl. Acad. Sci. U.S.A.,* 75, 1209, 1978.
313. **Varenne, S., Piovant, M., Pagès, J. M. and Lazdunski, C.,** Evidence for synthesis of alkaline phosphatase on membrane-bound polysomes in *Escherichia coli, Eur. J. Biochem.,* 86, 603, 1978.
314. **Halequoa, S., Sekizawa, J., and Inouye, M.,** A new form of structural lipoprotein of outer membrane of *Escherichia coli, J. Biol. Chem.,* 252, 2324, 1977.
315. **Inouye, S., Wang, S., Sekizawa, J., Haleqoua, S., and Iouye, M.,** Amino acid sequence for the peptide extension on the prolipoprotein of the *Escherichia coli* outer membrane, *Proc. Natl. Acad. Sci. U.S.A.,* 74, 1004, 1977.
316. **Inouye, H. and Beckwith, J.,** Synthesis and processing of an *Escherichia coli* alkaline phosphatase precursor *in vitro, Proc. Natl. Acad. Sci. U.S.A.,* 74, 1440, 1977.
317. **Sarvas, M., Hirth, K. P., Fuchs, E., and Simons, K.,** A precursor form of penicillinase from *Bacillus licheniformis, FEBS Lett.* 95, 76, 1978.
318. **Glenn, A. R.,** Production of extracellular proteins by bacteria, *Annu. Rev. Microbiol.* 30, 41, 1976.
319. **Priest, F. G.,** Extracellular enzyme synthesis in the genus *Bacillus, Bacteriol. Rev.,* 41, 711, 1977.
320. **Schlesinger, M. J.,** The reversible dissociation of the alkaline phosphatase of *Escherichia coli.* II. Properties of the subunit, *J. Biol. Chem.,* 240, 4293, 1965.
321. **Schlesinger, M. J.,** Activation of a mutationally altered form of the *Escherichia coli* alkaline phosphatase, *J. Biol. Chem.,* 241, 3181, 1966.
322. **Schlesinger, M. J.,** The reversible dissociation of the alkaline phosphatase of *Escherichia coli.* III. Properties of antibodies directed against the subunit, *J. Biol. Chem.,* 242, 1599, 1967.
323. **Schlesinger, M. J.,** Formation of a defective alkaline phosphatase subunit by a mutant of *Escherichia coli, J. Biol. Chem.,* 242, 1604, 1967.
324. **Schlesinger, S. and Schlesinger, M. J.,** The effect of amino acid analogues on alkaline phosphatase formation in *Escherichia coli* K12. I. Substitution of triazolealanine for histidine, *J. Biol. Chem.,* 242, 3369, 1967.
325. **Schlesinger, M. J.,** Secretion of alkaline phosphatase subunits by spheroplasts of *Escherichia coli, J. Bacteriol.* 96, 727, 1968.
326. **Schlesinger, S. and Schlesinger, M. J.,** The effect of amino acid analogues on alkaline phosphatase formation in *Escherichia coli* K12. III. Substitution of 2-methylhistidine for histidine, *J. Biol. Chem.,* 244, 3803, 1969.

327. **Attias, J., Schlesinger, M. J., and Schlesinger, S.,** The effect of amino acid analogues on alkaline phosphatase formation in *Escherichia coli* K12, IV. Substitution of canavanine for arginine, *J. Biol. Chem.*, 244, 3810, 1969.
328. **Morris, H. and Schlesinger, M. J.,** Effects of proline analogues on the formation of alkaline phosphatase in *Escherichia coli, J. Bacteriol.*, 111, 203, 1972.
329. **Halford, S. E., Lennette, D. A., Kelley, P. M., and Schlesinger, M. J,** A mutationally altered alkaline phosphatase from *Escherichia coli.* I. Formation of an active enzyme *in vitro* and phenotypic suppression *in vivo, J. Biol. Chem.*, 247, 2087, 1972.
330. **Pagès, J.-M., Louvard, D., and Lazdunski, C.,** Preparation of active iodinated specific antibodies, *FEBS Lett.* 59, 32, 1975.
331. **Lazdunski, C. J., Pagès, J.-M., and Louvard, D.,** Antibodies as probes for detection of conformational changes in proteins. A model study with the alkaline phosphatase of *Escherichia coli, J. Mol. Biol.*, 97, 309, 1975.
332. **Pagès, J.-M., Varenne, S., and Lazdunski, C.,** Effects of antibodies to various molecular forms of a mutationally altered *Escherichia coli* alkaline phosphatase on its activation by zinc, *Eur. J. Biochem.*, 67, 145, 1976.
333. **Pagès, J.-M., Piovant, M., Varenne, S., and Lazdunski, C.,** Mechanistic aspects of the transfer of nascent periplasmic proteins across the cytoplasmic membrane of *Escherichia coli, Eur. J. Biochem.* 86, 589, 1978.
334. **Rotman, M. B. and Celada, F.,** Antibody-mediated activation of a defective β-D-galactosidase extracted from an *Escherichia coli* mutant, *Proc. Natl. Acad. Sci. U.S.A.*, 60, 660, 1968.
335. **Inouye, M.,** Biosynthesis and assembly of the outer membrane proteins of *Escherichia coli,* in *Membrane Biogenesis,* **Tzagoloff, A., Ed.,** Plenum Press, New York, 1975, 351.
336. **Braun, V., Rehn, K., and Wolff, H.,** Supramolecular structure of the rigid layer of the cell wall of *Salmonella, Serratia, Proteus,* and *Pseudomonas fluorescens.* Number of lipoprotein molecules in a membrane layer, *Biochemistry,* 9, 5041, 1970.
337. **Halegoua, S., Hirashima, A., and Inouye, M.,** Existance of a free form of specific membrane lipoprotein in gram-negative bacteria, *J. Bacteriol.*, 120, 1204, 1974.
338. **Nakamura, K., Pirtle, R. M., and Inouye, M.,** Homology of the gene coding for outer membrane lipoprotein within various gram-negative bacteria, *J. Bacteriol.*, 137, 595, 1979.
339. **Braun, V., Bosch, V., Klumpp, E. R., Neff, I., Mayer, H., and Schlecht, S.,** Antigenic determinants of murein lipoprotein and its exposure at the surface of *Enterobacteriaceae, Eur. J. Biochem.*, 62, 555, 1976.
340. **Katz, E., Loring, D., Inouye, S., and Inouye, M.,** Lipoprotein from *Proteus mirabilis, J. Bacteriol.*, 134, 674, 1978.
341. **Inouye, M.,** A three-dimensional molecular assembly model of a lipoprotein from the *Escherichia coli* outer membrane, *Proc. Natl. Acad. Sci. U.S.A.*, 71, 2396, 1974.
342. **Lee, N., Cheng, E., and Inouye, M.,** Optical properties of an outer membrane lipoprotein from *Escherichia coli, Biochim. Biophys. Acta,* 465, 650, 1977.
343. **Hirashima, A., Wang, S., and Inouye, M.,** Cell-free synthesis of a specific lipoprotein of the *Escherichia coli* outer membrane directed by purified messenger RNA, *Proc. Natl. Acad. Sci. U.S.A.*, 71, 4149, 1974.
344. **Hirashima, A. and Inouye, M.,** Biosynthesis of a specific lipoprotein of the *Escherichia coli* outer membrane on polyribosomes, *Eur. J. Biochem.*, 60, 395, 1975.
345. **Wang, S., Marcu, K. B., and Inouye, M.,** Translation of a specific m-RNA from *Escherichia coli* in a eukaryotic cell-free system, *Biochem. Biophys. Res. Commun.*, 68, 1194, 1976.
346. **Wu., C. H. and Lin, J. J.-C.,** *Escherichia coli* mutants altered in murein lipoprotein, *J. Bacteriol.*, 126, 147, 1976.
347. **Inouye, S., Takeishi, K., Lee, N., De Martini, M., Hirashima, A., and Inouye, M.,** Lipoprotein from the outer membrane of *Escherichia coli:* purification, paracrystallization, and some properties of its free form, *J. Bacteriol.*, 127, 555, 1976.
348. **Hirota, Y., Suzaki, H., Nishimura, Y., and Yasuda, S.,** On the process of cellular division in *Escherichia coli:* a mutant of *E. coli* lacking murein-lipoprotein, *Proc. Natl. Acad. Sci. U.S.A.*, 74, 1417, 1977.
349. **Yem, D. W. and Wu, H. C.,** Physiological characterization of an *Escherichia coli* mutant altered in the structure of murein lipoprotein, *J. Bacteriol,* 133, 1419, 1978.
350. **Griffiths, E. K., Yoonessi, S., and Neter, E.,** Antibody response to enterobacterial lipoprotein of patients with varied infections due to *Enterobacteriaceae, Proc. Soc. Exp. Biol. Med.*, 154, 246, 1977.
351. **Smith, J. W.,** Local immune response to lipoprotein of the outer membrane of *Escherichia coli* in experimental pyelonphritis, *Infect. Immun.*, 17, 366, 1977.
352. **Melchers, F., Braun, V., and Galanos, C.,** The lipoprotein of the outer membrane of *Escherichia coli:* a B-lyphocyte mitogen, *J. Exp. Med.*, 142, 473, 1975.

353. **Bessler, W. G. and Ottenbreit, B. P.,** Studies on the mitogenic principle of the lipoprotein from the outer membrane of *Escherichia coli, Biochem. Biophys. Res. Commun.,* 76, 239, 1977.

354. **Bessler, W., Resch, K., Hancock, E., and Hantke, K.,** Induction of lymphocyte proliferation and membrane changes by lipopeptide derivatives of the lipoprotein from the outer membrane of *Escherichia coli, Z. Immunitaetsforsch.,* 153, 11, 1977.

355. **Andersson, J., Melchers, F., Galanos, C., and Lüderitz, O,** The mitogenic effect of lipopolysaccharide on bone marrow-derived mouse lymphocytes: Lipid A as the mitogenic part of the molecule, *J. Exp. Med.,* 137, 943, 1973.

356. **Dankert, J. and Hofstra, H.,** Antibodies against outer membrane proteins in rabbit antisera prepared against *Escherichia coli* 026K60, *J. Gen. Microbiol.,* 104, 311, 1978.

356a. **Lugtenberg, B., Bronstein, H., van Selm, N., and Peters, R.,** Peptidoglycan-associated outer membrane proteins in Gram negative bacteria, *Biochim. Biophys. Acta,* 465, 571, 1977.

356b. **Jann, B., and Jann, K.,** SDS polyacrylamide gel electrophoresis patterns of the outer membrane proteins from *E. coli* strains of different pathogenic origin, *FEMS Microbiol. Lett.,* 7, 19, 1980.

356c. **Paakkanen, J., Gotschlich, E. C. and Mäkelä, P. H.,** Protein K: a new major outer membrane protein found in encapsulated *Escherichia coli, J. Bacteriol.* 139, 835, 1979.

356d. **Achtman, M.,** Epidemiological analysis of major outer membrane proteins from encapsulated invasive *E. coli, FEMS Symposium on Microbial Envelopes,* Abstract no. 42, 1980.

357. **Hofstra, H. and Dankert, J.,** Antigenic cross-reactivity of major outer membrane proteins in *Enterobacteriaceae* species, *J. Gen. Microbiol.,* 111, 293, 1979.

357a. **Hofstra, H. and Dankert, J.,** Antigenic cross-reactivity of outer membrane proteins of *Escherichia coli* and *Proteus* species, *FEMS Microbiol. Lett.,* 7, 171, 1980.

357b. **Hofstra, H. and Dankert, J.,** Major outer membrane proteins: common antigens in *Enterobacteriaceae* species, *J. Gen. Microbiol.,* 119, 123, 1980.

357c. **Hofstra, H., van Tol, M. J. D. and Dankert, J.,** Cross-reactivity of major outer membrane proteins of *Enterobacteriaceae,* studied by crossed immunoelectrophoresis, *J. Bacteriol.,* 143, 328, 1980.

357d. **Datta, D. B., Krämer, C., and Henning, U.,** Diploidy for a structural gene specifying a major protein of the outer cell envelope membrane from *Escherichia coli* K-12, *J. Bacteriol.,* 128, 834, 1976.

357e. **Lee, D. R. and Schnaitman, C. A.,** Comparison of outer membrane porin proteins produced by *Escherichia coli* and *Salmonella typhimurium, J. Bacteriol.,* 142, 1019, 1980.

358. **Rosenbusch, J. P.,** Characterisation of the major envelope protein from *Escherichia coli:* regular arrangement on the peptidoglycan and unusual dodecyl sulphate binding, *J. Biol. Chem.* 249, 8019, 1974.

359. **Diedrich, D. L., Summers, A. O., and Schnaitman, C. A.,** Outer membrane proteins of *Escherichia coli.* V. Evidence that protein 1 and bacteriophage-directed protein 2 are different polypeptides, *J. Bacteriol.,* 131, 598, 1977.

360. **Foulds, J. and Chai, T.-J.,** New major outer membrane protein found in an *Escherichia coli tolF* mutant resistant to bacteriophage TuIb, *J. Bacteriol.,* 133, 1478, 1978.

361. **Benz, R., Boehler-Kohler, B. A., Dieterle, R., and Boos, W.,** Porin activity in the osmotic shock fluid of *Escherichia coli, J. Bacteriol.,* 135, 1080, 1978.

362. **Pugsley, A. P. and Schnaitman, C. A.,** Identification of three genes controlling production of new outer membrane pore proteins in *Escherichia coli* K-12, *J. Bacteriol.,* 135, 1118, 1978.

363. **van Tol, M. J. D., Hofstra, H., Dankert, J.,** Major outer membrane proteins of *Escherichia coli* analysed by crossed immunoelectrophoresis, *FEMS Microbiol. Lett.,* 5, 349, 1979.

363a. **Hofstra, H., van Tol., M. J. D., and Dankert, J.,** Immunofluorescent detection of the major outer membrane protein II* in *Escherichia coli* 026K60, *FEMS Microbiol. Lett.,* 6, 147, 1979.

364. **Randall-Hazelbauer, L. and Schwartz, M.,** Isolation of the bacteriophage lambda receptor from *Escherichia coli, J. Bacteriol.,* 116, 1436, 1973.

365. **Palva, E. T.,** Major outer membrane protein in *Salmonella typhimurium* induced by maltose, *J. Bacteriol.,* 136, 286, 1978.

366. **Pick, K.-H. and Wöber, G.,** Maltodextrin pore proteins in the outer membrane of *Escherichia coli and Klebsiella pneumoniae:* immunological comparison, *FEMS Microbiol. Lett.,* 5, 119, 1979.

366a. **Chai, T.-J. and Foulds, J.,** Isolation and partial characterization of protein E, a major protein found in certain *Escherichia coli* K-12 mutant strains: relationship to other outer membrane proteins, *J. Bacteriol.,* 139, 418, 1979.

367. **Nakamura, K. and Mizushima, S.,** Effects of heating in dodecyl sulfate solution on the conformation and electrophoretic mobility of isolated major outer membrane proteins from *Escherichia coli* K12, *J. Biochem. (Tokyo),* 80, 1411, 1976.

367a. **Manning, P. A., Beutin, L., and Achtman, M.,** Outer membrane of *Escherichia coli:* properties of the F sex factor *tra T* protein which is involved in surface exclusion, *J. Bacteriol.,* 142, 285, 1980.

367b. **Moll, A., Manning, P. A., and Timmis, K. N.,** Plasmid-determined resistance to serum bactericidal activity: a major outer membrane protein, the *tra T* gene product, is responsible for plasmid-specified serum resistance in *Escherichia coli, Infect. Immun.*, 28, 359, 1980.
368. **Swanson, J.,** Studies on gonococcus infection. XVIII. ^{125}I-Labeled peptide mapping of the major protein of the gonococcal cell wall outer membrane, *Infect. Immun.*, 23, 799, 1979.
369. **McDonough, M. W.,** Tryptic peptide maps of mutant *Salmonella* flagellins, *Biochem. J.*, 84, 114P, 1962.
370. **Heckels, J. E.,** The surface properties of *Neisseria gonorrhoeae:* isolation of the major components of the outer membrane, *J. Gen. Microbiol.*, 99, 333, 1977.
371. **Lamden, P. R. and Heckels, J. E.,** Outer membrane protein composition and colonial morphology of *Neisseria gonorrhoeae* strain P9, *FEMS Microbiol. Lett.*, 5, 263, 1979.
372. **Buchanan, T. M., Peare, W. A., Schoolnick, G. K., and Arko, R. J.,** Protection against infection with *Neisseria gonorrhoeae* by immunization with outer membrane complex and purified pili, *J. Infect. Dis.*, 136, S132, 1977.
373. **Hildebrandt, J. F., Mayer, L. W., Wang, S. P., and Buchanan, T. M.,** *Neisseria gonorrhoeae* acquire a new principal outer-membrane protein when transformed to resistance to serum bactericidal activity, *Infect. Immun.*, 20, 267, 1978.
374. **Chen, N. C., Kamel, K., Zuckerman, J., and Gaafar, H. A.,** Purification of *Neisseria gonorrhoeae* surface L-antigen, *Infect. Immun.*, 18, 230, 1977.
375. **Gaafar, H. A. and D'Arcangelis, D. C.,** Fluorescent antibody test for serological diagnosis of gonorrhea, *J. Clin. Microbiol.* 3, 438, 1976.
376. **King, G. J. and Swanson, J.,** Studies on gonococcus infection. XV. Identification of surface proteins of *Neisseria gonorrhoeae* correlated with leukocyte association, *Infect. Immun.*, 21, 575, 1978.
377. **Swanson, J.,** Studies on gonococcal infection. XII. Colony color and opacity variants of gonococci, *Infect. Immun.*, 19, 320, 1978.
378. **James, J. F. and Swanson, J.,** Studies on gonococcal infection. XIII. Occurrence of color/opacity colonial variants in clinical cultures, *Infect. Immun.*, 19, 332, 1978.
379. **Swanson, J.,** Studies on gonococcus infection. XIV. Cell wall protein differences among color/opacity colony variants of *Neisseria gonorrhoeae, Infect. Immun.*, 21, 292, 1978.
380. **Salit, I. E. and Gotschlich, E. C.,** Gonococcal color and opacity varients: virulence for chicken embryos, *Infect. Immun.*, 22, 359, 1978.
381. **Heckels, J. E. and Everson, J. S.,** The isolation of a new outer membrane protein from the parent strain of *Neisseria gonorrhoeae* P9, *J. Gen. Microbiol.*, 106, 179, 1978.
382. **Frasch, C. E. and Gotschlich, E. C.,** An outer membrane protein of *Neisseria meningitidis* Group B responsible for serotype specificity, *J. Exp. Med.*, 140, 87, 1974.
383. **Frasch, C. E., McNelis, R. M., and Gotschlich, E. C.,** Strain-specific variation in the protein and lipopolysaccharide composition of the Group B meningococcal outer membrane, *J. Bacteriol.*, 127, 973, 1976.
384. **Kasper, D. L. and Seiler, M. W.,** Immunochemical characterization of the outer membrane complex of *Bacteroides fragilis* subspecies *fragilis, J. Infect. Dis.*, 132, 440, 1975.
385. **Zollinger, W. D., Kasper, D. L., Veltri, B. J., and Artenstein, M. S.,** Isolation and characterization of a native cell wall complex from *Neisseria meningitidis, Infect. Immun.*, 6, 835, 1972.
386. **Hill, J. C. and Weiss, E.,** Protein fraction with immunogenic potential and low toxicity isolated from the cell wall of *Neisseria meningitidis* Group B, *Infect. Immun.*, 10, 605, 1974.
387. **Frash, C. E. and Robbins, J. D.,** Protection against Group B meningococcal disease. III. Immunogenicity of serotype 2 vaccines and specificity of protection in a guinea pig model, *J. Exp. Med.*, 147, 629, 1978.
388. **Frasch, C. E. and Chapman, S. S.,** Classification of *Neisseria meningitidis* group B into distinct serotypes. III. Application of a new bacteriocidal inhibition technique to distribution of serotypes among cases and carriers, *J. Infect. Dis.*, 127, 149, 1973.
389. **Munford, R. S., Patton, C. M., and Gorman, G. W.,** Epidemiologic studies of serotype antigens common to groups B and C *Neisseria meningitidis, J. Infect. Dis.*, 131, 286, 1975.
390. **Jones, D. M. and Tobin, B. M.,** Serotypes of group B meningococci *J. Clin. Pathol. (London)*, 29, 746, 1976,
390a. **Tsai, C.-M. and Frasch, C. E.,** Chemical analysis of major outer membrane proteins of *Neisseria meningitidis:* comparison of serotypes 2 and 11, *J. Bacteriol.*, 141, 169, 1980.
390b. **Hoff, G. E. and Frasch, C. E.,** Outer membrane antigens of *Neisseria meningitidis* group B serotype 2 studied by crossed immunoelectrophoresis, *Infect. Immun*, 25, 849, 1979.
390c. **Kuusi, N., Nurminen, M., Saxen, H., Valtonen, M., and Mäkelä, P. H.,** Immunization with major outer membrane proteins in experimental salmonellosis of mice, *Infect. Immun.*, 25, 857, 1979.

390d. **Svenson, S. B., Nurminen, M., and Lindberg, A. A.,** Artificial *Salmonella* vaccines: O-antigenic oligosaccharide-protein conjugates induce protection against infection with *Salmonella typhimurium, Infect. Immun.,* 25, 863, 1979.

390e. **Geyer, R., Galanos, C., Westphal, O., and Golecki, J. R.,** A lipopolysaccharide-binding cell-surface protein from *Salmonella minnesota.* Isolation, partial characterization and occurrence in different *Enterobacteriaceae, Eur. J. Biochem.,* 98, 27, 1979.

390f. **Bub, F., Bieker, P., Martin, H. H., and Nixdorff, K.,** Immunological characterization of two major proteins isolated from the outer membrane of *Proteus mirabilis, Infect. Immun.,* 27, 315, 1980.

390g. **Chen, Y.-H. U., Hancock, R. E. W., and Mishell, R. I.,** Mitogenic effects of purified outer membrane proteins from *Pseudomonas aeruginosa, Infect. Immun,* 28, 178, 1980.

390h. **Mizuno, T. and Kageyama, M.,** Isolation and characterization of major outer membrane proteins of *Pseudomonas aeruginosa* strain PAO with special reference to peptidoglycan-associated protein, *J. Biochem. (Tokyo),* 86, 979, 1979.

390i. **Mizuno, T. and Kageyama, M.,** Isolation and characterization of a major outer membrane protein of *Pseudomonas aeruginosa.* Evidence for the occurrence of a lipoprotein, *J. Biochem. (Tokyo),* 85, 115, 1979.

390j. **Mizuno, T.,** A novel peptidoglycan-associated lipoprotein found in the cell envelope of *Pseudomonas aeruginosa* and *Escherichia coli J. Biochem. (Tokyo),* 86, 991, 1979.

390k. **Wicken, A. J. and Knox, K. W.,** Bacterial cell surface amphiphiles, *Biochim. Biophys. Acta,* 604, 1, 1980.

391. **Inoue, K. and Nojima, S.,** Immunochemical studies of phospholipids. I. Reactivity of various synthetic cardiolipid derivatives with Wassermann antibody, *Chem. Phys. Lipids,* 1, 360, 1967.

392. **Inoue, K. and Nojima, S.,** Immunochemical studies of phospholipids. III. Production of antibody to cardiolipid, *Biochim. Biophys. Acta,* 144, 409, 1967.

393. **Inoue, K. and Nojima, S.,** Immunochemical studies of phospholipids. IV. Reactivities of antiserum against natural cardiolipid and synthetic cardiolipin analog-containing antigens, *Chem. Phys. Lipids,* 3, 70, 1969.

394. **Rapport, M. M. and Graf, L.,** Immunochemical reactions of lipids, *Progr. Allergy,* 13, 273, 1969.

395. **Kataoka, T. and Nojima, S.,** Immunochemical studies of phospholipids, VI. Haptenic activity of phosphatidylinositol and the role of lecithin as an auxiliary lipid, *J. Immunol.,* 105, 502, 1970.

396. **DeSiervo, A. J.,** Anti-cardiolipin and anti-phosphatidylglycerol antibodies prepared against bacterial phospholipids, *Infect. Immun.,* 9, 835, 1974.

397. **Schiefer, H.-G., Gerhardt, J., and Brunner, H.,** Immunological studies on the localization of phosphatidylglycerol in the membranes of *Mycoplasma hominis, Hoppe-Seyler's Z. Physiol. Chem.,* 356, 559, 1975.

398. **Marcus, D. M. and Schwarting, G. A.,** Immunochemical properties of glycolipids and phospholipids, in *Advances in Immunology,* Vol. 23, **Kunkel, H. G. and Dixon, F. J., Eds.,** Academic Press, New York, 1976, 203.

399. **Dörner, I., Brunner, H., Schiefer, H.-G., and Wellensiek, H.-J.,** Complement-mediated killing of *Acholeplasma laidlawii* by antibodies to various membrane components, *Infect. Immun.,* 13, 1663, 1976.

400. **Dörner, I., Brunner, H., Schiefer, H.-G., Loos, M., and Wellensiek, H.-J.,** Antibodies to *Acholeplasma laidlawii* membrane lipids in normal guinea pig serum, *Infect. Immun.,* 18, 1, 1977.

401. **Kenny, G. E. and Grayston, J. T.,** Eaton pleuropneumonia-like organism *(Mycoplasma pneumoniae)* complement-fixation antigen: extraction with organic solvents, *J. Immunol.,* 95, 19, 1965.

402. **Deeb, B. J. and Kenny, G. E.,** Characterisation of *Mycoplasma pulmonis* varients isolated from rabbits. II. Basis for differentiation of antigenic subtypes, *J. Bacteriol.,* 93, 1425, 1967.

403. **Lemcke, R. M. Marmion, B. P., and Plackett, P.,** Immunochemical analysis of *Mycoplasma pneumoniae, Ann. N. Y. Acad. Sci.,* 143, 691, 1967.

404. **Marmion, B. P., Plackett, P., and Lemcke, R. M.,** Immunochemical analysis of *Mycoplasma pneumoniae.* I. Methods of extraction and reaction of fractions from *M. pneumoniae* and *M. mycoides* with homologous antisera and with antisera against *Streptococcus* MG, *Aust. J. Exp. Biol. Med. Sci.,* 45, 163, 1967.

405. **Plackett, P. and Shaw, E. J.,** Glycolipids from *Mycoplasma laidlawii* and *Streptococcus* MG, *Biochem. J.,* 104, 61c, 1967.

406. **Beckman, B., and Kenny, G. E.,** Immunochemical analysis of serologically active lipids of *Mycoplasma pneumoniae, J. Bacteriol.,* 96, 1171, 1968.

407. **Lemcke, R. M., Plackett, P., Shaw, E. J., and Marmion, B. P.,** Immunochemical analysis of *Mycoplasma pneumoniae.* 2. Properties of chloroform-methanol extract from *M. pneumoniae. Aust. J. Exp. Biol. Med. Sci.,* 46, 123, 1968.

408. **Plackett, P., Marmion, B. P., Shaw, E. J., and Lemcke, R. M.,** Immunochemical analysis of *Mycoplasma pneumoniae.* 3. Separation and chemical identification of serologically active lipids, *Aust. J. Exp. Biol. Med. Sci.,* 47, 171, 1969.
409. **Razin, S., Prescott, B., Caldes, G., James, W. D., and Chanock, R. M.,** Role of glycolipids and phosphatidylglycerol in the serological activity of *Mycoplasma pneumoniae, Infect. Immun.,* 1, 408, 1970.
410. **Razin, S., Prescott, B., James, W. D., Caldes, G., Valdesuso, J., and Chanock, R. M.,** Production and properties of antisera to membrane glycolipids of *Mycoplasma pneumoniae, Infect. Immun.,* 3, 420, 1971.
411. **Kenny, G. E.,** Immunogenicity of *Mycoplasma pneumoniae, Infect. Immun.,* 3, 510, 1971.
412. **Kenny, G. E.,** Serological cross-reaction between lipids of *Mycoplasma pneumoniae* and *Mycoplasma neurolyticum, Infect. Immun.,* 4, 149, 1971.
413. **Dupouey, P.,** Role of cerebrosides and a galactodiglyceride in the antigenic cross-reaction between nerve tissue and treponema, *J. Immunol.,* 109, 146, 1972.
414. **Feinman, S. B., Prescott, B., and Cole, R. M.,** Serological reactions of glycolipids from streptococcal L-forms, *Infect. Immun.,* 8, 752, 1973.
415. **Romano, N. and Scarlata, G.,** Serological activity of lipids of a T strain of *Mycoplasma, Infect. Immun.,* 9, 1062, 1974.
416. **Sugiyama, T., Smith, P. F., Langworthy, T. A., and Mayberry, W. R.,** Immunological analysis of glycolipids and lipopolysaccharides derived from various *Mycoplasmas, Infect. Immun.,* 10, 1273, 1974.
417. **Kenny, G. E.,** Antigens of the *Mycoplasmatales* and *Chlamydiae,* in *The Antigens,* Vol. 3, **Sela, M., Ed.,** Academic Press, New York, 1975, 449.
418. **Ryan, M. D., Noker, P., and Matz, L. L.,** Immunological properties of glycolipids from membranes of *Acholeplasma laidlawii, Infect. Immun.,* 12, 799, 1975.
419. **Schiefer, H.-G., Gerhardt, U., and Brunner, H.,** Localization of a phosphoglycolipid in mycoplasma membranes using specific antilipid-antibodies, *Zentralbl. Bakteriol., Parasitenkd., Infektionskr. Hyg., Abt. 1: Orig., Reihe A,* 239, 262, 1977.
420. **Pollack, J. D., Somerson, N .L. and Senterfit, L. B.,** Isolation, characterization, and immunogenicity of *Mycoplasma pneumoniae* membranes, *Infect. Immun.,* 2, 326, 1970.
421. **Romano, N., LaLicata, R., and Scarlata, G.,** Immunological analysis of plasma membranes of a T-strain of Mycoplasma *(Ureaplasma urealyticum), Infect. Immun.,* 16, 734, 1977.
422. **Stone, S. S. and Razin, S.,** Immunoelectrophoretic analysis of *Mycoplasma mycoides* var. *mycoides, Infect. Immun.,* 7, 922, 1973.
423. **Razin, S., Prescott, B., and Chanock, R. M.,** Immunogenicity of *Mycoplasma pneumoniae* glycolipids. A novel approach to the production of antisera to membrane lipids, *Proc. Natl. Acad. Sci. U.S.A,* 67, 590, 1970.
424. **Guarneri, M.,** Reaction of cardiolipin and phosphatidylinositol antisera with phospholipid antigens, *Lipids,* 9, 692, 1974.
425. **Shaw, N.,** Bacterial glycolipids and glycophospholipids, in *Advances in Microbial Physiology,* Vol. 12, **Rose, A. H. and Tempest, D. W., Eds.,** Academic Press, London, 1975, 141.
426. **Brunner, H., Razin, S., Kalica, A. R., and Chanock, R. M.,** Lysis and death of *Mycoplasma pneumoniae* by antibody and complement, *J. Immunol.,* 106, 907, 1971.
427. **Brunner, H., Dörner, I., Schiefer, H.-G., Krauss, H., and Wellensiek, H.-J.,** Lysis of *Acholeplasma laidlawii* by antibodies and complement, *Infect. Immun.,* 13, 1671, 1976.
428. **Kenny, G. E.,** Heat-lability and organic solvent-solubility of *Mycoplasma* antigens, *Ann. N. Y. Acad. Sci.,* 143, 676, 1967.
429. **Greenberg, H., Prescott, B., Brunner, H., James, W., and Chanock, R. M.,** Sharing of glycolipid antigenic determinants by *Mycoplasma pneumoniae,* vegetables and certain bacteria, in *New Approaches for Inducing Natural Immunity to Pyogenic Organisms,* **Robbins, J. B., Horton, R. E. and Krause, R. M., Eds.,** Winter Park, Florida, 1973, 151.
430. **Dupouey, P., Billecocq, A., and Lefroit M.,** Comparative study of the immunological properties of galactosyldiglyceride and galactosylceramide included within natural membranes, *Immunochemistry,* 13, 289, 1976.
431. **Dupouey, P., Coulon-Morelec, M. J., and Maréchal, J.,** Commonauté antigénique entre le galactodiglycéride de *Treponema reiteri* et les cérébrosides, *C. R. Seances Acad. Sci., Ser. D,* 270, 1541, 1970.
432. **Langworthy, T. A.,** Long-chain diglycerol tetraethers from *Thermoplasma acidophilum, Biochim. Biophys. Acta,* 487, 37, 1977.
433. **Razin, S.,** The mycoplasmas, *Microbiol. Rev.,* 42, 414, 1978.

434. **Mayberry-Carson, K. J., Langworthy, T. A., Mayberry, W. R., and Smith, P. F.,** A new class of lipopolysaccharide from *Thermoplasma acidophilum, Biochim. Biophys. Acta* 360, 217, 1974.
435. **Smith, P. F.,** Homogeniety of lipopolysaccharide from *Acholeplasma, J. Bacteriol.* 130, 393, 1977.
436. **Goren, M. B.,** Mycobacterial lipids: selected topics, *Bacteriol. Rev.*, 36, 33, 1972.
437. **Kabat, E. A.,** *Structural Concepts in Immunology and Immunochemistry,* **Ebert, J. D., Loewy, A. G., and Schneiderman, H. A. Eds.,** Holt, Rinehart & Winston, New York, 1968, chap. 4.
438. **Kaback, H. R., Ramos, S., Robertson, D. E., Stroobant, P., and Tokuda, H.,** Energetics and molecular biology of active transport in bacterial membrane vesicles, *J. Supramol. Struct.*, 7, 443, 1977.
439. **Boman, H. G. and Monner, D. A.,** Characterization of lipopolysaccharides from *Escherichia coli* K-12 mutants, *J. Bacteriol.*, 121, 455, 1975.
440. **McLachlan, A. D.,** The double helix coiled coil structure of murein lipoprotein from *Escherichia coli, J. Molec. Biol.*, 121, 493, 1978.
441. **Rothman, J. E. and Lenard, J.,** Membrane asymmetry, *Science,* 195, 743, 1977.
442. **Bergelson, L. D. and Barsukov, L. I.,** Topological asymmetry of phospholipids in membranes, *Science,* 197, 224, 1977.
443. **Coons, A. H., Creech, H. J., Jones, R. N., and Berliner, E.,** The demonstration of pneumococcal antigen in tissues by the use of fluorescent antibody, *J. Immunol.*, 45, 159, 1942.
444. **Raff, M. C.,** Cell-surface immunology, *Sci. Am.*, 234 (5), 30, 1976.
445. **Cole, R. M.,** Bacterial cell-wall replication followed by immunofluorescence, *Bacteriol. Rev.*, 29, 326, 1965.
446. **Hunsley, D. and Kay, D.,** Wall structure of the *Neurospora* hyphal apex: immunofluorescent localization of wall surface antigens, *J. Gen. Microbiol.*, 95, 233, 1976.
447. **Nakane, P. K., SriRam, J., and Pierce, G. B.,** Enzyme-labeled antibodies for light and electron microscopic localization of antigens, *J. Histochem. Cytochem.*, 14, 789, 1966.
448. **Nakane, P. K. and Pierce, G. B.,** Enzyme-labeled antibodies: preparation and application for the localization of antigens, *J. Histochem. Cytochem.*, 14, 929, 1966.
449. **Avrameas, S., and Uriel, J.,** Méthode de marguage d'antigènes et d'anticorps avec des enzymes et son application en immunodiffusion, *C. R. Seances Acad. Sci.*, Ser. D, 262, 2543, 1966.
450. **Nakane, P. K., Nichoalds, G. E., and Oxender, D. L.,** Cellular location of leacine-binding protein from *Escherichia coli, Science,* 161, 182, 1968.
451. **Cox, J. C., Pihl, E., Read, R. S. D., and Nairn, R. C.,** Rapid localization of bacterial surface antigens by whole-mount immunoperoxidase technique, *J. Gen. Microbiol.*, 70, 385, 1972.
452. **Lai, C.-H., Listgarten, M. A. and Rosan, B.,** Immunoelectron microscopic identification and localization of *Streptococcus sanguis* with peroxidase-labeled antibody: localization of surface antigens in pure culture, *Infect. Immun.*, 11, 193, 1975.
453. **Short, J. A. and Walker, P. D.,** The location of bacterial antigens on sections of *Bacillus cereus* by use of the soluble peroxidase-anti-peroxidase complex and unlabeled antibody, *J. Gen. Microbiol.*, 89, 93, 1975.
454. **Dahlén, G., Nygren, H., and Hansson, H.-A.,** Immunoelectron microscopic localization of lipopolysaccharides in the cell wall of *Bacteroides oralis* and *Fusobacterium nucleatum, Infect. Immun.*, 19, 265, 1978.
455. **Shands, J. W.,** Localization of somatic antigen on gram-negative bacteria by electron microscopy, *J. Bacteriol.*, 90, 266, 1965.
456. **Shands, J. W.,** Localization of somatic antigen on gram-negative bacteria using ferritin antibody conjugates, *Ann. N. Y. Acad. Sci.*, 133, 292, 1966.
457. **Fukuski, K., Ariji, F., Yamaguchi, J., and Oka, S.,** Electron microscopic localization of endotoxin of *Escherichia coli* by the use of ferritinconjugated antibody, *J. Electron Microsc.*, 15, 137, 1966.
458. **Mühlradt, P. F., Menzel, J., Golecki, J. R., and Speth, V.,** Outer membrane of *Salmonella:* sites of export of newly synthesised lipopolysaccharide on the bacterial surface, *Eur. J. Biochem.*, 35, 471, 1973.
459. **Bayer, M. E.,** Ultrastructure and organization of the bacterial envelope, *Ann. N. Y. Acad. Sci.*, 235, 6, 1974.
460. **Mühlradt, P. F., Menzel, J., Golecki, J. R, and Speth, V.,** Lateral mobility and surface density of lipopolysaccharide in the outer membrane of *Salmonella typhimurium, Eur. J. Biochem.*, 43, 533, 1974.
461. **Mühlradt, P. F. and Golecki, J. R.,** Asymmetrical distribution and artifactual reorientation of lipopolysaccharide in the outer membrane of *Salmonella typhimurium, Eur. J. Biochem.*, 51, 343, 1975.
462. **Mühlradt, P. F.,** Topology of outer membrane assembly in *Salmonella, J. Supramol. Struct.*, 5, 103, 1976.

463. **Morgan, C.,** The use of ferritin-conjugated antibodies in electronmicroscopy, *Int. Rev. Cytol.* 32, 291, 1972.
464. **Wagner, M.,** Methods of labeling antibodies for electron microscopic localization of antigens, in *Research in Immunochemistry and Immunobiology,* Vol. 3, **Kwapinski, J. B. G., Ed.,** University Park Press, Baltimore, 1973, 185.
465. **Sternberger, L. A.,** *Immunochemistry,* Prentice-Hall, Englewood Cliffs, N. J., 1974.
466. **Huis in 'T Veld, J. H. J., and Linssen, W. H.,** The localization of streptococcal group and type antigens: an electron-microscopic study using ferritin-labelled antisera, *J. Gen. Microbiol.,* 74, 315, 1973.
467. **Walker, P. D., Short, J., Thomson, R. O. and Roberts, D. S.,** The fine structure of *Fusiformis nodosus* with special reference to the location of antigens associated with immunogenicity, *J. Gen. Microbiol.,* 77, 351, 1973.
468. **Fox, E. N.,** M proteins of group A streptococci, *Bacteriol. Rev.,* 38, 57, 1974.
469. **Elliot, T. S. J., Ward, J. B., Wyrick, P. B., and Rogers, H. J,** Ultrastructure study of the reversion of protoplasts of *Bacillus licheniformis* to bacilli, *J. Bacteriol.,* 124, 905, 1975.
470. **McCoy, E. C., Doyle, D., Burda, K., Corbeil, L. B., and Winter, A. J.,** Superficial antigens of *Campylobacter (Vibrio) fetus:* characterization of an antiphagocytic component, *Infect. Immun.,* 11, 517, 1975.
471. **Ward, J. B.,** The reversion of bacterial protoplasts and L-forms, in *Relations between Structure and Function in the Prokaryotic Cell,* **Stanier, R. Y., Rogers, H. J., and Ward, J. B., Eds.,** Cambridge University Press, Cambridge, England, 1978, 249.
472. **Wagner, M., Wagner, B., and Rýc, M.,** An electron microscope study of the localization of peptidoglycan in group A and C streptococcal cell walls, *J. Gen. Microbiol.,* 108, 283, 1978.
473. **Muñoz, E., Freer, J. H., Ellar, D. J., and Salton, M. R. J.,** Membrane-associated ATPase activity from *Micrococcus lysodeikticus, Biochim. Biophys. Acta,* 150, 531, 1968.
474. **van Driel, D., Wicken, A. J., Dickson, M. R., and Knox, K. W.,** Cellular location of the lipoteichoic acids of *Lactobacillus fermenti* NCTC 6991 and *Lactobacillus casei* NCTC 6375, *J. Ultrastruct. Res.,* 43, 483, 1973.
475. **Coulter, J. R and Mukherjee, T. M.,** Electron microscopic localization of alpha toxin within the staphylococcal cell by ferritin-labeled antibody, *Infect. Immun.,* 4, 650, 1971.
476. **MacAlistair, T. J., Irvin, R. T., and Costerton, J. W.,** Cell surface-localized alkaline phosphatase of *Escherichia coli* as visualized by reaction product deposition and ferritin-labeled antibody, *J. Bacteriol.,* 130, 318, 1977.
477. **MacAlister, T. J., Irvin, R. T., and Costerton, J. W.,** Immunocytological investigation of protein synthesis in *Escherichia coli, J. Bacteriol.,* 130, 329, 1977.
478. **Ghosh, A. and Ghosh, B. K.,** Immunoelectron microscopic localization of penicillinase in *Bacillus licheniformis, J. Bacteriol.* 137, 1374, 1979.
479. **Smirnova, T. A., Kushnarev, V. M., and Tshaikovskaja, S. M.,** Electron microscopic study of the secretion of some bacterial enzymes and toxins, *J. Ultrastruct. Res.,* 37, 269, 1971.
480. **de Leij, L., Kingma, J., and Witholt, B.,** Nature of the regions involved in the insertion of newly synthesized protein into the outer membrane of *Escherichia coli, Biochim. Biophys. Acta,* 553, 224, 1979.
481. **DiRienzo, J. M. and Inouye, M.,** Lipid fluidity-dependent biosynthesis and assembly of the outer membrane proteins of *E. coli, Cell,* 17, 155, 1979.
481a. **Takamiya, H., Batsford, S. Gelderblom, H., and Vogt, A.,** Immuno-electron microscopic localization of lipopolysaccharide antigens on ultrathin sections of *Salmonella typhimurium, J. Bacteriol.,* 140, 261, 1979.
481b. **Funahara, Y. and Nikaido, H.,** Asymmetric localization of lipopolysaccharide on the outer membrane of *Salmonella typhimurium, J. Bacteriol.,* 141, 1463, 1980.
482. **Lee, A. G., Birdsell, N. J. M., and Metcalfe, J. C.,** Measurement of fast lateral diffusion of lipids in vesicles and in biological membranes by ^{1}H-nuclear magnetic resonance, *Biochemistry,* 12, 1650, 1973.
483. **Bayer, M. E. and Thurow, H.,** Polysaccharide capsule of *Escherichia coli:* microscopic study of its size, structure and sites of synthesis, *J. Bacteriol.,* 130, 911, 1977.
484. **Smit, J. and Nikaido, H.,** Outer membrane of gram-negative bacteria. XVIII. Electron microscopic studies on porin insertion sites and growth of cell surface of *Salmonella typhimurium, J. Bacteriol.,* 135, 687, 1978.
485. **Begg, K. J. and Donachie, W. D.,** Growth of the *Escherichia coli* cell surface, *J. Bacteriol.,* 129, 1524, 1977.
486. **Pendleton, I. R., Kim, K. S., and Bernheimer, A. W.,** Detection of cholesterol in cell membranes by use of bacterial toxins, *J. Bacteriol.,* 110, 722, 1972.
487. **Bernheimer, A. W.,** Interactions between membranes and cytolytic bacterial toxins, *Biochim. Biophys. Acta,* 344, 27, 1974.

488. **Schiefer, H.-G., Gerhardt, U., Brunner, H., and Krüpe, M.,** Studies with lectins on the surface carbohydrate structures of *Mycoplasma* membranes, *J. Bacteriol.,* 120, 81, 1974.
489. **Sharon, N.,** Lectins, *Sci. Am.,* 236 (6), 108, 1977.
490. **Birdsell, D. C., Doyle, R. J., and Morgenstern, M.,** Organization of teichoic acid in the cell wall of *Bacillus subtilis, J. Bacteriol.,* 121, 726, 1975.
491. **Schiefer, H.-G., Krauss, H., Schummer, U., Brunner, H., and Gerhardt, U.,** Studies with ferritin-conjugated concanavalin A on carbohydrate structures of *Mycoplasma* membranes, *FEMS Microbiol. Lett.,* 3, 183, 1978.
492. **Mayberry-Carson, K. J., Jewell, M. J., and Smith, P. F.,** Ultrastructural localization of *Thermoplasma acidophilum* surface carbohydrate by using concanavalin A, *J. Bacteriol.,* 133, 1510, 1978.
493. **Hubbard, A. L. and Cohn, Z. A.,** Specific labels for cell surfaces, in *Biochemical Analysis of Membranes,* **Maddy, A. H., Ed.,** Chapman and Hall, London, 1976, 427.
494. **Bellanti, J. A.,** *Immunology,* Vol. II. W. B. Saunders, Philadelphia, 1978, 133.
495. **Amzel, L. M. and Poljak, R. J.,** Three-dimensional structure of immunoglobulin, *Annu. Rev. Biochem.,* 48, 961, 1979.
496. **Wicken, A. J., Gibbens, J. W. and Knox, K. W.,** Anti-teichoic acid antibodies and non-treponemal serological tests for syphilis, *Infect. Immun.,* 5, 982, 1972.
497. **Schiefer, H.-G.,** Orientation of serologically active phosphilipids in mitochondrial membranes. I, *Hoppe-Seyler's Z. Physiol. Chem.,* 354, 722, 1973.
498. **Schiefer, H.-G.,** Orientation of serologically active phospholipids in mitochondrial membranes. II. Studies *in vitro* with a complement fixation reaction, *Hoppe-Seyler's Z. Physiol. Chem.,* 354, 725, 1973.
499. **Guarneri, M., Stechmiller, B., and Lehninger, A. L.,** Use of an antibody to study the location of cardiolipin in mitochondrial membranes, *J. Biol. Chem.,* 246, 7526, 1971.
500. **Guarnieri, M. and Eisner, D.,** A DNA antigen that reacts with antisera to cardiolipin, *Biochem. Biophys. Res. Commun.,* 58, 347, 1974.
501. **van Tol, M. J. D., Hofstra, H., and Dankert, J.,** Major outer membrane proteins of *Escherichia coli* analyzed by crossed immunoelectrophoresis, *FEMS Microbiol. Lett.,* 5, 349, 1979.
502. **Loft, H.,** Identification of human serum proteins by polyacrylamide gel electrophoresis combined with quantitative immunoelectrophoresis methods, *Scand J. Immunol.,* 4 (Suppl. 2), 115, 1975.
503. **Webb, K. S., Mickey, D. D., Stone, K. R., and Paulson, D. F.,** Correlation of apparent molecular weight and antigenicity of viral proteins: an SDS-PAGE separation followed by acrylamide-agarose electrophoresis and immunoprecipitation, *J. Immunol. Methods,* 14, 343, 1977.
504. **Chua, N.-H. and Blomberg, F.,** Immunochemical studies of thylakoid membrane polypeptides from spinach and *Chlamydomonas reinhardtii:* a modified procedure for crossed immunoelectrophoresis of dodecyl sulphate • protein complexes, *J. Biol. Chem.,* 254, 215, 1979.
505. **Olden, K. and Yamada, K. M.,** Direct detection of antigens in sodium dodecyl sulphate-polyacrylamide gels, *Anal. Biochem.,* 78, 483, 1977.
506. **Poolman, J. T., Hopman, C. Th. P., and Zanen, H. C.,** Immunochemical characterization of outer membrane complexes from *Neisseria meningitidis* by the SDS-polyacrylamide-gel-electrophoresis-immunoperoxidase technique (SGIP), *FEMS Microbiol. Lett.,* 4, 245, 1978.
507. **Adair, W. S., Jurivich, D., and Goodenough, U. W.,** Localization of cellular antigens in sodium dodecyl sulphate-polyacrylamide gels, *J. Cell. Biol.,* 79, 281, 1978.
508. **Broome, S., and Gilbert, W.,** Immunochemical screening method to detect specific translation products, *Proc. Natl. Acad. Sci. U.S.A.,* 75, 2746, 1978.
509. **Renart, J., Reiser, J., and Stark, G. R.,** Transfer of proteins from gels to diazobenzyloxymethyl-paper and detection with antisera: a method for studying antibody specificity and antigen structure, *Proc. Natl. Acad. Sci. U.S.A.,* 76, 3116, 1979.
510. **Norén, O. and Sjöström, H.,** Fluorography of tritum-labelled proteins in immunoelectrophoresis, *J. Biochem. Biophys. Methods,* 1, 59, 1979.
511. **Spratt, B. G.,** Properties of penicillin-binding proteins of *Escherichia coli, Eur. J. Biochem.,* 72, 341, 1977.
512. **Proia, R. L., Hart, D. A., Holmes, R. K., Holmes, K. V., and Eidels, L.,** Immunoprecipitation and partial characterization of diphtheria toxin-binding glycoproteins from surfaces of guinea pig cells, *Proc. Natl. Acad. Sci. U.S.A.,* 76, 685, 1979.
513. **Caron, M. G., Srinivasan, Y., Snyderman, R., and Lefkowitz, R. J.,** Antibodies raised against purified β-adrenergic receptors specifically bind β-adrenergic ligands, *Proc. Natl. Acad. Sci. U.S.A.,* 76, 2263, 1979.
514. **Amar, A., Rottem, S., and Razin, S.,** Disposition of membrane proteins as affected by changes in the electrochemical gradient across mycoplasma membranes, *Biochem. Biophys. Res. Commun.,* 84, 306, 1978.

515. **Klingenberg, M., Riccio, P., and Aquila, H.,** Isolation of the ADP, ATP carrier as the carboxyatractylate•protein complex from mitochondria, *Biochim. Biophys. Acta,* 503, 193, 1978.
516. **Buchanan, B. B., Eiermann, W., Riccio, P., Aquila, H., and Klingenberg, M.,** Antibody evidence for different conformational states of ADP, ATP translocator protein isolated from mitochondria, *Proc. Natl. Acad. Sci. U.S.A.,* 73, 2280, 1976.
517. **McCans, J. L., Lane, L. K., Lindenmayer, G. E., Butler, V. P., Jr., and Schwartz, A.,** Effects of an antibody to highly purified Na^+, K^+-ATPase from canine renal medula: Separation of the "holoenzyme antibody" into catalytic and cardiac glycoside receptor-specific components, *Proc. Natl. Acad. Sci. U.S.A.,* 71, 2449, 1974.
518. **Koepsell, H.,** Conformational changes of membrane-bound (Na^+-K^+)-ATPase as revealed by antibody inhibition, *J. Membr. Biol.* 45, 1, 1979.
519. **Arnon, R.,** Immunochemistry of lysozyme, in *Immunochemistry of Enzymes and their Antibodies,* **Salton, M. R. J., Ed.,** John Wiley & Sons, 1977, 1.
520. **Benjamin, D. C. and Teale, J. M.,** The antigenic structure of bovine serum albumin: evidence for multiple, different, domain-specific antigenic determinants, *J. Biol. Chem.,* 253, 8087, 1978.
521. **Celada, F., Fowler, A. V., and Zabin, I.,** Probes of β-galactosidase structure with antibodies. Reaction of anti-peptide antibodies against native enzyme, *Biochemistry,* 17, 5156, 1978.
522. **Henderson, R.,** The purple membrane from *Halobacterium halobium, Annu. Rev. Biophys. Bioeng.,* 6, 87, 1977.
523. **Dowham, W., Wickner, W. T., and Kennedy, E. P.,** Purification and properties of phosphatidylserine decarboxylase from *Escherichia coli, J. Biol. Chem.,* 249, 3079, 1974.
524. **Hirabayashi, T., Larson, T. J., and Dowham, W.,** Membrane-associated phosphatidylglycerophosphate synthetase from *Escherichia coli:* purification by substrate affinity chromatography on cytidine 5'-diphospho-1, 2-diacyl-*sn*-glycerol sepharose, *Biochemistry, 15,* 5205, 1976.
525. **Fillingame, R. H.,** Purification of the carbodiimide-reactive protein component of the ATP energy-transducing system of *Escherichia coli, J. Biol. Chem.,* 251, 6630, 1976.
526. **Wilson, D. B.,** Cellular transport mechanisms, *Annu. Rev. Biochem.,* 47, 933, 1978.
527. **Kusaka, I. and Kanai, K.,** Purification and characterization of alanine carrier isolated from H-proteins of *Bacillus subtilis, Eur. J. Biochem.,* 83, 307, 1978.
528. **Futai, M. and Kimura H.,** Inducible membrane-bound L-lactate dehydrogenase from *Escherichia coli, J. Biol. Chem.,* 252, 5820, 1977.
529. **Sandermann, H., Jr., Bavoil, P., and Nikaido, H.,** Phage lambda receptor protein from *Escherichia coli:* solubilization and purification in an aprotic solvent, *FEBS Lett.,* 95, 107, 1978.
530. **Erlich, H. A., Cohent, S. N., and McDevitt, H. O.,** A sensitive radioimmunoassay for detecting products translated from cloned DNA fragments, *Cell,* 13, 681, 1978.
531. **Henning, U., Schwarz, H., and Chen, R.,** Radioimmunological screening method for specific membrane proteins, *Anal. Biochem.,* 97, 153, 1979.
531a. **Van de Rijn, I. and Kessler, R. E.,** Chemical analysis of changes in membrane composition during growth of *Streptococcus pyogenes, Infect. Immun.,* 26, 883, 1979.
531b. **Kessler, R. E. and van de Rijn, I.,** Quantitative immunoelectrophoretic analysis of *Streptococcus pyogenes* membrane, *Infect. Immun.,* 26, 892, 1979.
532. **Barka, T. and Anderson, P. J.,** *Histochemistry: Theory, Practice and Bibliography,* Harper & Row, New York, 1963, chap. 13.
533. **Baptist, J. N., Shaw, C. R., and Mandel, M.,** Zone electrophoresis of enzymes in bacterial taxonomy, *J. Bacteriol.,* 99, 180, 1969.
534. **Weinbaum, G., and Markman, R.,** A rapid technique for distinguishing enzymatically active proteins in the cell "envelope" of *Escherichia coli* B, *Biochim. Biophys. Acta,* 124, 207, 1966.
535. **Linke, R., Zwilling, R., Herbold, D., and Phleiderer, G.,** Zur Evolution der Endopeptidasen. IX. Immunologische Untersuchungen des Trypsins und der niedermolekuleren Protease aus drei decapod Crustaceen, *Hoppe-Seyler's Z. Physiol. Chem.,* 350, 877, 1969.
536. **Herbold, D., Zwilling, R., and Pfleiderer, G.,** Zur Evolution der Endopeptidasen. XIII. Biochemische und immunologische Untersuchungen über Trypsin und niedermolekulare Protease aus der Strandkrabbe *Carcinus maenus* L., *Hoppe-Seyler's Z. Physiol. Chem.,* 352, 583, 1971.
537. **Johansson, K.-E., Pertoft, H., and Hjertén, S.,** Characterization of the Tween 20-soluble membrane proteins of *Acholeplasma laidlawii, Int. J. Biol. Macromol.,* 1, 111, 1979.
538. **Johansson, K.-E.,** personal communication, 1979.
539. **Capaldi, R. A. and Vanderkooi, G.,** The low polarity of many membrane proteins, *Proc. Natl. Acad. Sci. U.S.A.,* 69, 930, 1972.
540. **Fornstedt, N. and Porath, J.** Characterization studies on a new lectin found in seeds of *Vicia Ervilia, FEBS Lett.,* 57, 187, 1975.

Chapter 4

THE MYCOPLASMA MEMBRANE

Shmuel Razin

TABLE OF CONTENTS

I. INTRODUCTION

The single most important characteristic which distinguishes mycoplasmas from all other prokaryotes is their complete lack of a cell wall. In a recent classification of the kingdom *Procaryotae,* the mycoplasmas are given the status of a major division named *Mollicutes* (mollis, soft; cutes, skin) based on the nature of the cell envelope.[112] The three other divisions are: *Gracilicutes,* the Gram-negative bacteria, *Firmacutes,* the Gram-positive bacteria, and *Mendocutes,* bacteria with incomplete cell walls. Since the late 1950s, evidence has been accumulating for the lack of a cell wall in mycoplasmas. That the unique components of peptidoglycan, muramic and diaminopimelic acids, could not be detected in mycoplasmas,[168,215,249] explains the resistance of these organisms to lysis by lysozyme and to growth inhibition by penicillin.[249,277] The susceptibility of mycoplasmas to lysis by osmotic shock and by various agents causing the lysis of bacterial protoplasts[276] also supported the notion that mycoplasmas lack cell walls. However, the first direct evidence for the absence of cell walls was provided by electron micrographs of thin-sectioned mycoplasmas, showing the cells to be bounded by a single membrane resembling the bacterial plasma membrane in its trilaminar shape in osmium-fixed sections.[426] Subsequently, demonstration of a single membrane in cell sections has become essential to define a new isolate as a mycoplasma.[403]

The total absence of a cell wall is thus a unique property of mycoplasmas. Yet, wall-covered bacteria may be transformed into protoplasts by various means, and these in turn may be cultivated to produce the so-called bacterial L-forms or L-phase. Although most of the L-forms retain the ability to synthesize a cell wall, some have permanently lost it and have become stable L-forms. They resemble mycoplasmas in many ways, which include a similar ultrastructure in thin sections, a "fried-egg" colony shape, and total resistance to lysozyme and penicillin.[240] This resemblance caused some confusion in the past as to the taxonomic status of the mycoplasmas. However, the much smaller size of the mycoplasma genome, the failure to hybridize it with genomes of wall-covered bacteria and bacterial L-forms[398] and the mycoplasmas' singular requirement for cholesterol[293] appear to exclude the possibility that the mycoplasmas represent stable L-forms of existing wall-covered bacteria. Furthermore, the mycoplasmas are incapable of synthesizing the soluble nucleotide precursors of the peptidoglycan,[25,430] a property retained by stable L-forms;[300] neither can any mycoplasma synthesize the characteristic lipopolysaccharides of gram-negative bacteria.[386]

The number of newly recognized species of Mollicutes is continuously increasing due to improvement in cultivation and identification techniques. It is now clear that the mycoplasmas comprise an extremely wide and prevalent group of microorganisms. As can be seen in Table 1, there are currently about 70 established species classified in several genera and families, but there is little doubt that we are still far from having cultivated and identified all the wall-less procaryotes. Mycoplasmas have been cultivated from only a few diseased plants and insects, though microscopical evidence indicates their presence in many more.[346] Moreover, electron microscopy suggests the presence of wall-less prokaryotes resembling mycoplasmas in organisms as diverse as amphibians, mollusks, trematode worms, and fungi.[273]

Thin sections of mycoplasmas reveal an extremely simple ultrastructure consisting of a plasma membrane, ribosomes, and a prokaryotic nucleoid. No mesosomes or other intracytoplasmic membranes can be seen. Vacuole-like structures bounded by a membrane are shown by serial sections to represent deep invaginations of the plasma membrane.[41,266] Hence the mycoplasmas are unique in being the only self-replicating organisms with a single membrane. This can be regarded as a most useful property for membrane studies, for it facilitates the isolation of pure plasma membranes from mycoplasmas, uncontaminated by other membrane types.[271]

As would be expected in a wall-less organism, the coccus is the basic form in mycoplasma cultures (Figure 1). However, under adequate growth conditions many mycoplasma species produce long filaments at the early logarithmic phase of growth. Once a filament reaches a certain length, it transforms into a chain of cocci which subsequently fragments into single coccoid cells.[46,103,280] It is now widely accepted that the mode of reproduction of mycoplasmas is not essentially different from that of other prokaryotes dividing by binary fission.[268] For typical binary fission to occur, cytoplasmic division must be fully synchronized with genome replication, resulting in the formation of multinucleate filaments (Figure 2). In some mycoplasmas isolated from plants and insects, the filaments assume a helical shape and are capable of flexional and rotatory motion. These mycoplasmas have been accordingly named spiroplasma. The factors responsible for the helicity and contractility of these filaments are still unclear (Section IV.A.5.).

In addition to structural simplicity, the mycoplasmas are relatively simple biochemically. These minute parasites lack many of the biosynthetic pathways found in

Table 1
TAXONOMY AND PROPERTIES OF ORGANISMS INCLUDED IN THE CLASS MOLLICUTES

Classification	Current number of established species	Genome size ($\times 10^8$ daltons)[a]	Cholesterol requirement	NADH oxidase localization[a]	Characteristic properties	Habitat
Mycoplasmataceae						
Mycoplasma	About 60	5	+	Cytoplasm	—	Animals
Ureaplasma	1	5	+	N.D.	Urease activity	Animals
Acholeplasmataceae						
Acholeplasma	8	10	–	Membrane	Synthesize carotenoids	Animals
Spiroplasmataceae						
Spiroplasma	1	10	+	Cytoplasm	Helical filaments	Insects and plants
Genera of uncertain taxonomic position						
Anaeroplasma	2	N.D.	Some + some –	N.D.	Anaerobic; some digest bacteria	Rumens of cattle and sheep
Thermoplasma	1	10	–	Membrane	Thermophilic (optimum 59°C) and acidophilic (optimum pH 1.0—2.0)	Burning coal-refuse piles

[a] N.D., not determined.

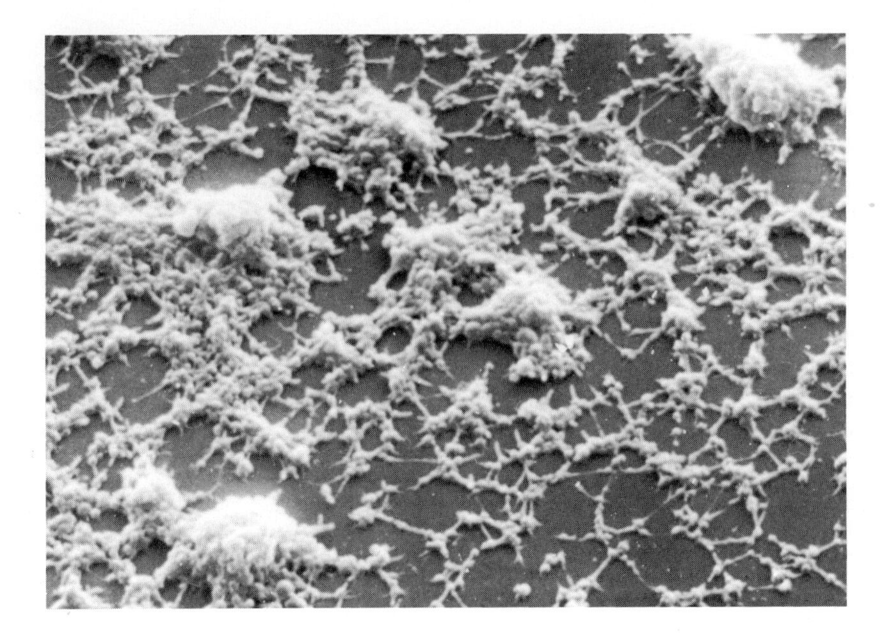

FIGURE 1. Scanning electron micrograph of a *Mycoplasma pneumoniae* culture grown on glass, showing the coccoid and filamentous shapes of the organisms. (Magnification × 40,000.) (From Razin, S. et al., *Infect. Immun.,* 30, 538, 1980. With permission.)

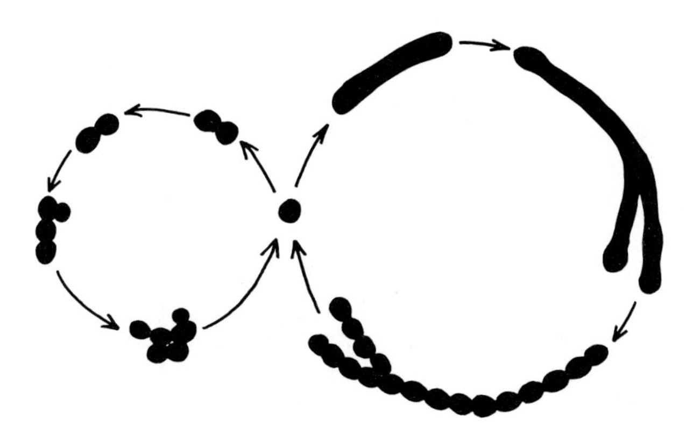

FIGURE 2. Schematic presentation of the mode of mycoplasma replication. Cells may either divide by regular binary fission or elongate first to multinucleate filaments which subsequently break up into coccoid bodies.

the wall-covered bacteria. Therefore they require complex growth media to provide the numerous essential nutrients, including membrane lipid components or their precursors, which cannot be synthesized by mycoplasmas.[268,271] This dependence on an external supply of lipids provides a powerful tool for the introduction of controlled changes in their membrane lipids, enabling the analysis of effects of specific lipid components on membrane structure and function.[271]

For a complete and systematic treatment of all aspects of mycoplasmology, the reader is referred to a three-volume treatise on mycoplasmas published recently.[23] The development of mycoplasma membrane research can be followed in a series of previous reviews on this subject.[260,264,265,266,268,271,273] Specific problems concerning the organization and role of lipids in mycoplasma membranes are discussed in a recent review by Rottem.[319] The purpose of the present contribution is to provide a systematic and up-to-date coverage of the various aspects of mycoplasma membrane research, emphasizing similarities and differences between the mycoplasma membrane and membranous structures in other prokaryotes.

II. GENERAL CHARACTERISTICS OF MYCOPLASMA MEMBRANES

A. Cell Lysis and Membrane Isolation

The lack of a cell wall in mycoplasmas facilitates the isolation of their cytoplasmic membranes by eliminating the need for the tedious procedures usually involved in the separation of the bacterial cell wall from the cytoplasmic membrane. Osmotic lysis is the preferred technique for isolating mycoplasma membranes, as it is gentle enough to keep the membrane from disintegrating into small pieces, so that the cytoplasmic fluid can be separated without contamination by minute membrane fragments. In addition, no foreign substance is introduced into the preparation during this procedure.[255,256,262,263] Osmotic lysis is not always effective, however, because the sensitivity of the mycoplasmas to osmotic shock decreases most markedly with the aging of the culture[262,263] and the presence of even traces of divalent cations may provide complete protection from lysis.[263,303] Thus, osmotic lysis is most effective when the organisms are harvested at the logarithmic phase of growth, with the divalent cation concentration in the lysis medium kept to a minimum. Good control of the growth rates is absolutely essential and can be achieved only when the strain is well adapted to the growth medium. This is relatively easy in the case of the fast-growing mycoplasmas, such as *Acholeplasma laidlawii* or *Mycoplasma mycoides* subsp. *capri,* but may prove extremely difficult with the more exacting, slow-growing mycoplasmas, such as *Mycoplasma pneumoniae.*[257]

It is not clear why the susceptibility of mycoplasmas to osmotic lysis decreases upon aging of cultures. Considerable changes in membrane lipid composition, in the lipid-to-protein ratio, and consequently in membrane fluidity, are known to accompany aging in mycoplasma cultures.[8,9,12,269,323,328] The dependence of membrane elasticity on the fluidity of membrane lipids has been suggested by the studies of de Kruyff et al.[78] and McElhaney et al.[205] Transformation of the membrane lipid bilayer into the viscous gel state reduces membrane elasticity, so that the cells lyse rather than swell when placed in hypotonic solutions. This idea fits in very well with the extreme sensitivity to osmotic lysis of *A. laidlawii* cells enriched with palmitate or stearate,[295] since at room temperature, or even at 37°C, their membrane lipids are in the gel state.[203,400] However, upon aging of mycoplasma cultures the reverse actually happens: the membranes become more viscous while the cells show higher resistance to osmotic lysis. The much higher protein content of these membranes[9,269] may perhaps contribute to their marked resistance to brittle fracture.

Interestingly, despite their structural similarity to bacterial protoplasts and bacterial L-forms, mycoplasmas are generally more resistant to osmotic lysis (Figure 3). One possible explanation for the greater resistance of mycoplasmas can be based on the higher tensile strength of their membranes. Thus, *A. laidlawii* membranes resisted fragmentation by sonication far better than did *Micrococcus lysodeikticus* protoplast membranes.[264] The possibility that the cholesterol component of mycoplasma membranes contributes to their tensile strength is supported by the higher susceptibility to lysis of the cholesterol-poor *M. mycoides* subsp. *capri* membranes.[264,243] Another explanation is that due to the high surface-to-volume ratio of the minute mycoplasma

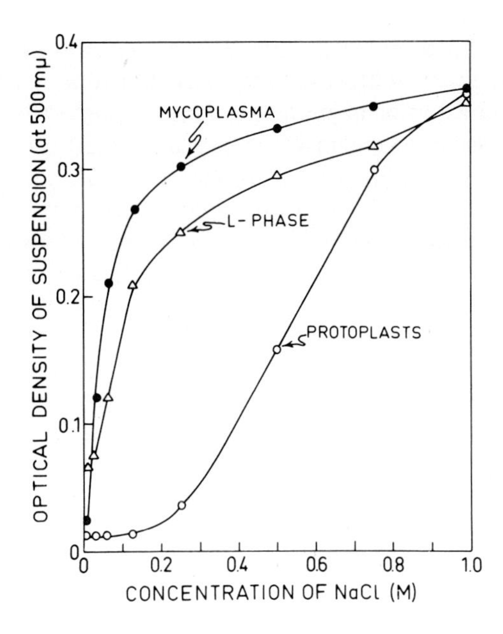

FIGURE 3. Osmotic fragility of *A. laidlawii*, L-phase of *Streptobacillus moniliformis*, and protoplasts of *Micrococcus lysodeikticus*. (From Razin, S. in *The Mycoplasmatales and the L-Phase of Bacteria*, Hayflick, L., Ed., Appleton-Century-Crofts, New York, 1969, 317. With permission.

cells, rapid liberation of the internal solutes may occur upon transfer to a hypotonic solution, quickly lowering the internal osmotic pressure so as to cushion the osmotic shock. The marked leakiness of mycoplasmas in nonnutrient solutions is well known.[51,265,293] Moreover, filamentous growth, common in mycoplasmas, is likely to augment these effects, since the surface-to-volume ratio is much higher in a filament than in a spherical cell, and filaments are able to absorb large amounts of water and turn into large spherical bodies without any stretching of the membrane.[280,303]

Mycoplasmas, in particular the sterol-requiring species, contain considerable quantities of cholesterol in their cell membrane (Section III.A.6.a.) and are therefore sensitive to lysis by digitonin.[277,294,392] Digitonin forms a complex with cholesterol in the membrane, withdrawing the cholesterol molecules from their interaction with membrane phospholipids. As a result of the extensive rearrangements in the lipid bilayer, membrane permeability increases, leading to cell lysis.[337] The advantage of digitonin-induced lysis over osmotic lysis is that it is less dependent on the age of culture and can take place in the presence of divalent cations. Elimination of the need to reduce the ionic strength and divalent cation concentration of the lysis medium may prevent the loosely bound membrane proteins from detaching during membrane isolation (Section IV.A.2.) On the other hand, membranes isolated by digitonin-induced lysis are different from native membranes in that they contain the cholesterol-digitonide complexes.[337] Hence, the use of digitonin is recommended only when osmotic shock fails to lyse the mycoplasmas, as is the case with *Mycoplasma gallisepticum*[185] and *Ureaplasma urealyticum*.[195,196,311]

A simple means for lysing the wall-less mycoplasmas is by the use of alkali.[277] The osmotically resistant *M. gallisepticum* and *Thermoplasma acidophilum* can be lysed by raising the pH to about 9.5[115,345] However, *T. acidophilum* membranes isolated in this way may be deficient in protein[177,345] as they depend on protons for stability.[176,177] The procedure recommended by Smith et al.[390] consists of disruption of the thermoplasmas by sonic oscillations in a buffer of moderate ionic strength (0.05) and at pH 5.0.

Sonic or ultrasonic oscillators and mechanical presses rupture mycoplasma cells most effectively. However, the membranes usually disintegrate during these treatments into minute fragments, some of which cannot be collected even at very high gravitational forces.[163,256] Nevertheless, in cases in which lysis by osmotic shock or digitonin fails to give satisfactory results, one can resort to these mechanical means.[311]

B. Gross Chemical Composition

Isolated mycoplasma membranes resemble plasma membranes of other prokaryotes in gross chemical composition, being composed mainly of proteins and lipids. The protein comprises roughly two-thirds of the mass of the membrane, the balance being largely lipid.[278] Tabulated information on composition of membranes isolated from different mycoplasmas can be found in previous reviews.[271,274] Criteria for checking the purity of membrane preparations, including electron microscopy, nucleic acid content, density-gradient analysis, and enzymatic activities as markers have been described and evaluated in detail by Razin and Rottem.[274,292]

Contamination of mycoplasma membrane preparations with precipitated components of the growth medium may seriously hamper their chemical, enzymatic and antigenic characterization. This problem is usually encountered with the fast-growing acid-producing mycoplasmas (such as *M. mycoides* subsp. *capri*) when allowed to grow long enough to decrease the pH of the culture medium to less than pH 6.0, or with mycoplasmas (such as ureaplasmas) which grow very poorly in the serum-containing culture media. In the first instance, the low pH causes the precipitation of proteins and lipoproteins from the serum component of the growth medium,[43,55,448] which cosediment with the cells during harvesting and washing, and may consequently cosediment with[448] or absorb to the isolated membranes.[325] One- or two-dimensional polyacrylamide gel electrophoresis of *Spiroplasma citri* proteins labeled during growth with radioactive amino acids revealed several major unlabeled protein bands or spots, representing contaminating serum proteins.[216]

In the case of the ureaplasmas, cell yield is so low that most of the pellet obtained by centrifugation of the culture consists of noncellular components which subsequently contaminate the isolated membrane preparations. Thus, Masover et al.[195] reported that the protein content of the "membrane" fraction isolated by digitonin treatment of *U. urealyticum* grown in unfiltered Hayflick's medium was about 25 times higher than the amount of protein in the soluble cytoplasmic fraction. Reduction of the serum content and prefiltration of the growth medium[194] reduced the "membrane"-to-cytoplasmic protein ratio to about 10, a value still much higher than the 0.5 ratio reported for similar fractions from classical mycoplasmas.[278] Centrifugation on sucrose-density gradients has been proposed as a means of separating the ureaplasma cells from nonspecific precipitates,[333] but the great losses of precious cell material occurring in this procedure considerably limit its usefulness.

C. Slime Layers

Electron microscopy of thin sections prepared from a large number of *Mycoplasma, Ureaplasma, Acholeplasma* and *Spiroplasma* species reveals the presence of "fuzzy" layers on the cell surface.[38,61,66,133,142,214] The application of the Ruthenium-red staining techniques intensifies the image of mycoplasmal slime layers,[2,5,121,134,439] but in most of the cases the chemical nature of the capsular material has not been determined. Ruthenium red reacts with a variety of polyanions and has been used to demonstrate the polysaccharide glycocalyxes of eucaryotic cells and slime layers of prokaryotes. Hence it is probable that in mycoplasmas also the extracellular material consists of or contains carbohydrate. In fact, the *U. urealyticum* slime layer reacted with concanavalin A-iron dextran, indicating the presence of glycosyl residues.[301] The possibility that the ruthenium red-positive material consists of growth medium components adsorbed to the cell surface[301,357] rather than slime layers cannot be ruled out. However, the recent demonstration of a ruthenium red-positive slim layer in *M. dispar* in sections of lung tissue from infected animals[5] supports the mycoplasmal origin of this layer, at least in this instance.

Direct evidence for the carbohydrate nature of mycoplasma slime layers is available for only two mycoplasmas: *M. mycoides* subsp. *mycoides,* which synthesizes a slime layer made of galactan,[119] and *A. laidlawii,* in which the slime layer is made of a hexosamine polymer consisting of *N*-acetylgalactosamine and *N*-acetylglucosamine.[114] The galactan slime layer of *M. mycoides* stains heavily with ruthenium red[134] and plays a role in the pathogenicity of this mycoplasma.[52] The hexosamine polymer produced by *A. laidlawii* has not been detected in any of the other *Acholeplasma* species, but even in *A. laidlawii* its quantity may vary significantly, depending on the strain and growth conditions. The hexosamine polymer is tightly bound to the membrane, as the bulk of it remains membrane-bound even after prolonged washing.[453]

III. MEMBRANE LIPIDS

A. Lipid Composition

An intriguing question is whether or not the lack of a cell wall in mycoplasmas is reflected in their membrane lipid composition. The permanent loss of the cell wall in streptococci transformed into stable L-forms is associated with an increase in the membrane lipid-to-protein ratio, alterations in fatty acid composition, and an increase in the membrane glycolipid content.[240] It appears that membrane lipid composition in mycoplasmas also differs from that of wall-covered bacteria. Thus, cholesterol or related sterols are important constituents of mycoplasma membranes,[293] and some mycoplasmas produce lipopolysaccharides very different in composition from those of the Gram-negative bacteria.[386] The thermophilic and acidophilic *Thermoplasma acidophilum* synthesizes peculiar phospho- and glycolipids which are well adapted to resist the harsh ecological niche of this organism.[176,177] Nevertheless, the major lipid constituents synthesized *de novo* by mycoplasmas are typical prokaryotic phospho- and glycolipids.

1. Phospholipids

The acidic phospholipids, phosphatidylglycerol, and diphosphatidylglycerol, are ubiquitous in mycoplasmas as in other prokaryotes.[6,102,268,245,281,329,338,383,386] Phosphatidic acid and lyso compounds may also be present, as either biosynthetic precursors or breakdown products.[323,386] Fully acylated derivatives of the acidic glycerophospholipids have been detected in some mycoplasmas.[245,250,319] Phosphatidylethanolamine, a very common phospholipid in bacteria, has been reported only once in mycoplasmas.[313] On the other hand ceramides, rare in bacteria, were found in *A. axanthum.* This organism synthesizes an hydroxyceramide phosphoryl glycerol containing an O-acyl group.[251] The predominant fatty acid in N-acyl linkage to the amino group of the long-chain base has been identified as D(−)-3-hydroxyhexadecanoate.[198] The phospholipids of the anaerobic *Anaeroplasma* species contain plasmalogens, as in other obligate anaerobic bacteria[178].

Although phosphatidylcholine and sphingomyelin are not synthesized by any of the

Table 2
CHOLESTEROL AND PHOSPHOLIPID UPTAKE FROM SERUM BY GROWING *ACHOLEPLASMA* AND *MYCOPLASMA* SPECIES

	Lipids in cells (μmol/g cell protein)			
Organism[a]	Free cholesterol	Esterfied cholesterol	Sphingomyelin	Phosphatidylcholine
Acholeplasma				
laidlawii	10.2	0	0	0
axanthum	3.7	0	0	0
granularum	28.5	0	0	0
Mycoplasma				
gallisepticum	76.0	4.7	15.5	37.8
hominis	76.4	30.6	32.7	20.3
arginini	58.2	27.5	37.8	31.9
pneumoniae	85.9	58.0	40.3	16.6
capricolum	67.2	67.7	20.8	40.1

[a] The organisms were grown with 5% (v/v) horse serum for 20—24 hr at 37°C. *M. pneumoniae* was grown with 10% (v/v) horse serum for 4 days at 37°C. Data of Razin et al.[285]

mycoplasmas tested,[386] significant quantities of these animal phospholipids can be found in membranes of mycoplasmas grown in serum-containing media[26,102,250,281,285,329] (Table 2). The ability to take up phospholipids from serum during growth appears to be a property common to *Mycoplasma* species[285] and *Spiroplasma citri*[102] but is lacking in *Acholeplasma* species.[285] *Mycoplasma* species took up phosphatidylcholine during growth in a medium supplemented with vesicles comprised of this phospholipid and cholesterol. Again, no phosphatidylcholine was taken up by *A. laidlawii* under these conditions.[165,285] *M. gallisepticum* is distinguished by its ability to modify the exogenous phosphatidylcholine taken up by converting it to a disaturated type. Apparently this mycoplasma inserts a saturated fatty acid at position 2 of lysophosphatidylcholine derived from the exogenous phospholipid by a deacylation-reacylation enzymatic sequence. However, attempts to demonstrate phospholipase activity in *M. gallisepticum* failed, and the modification of phosphatidylcholine also occurred in a serum-free medium.[329] Several findings indicate that the exogenous phospholipids taken up become integrated into the membrane and are not just adsorbed to its surface. Thus, the exogenous phospholipids resemble the endogenous phospholipids in extractibility by ether and susceptibility to phospholipase A_2 digestion. Incorporation of exogenous phospholipids in *M. gallisepticum* also inhibited the endogenous phosphatidylglycerol synthesis.[454]

A somewhat unexpected observation is that isolated *M. capricolum* membranes are incapable of taking up exogenous phospholipids.[285] A possible explanation can be based on the assumption that receptors for serum lipoproteins or lipid vesicles are lost during membrane isolation. The available evidence neither supports nor discounts the presence of specific receptors for lipoproteins and lipid vesicles on mycoplasma membranes. The finding that phospholipid uptake by *M. capricolum* cells was affected by inhibiting cell growth[285] suggests that phospholipid uptake is enhanced when the cell membrane is in a dynamic growing state. The exposure of proteins on the mycoplasma cell surface was recently shown to be affected by changes in the electrochemical-ion gradient across the cell membrane. Ionophores and growth inhibitors markedly decreased the exposure of proteins on the mycoplasma cell surface.[11] Hence, if proteins on the mycoplasma cell surface participate in the binding of lipoproteins and lipid vesicles, as was found for various eukaryotic cells,[24,238] then the higher degree of exposure of these proteins in metabolically active cells may facilitate contact with the lipid donors and enhance phospholipid transfer.

2. Glycolipids

Glycolipids constitute a significant portion of membrane lipids in *Acholeplasma* species and in *T. acidophilum,* and are found in much lesser amounts in some *Mycoplasma, Ureaplasma,* and *Spiroplasma* species.[386] The typical mycoplasmal glycolipids are glycosyl diglycerides containing one to five sugar residues; the different linkages of the sugars provide for the diversity of these lipids.[386] Small amounts of steryl glycosides and acylated glucose were found in some mycoplasmas[386] and in *S. citri,*[245] and carotenyl glycosides were detected in *Acholeplasma* species[386] (Figure 4).

The glycolipids in *A. laidlawii* and *A. oculi* are primarily of the monoglucosyl diglyceride (MGDG) and diglucosyl diglyceride (DGDG) types.[6,73,241,366,369,435] A diglyceride monoglucosyl diglyceride, has also been identified in *A. laidlawii.*[435] *A. modicum* synthesizes a pentaglycosyl diglyceride containing galactose, glucose, and mannoheptose,[196] and in *M. penumoniae,* glycolipids containing both glucose and galactose were described.[170,250,288]

The role of glycolipids in mycoplasma membranes is not clear. The recognition of the role of polyisoprenoid lipid intermediates in the synthesis of carbohydrate-containing polymers on the bacterial surface rules out such a function for glycosyl diglycerides.[244] Polyprenol phosphate has recently been detected in *Acholeplasma.*[393a] The apparent absence of glycolipid turnover[206,382] speaks in favor of a structural role for the glycolipids. Pask-Hughes et al.[244] suggest that glycolipids may act as membrane reinforcers in wall-less and thermophilic bacteria. They base their proposal on the finding of glycolipids as major membrane components in many mycoplasmas, L-forms and thermophiles, and on the demonstration of a rigidifying effect of diacylglycotetraosylglycerol in the membrane of the thermophile *Thermus aquaticus.* Accordingly, the large quantities of glycolipids in *Acholeplasma, T. acidophilum,* and bacterial L-forms fulfil a role similar to that of cholesterol in membranes of the sterol-requiring mycoplasmas (Section III.C.3.). In fact, glycolipids are completely absent, or constitute only minor components, in the cholesterol-rich membranes of the *Mycoplasma, Ureaplasma,* and *Spiroplasma* species tested so far.[386] However, it should be mentioned that carotenoids have also been proposed as membrane reinforcers in *Acholeplasma*[140,331] (Section III.C.4.), and the high rigidity of the *T. acidophilum* membrane can be explained by the peculiar structural properties of its lipids[175-177] (Section III. A. 5.). The notion that glycolipids may influence the physical properties of the membrane is also raised by Wieslander et al.[436] They claim that monoglucosyl diglyceride (MGDG) and diglucosyl diglyceride (DGDG) share several physical properties with phosphatidylethanolamine, a phospholipid absent in mycoplasmas. These properties include a low hydration capacity and a transition temperature higher than that of the acidic phospholipids (see also Reference 369). However, MGDG has an ability to transform into a nonlamellar (possibly hexagonal) phase in the membrane. This may be associated with the finding that a certain excess of monoglucosyl diglyceride (MGDG) over DGDG makes the cell membrane of an *A. laidlawii* mutant resistant to lysis by complement.[73]

The role of membrane glycolipids as major antigenic determinants has been well established for *M. pneumoniae.*[26,170,250,288] The purified glycolipids are haptens, which become immunogenic on association with membrane proteins in the native or reconstituted membranes.[289] The antisera to glycolipids can distinguish structural differences among closely related glycolipids. Thus, DGDG of *M. neurolyticum* will not

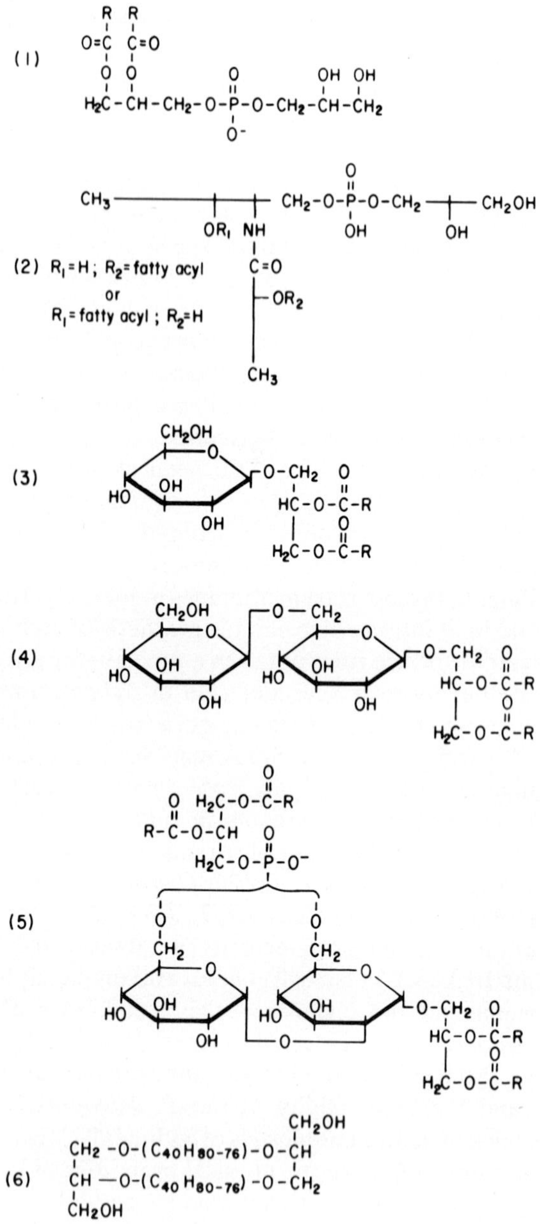

FIGURE 4. Mycoplasmal lipids: (1) phosphatidyl, (2) *O*-acyl ceramide, phosphoryl glycerol, (3) monoglucosyl diglyceride, (4) diglucosyl diglyceride, (5) glycerophosphoryl diglucosyl diglyceride, (6) dialkyl diglycerol tetraether.

react with antibodies to *A. laidlawii* DGDG, as the linkages of the glucose units in *M. neurolyticum* are β (1→6) and β (1→1), whereas in *A. laidlawii* the linkages are α (1→2) and α (1 → 1). Hence, the positive serological cross-reactions of the MGDG and DGDG isolated from *A. laidlawii, A. modicum,* and *A. oculi* are indicative of their identical structure.[6,404]

A. laidlawii membranes also contain large quantities of phosphoglucolipids, "hybrids" of diglucosyl diglyceride with glycerophosphate or phosphatidic acid, which were accordingly named glycerophosphoryl diglucosyl diglyceride and phosphatidyl diglucosyl diglyceride[384] (Figure 4). A glycerophosphoryl monoglucosyl diglyceride has been identified recently in *A. laidlawii*[435] and a phosphoglucolipid, possibly glycerophosphoryl diglucosyl diglyceride was detected in *M. mycoides* subsp. *capri.*[356] Neither glycolipids nor phosphoglycolipids could be detected in the closely related *M. capricolum.*[65]

The lipid composition of all the *Acholeplasma* species examined shows a striking resemblance: phospholipids comprise less than one half of the lipid and all the rest consists of glycolipids, with small amounts of neutral lipids (carotenoids, triglycerides).[6] The lipid composition of the *A. laidlawii* membrane has been established in several laboratories with strikingly close results: about 30% phosphatidylglycerol and diphosphatidylglycerol, about 60% glucolipids and phosphoglucolipids, and much less than 10% neutral lipids.[31,123,241,348,383,435] Hence, most membrane lipids in *A. laidlawii* contain carbohydrate residues.

3. *Lipopolysaccharides (Lipoglycans)*

As mentioned above, glycosyl diglycerides containing from one to five sugar residues have so far been detected in the *Acholeplasma* species, *M. neurolyticum,* and *T. acidophilum.* The two *Anaeroplasma* species contain glycolipids, but these have not been identified as yet.[178,386] Related compounds with more extended oligosaccharide chains were found in all of these mycoplasmas. They were named lipopolysaccharides, because the presence of oligosaccharide moieties conferred solubility properties resembling those of the classical lipopolysaccharides of Gram-negative bacteria.[389] Thus, the mycoplasmal lipopolysaccharides are extractable in hot aqueous phenol, but not in the conventional lipid solvents. The extracted lipopolysaccharides form long, ribbon-like micellar structures in negatively stained preparations, resembling the structure of the classical bacterial lipopolysaccharides.[201] It should be pointed out, however, that in chemical composition the mycoplasmal lipopolysaccharides are totally unrelated to those of the Gram-negative bacteria, as they lack heptoses, ketodeoxyoctonate, and phosphoryl ethanolamine.[386] The mycoplasmal lipopolysaccharides can be viewed as analogs of glycosyl diglycerides possessing a long linear polysaccharide chain, glycosidically linked to the diglyceride. They were recently named lipoglycans to distinguish them from the lipopolysaccharides of the Gram-negative bacteria.[365a]

Lipopolysaccharides account for 2 to 3% of the cell dry weight in *T. acidophilum* and for about 2% in the two *Anaeroplasma* species and in *M. neurolyticum.*[386] In acholeplasmas, lipopolysaccharides account for 0.7 to 2.3% of the cell dry weight. The greatest amount was found in *A. oculi* and the least in *A. laidlawii* and *A. modicum.*[6] No lipopolysaccharides were detected in *M. capricolum, M. gallisepticum, M. gallinarum, M. arthritidis,* and *S. citri.*[389] The *T. acidophilum* lipopolysaccharide is composed of 24 mannose residues, all in the α configuration. The tentative structure of the lipopolysaccharide is: [man-(1 $\overset{\alpha}{\rightarrow}$ 2)-man-(1 $\overset{\alpha}{\rightarrow}$ 2)-man-(1 $\overset{\alpha}{\rightarrow}$ 3)]$_8$-glc-diglycerol tetraether.[177,200] The other mycoplasmal lipopolysaccharides contain both neutral and amino sugars, apart from that of *A. oculi* which contains neutral sugars only.[6] The neutral sugars are glucose, galactose, mannose, and fucose, while the amino sugars include glucosamine, galactosamine, fucosamine, and quinovosamine.[386,389]

While purified mycoplasmal glycolipids are unable to elicit an antibody response in rabbits,[289] some of the purified mycoplasmal lipopolysaccharides are immunogenic, probably due to the large molecular size of these polymers in aqueous solutions (reaching 150,000 for the *A. laidlawii* lipopolysaccharide).[385] The lipopolysaccharide of *A. granularum* and *A. oculi* failed to elicit an antibody response, probably due to the low molecular weight (about 20,000) of the former and to the absence of amino sugars in the latter.[6]

The fact that the lipopolysaccharides are exclusively associated with the cell membrane, and the numerous indications for the localization of their hydrophilic carbohydrate chains on the outer cell surface[386] (Section III. B. 2.), support the idea that these membrane components play an important role in the interactions of the parasitic mycoplasmas with their hosts.[386,391] Whether in addition to antigenic stimulation the mycoplasmal lipopolysaccharides participate in adherence of the parasite to its host and also have pharmacological effects resembling those of lipid A of the classical lipopolysaccharides remains to be clarified.

4. Fatty Acids

The fatty acid residues of membrane phospholipids and glycolipids constitute the major portion of the hydrophobic core of the membrane, so that the physical properties of this core are largely determined by the composition of these residues. One of the greatest advantages in using mycoplasmas as models for membrane studies stems from the fact that these organisms are partially or totally incapable of synthesizing long-chain fatty acids and depend on the growth medium for their supply. This has been exploited most effectively to facilitate controlled alterations in the fatty acid composition of mycoplasma membranes and to study the effects of these alterations on the biophysical and biochemical properties of the membrane.[271,295,370,371,400] The most widely used mycoplasma for this purpose is *A. laidlawii.* This organism is dependent on an external supply of unsaturated fatty acids[290,295] but can readily synthesize saturated fatty acids from acetate,[213,258,335] a property shared by all known *Acholeplasma* species.[253] All the sterol-requiring *Mycoplasma*[129,253] and *Spiroplasma*[102,122] species tested are totally incapable of fatty acid synthesis. The finding of a *U. urealyticum* strain capable of synthesizing both saturated and unsaturated fatty acids from acetate[312] is somewhat unexpected, and more *Ureaplasma* strains should be tested before any generalizations can be made.

The synthesis of saturated fatty acids in *A. laidlawii* proceeds through the malonyl-coenzyme A pathway.[332] The inability of this organism to synthesize unsaturated fatty acids is attributed to the lack of β-hydroxydecanoyl thioester dehydrase at the point where the pathway branches for the synthesis of unsaturated fatty acids in bacteria.[332] Nevertheless, *A. laidlawii* is capable of elongating both *cis*- and *trans*-unsaturated fatty acids to the corresponding hexadecenoic and octadecenoic acids,[242,352] and the cyclopropane fatty acid *cis*-9, 10 methylenehexadecanoic acid to *cis*-11, 12-methyleneoctadecanoic acid.[241]

The ability of *A. laidlawii* to synthesize saturated fatty acids hampers attempts to achieve perfect control of the fatty acid composition of its membrane lipids. Although exogenous fatty acids are incorporated into the *A. laidlawii* membrane lipids, forming as much as 50 to 85% of the fatty acid in the membrane, a biosynthetic background of lauric, myristic, and palmitic acids always exists.[129,351] Recent attempts to abolish the biosynthetic background by inhibitors of fatty acid synthesis have succeeded. Of the three antilipogenic substances tested — avidin, *N,N*-dimethyl-4-oxo-2-*trans* dodecenamide (CM-55), and cerulenin — avidin was the best, as it abolished *de novo* fatty acid biosynthesis, and greatly reduced the chain elongation of exogenous fatty acids.[320,370,371] Under these conditions the exogenously supplied fatty acids constitute the only source of fatty acids for lipid biosynthesis. If the growth medium is carefully delipidated, a single added fatty acid may comprise over 98% of the total acyl groups in membrane lipids, resulting in a "fatty-acid homogenous" membrane (Table 3). As the sterol-requiring *Mycoplasma* species cannot synthesize and elongate long-chain fatty acids, there is no need to add an antilipogenic compound in order to obtain fatty-acid

Table 3
GROWTH YIELDS, MEMBRANE LIPID ENRICHMENT IN EXOGENOUS FATTY ACID SPECIES, AND LIPID PHASE TRANSITION TEMPERATURES FOR CULTURES OF *A. LAIDLAWII* B GROWN WITH AVIDIN PLUS VARIOUS EXOGENOUS FATTY ACIDS

Exogenous fatty acid	Growth (% of control)	Fatty acid in membrane (% total lipid acyl chains)	Phase transition temperature (T_c, °C)
14:0*ai*	61	95	−14.5
18:1*c*Δ^{11}	75	96	−8.3
16:1*t*Δ^{9}	87	98	7.6
17:0*ai*	95	96	8.2
14:0*i*	103	96	10.1
18:1*t*Δ^{11}	99	99	20.0
17:0*i*	98	98	28.0
14:0/16:0	91	49/50	32.9
Control[a]	100	—[a]	30.8

[a] Grown in the absence of avidin or fatty acid; the major membrane fatty acyl chains were ~45% palmitate, 30 to 35% myristate, and lesser amounts of lauric, stearic, and oleic acids. Data of Silvius et al.[369]

homogeneous membranes. However, this requires a growth medium with a well-defined lipid composition, difficult to achieve with the nutritionally exacting parasitic mycoplasmas. Strain Y of *M. mycoides* subsp. *mycoides,* which is somewhat less exacting than other sterol-requiring mycoplasmas, proved adequate for this purpose. When grown in a semidefined medium with elaidate as the only fatty acid, this acid comprised over 97% of the fatty acyl residues of membrane lipids.[305,306]

A rather wide variety of fatty acids supported good growth of *A. laidlawii* in the avidin-containing medium, or of strain Y in the semidefined medium, when added in pairs.[304,306,371] The selection of single fatty acids capable of supporting growth is, however, much more restricted. It includes fatty acids with 14 to 19 carbon atoms and no more than one double bond, acids which contribute intermediate fluidity to the membrane diacylglycerolipids.[306,371] The application of the fatty-acid homogeneous membranes to studies on membrane fluidity will be discussed in Section III.C.1.

Not all exogenous fatty acids are incorporated by the mycoplasmas. *A. laidlawii* selectively incorporates the exogenous acids which act to moderate the level of membrane fluidity,[348] while *S. citri* was found to incorporate palmitic acid preferentially and to discriminate against linoleic acid present in the growth medium. Stearic and oleic acids were incorporated into the *S. citri* membrane in about the same relative proportion as existed in the medium.[102,217] A preferential incorporation of palmitic over oleic acid was also noted for *M. hominis.* The much higher activity of coenzyme A:α-glycerophosphate transacylase with palmitoyl-CoA than with oleyl-CoA probably provides an answer for the preferential incorporation of palmitate by this organism.[323]

Rottem and Markowitz[329,330] have recently found that the positional distribution of the fatty acids in phospho- and glycolipids of *Mycoplasma* species is the reverse of that found in membrane lipids in nature. In the mycoplasmal lipids the fatty acids with the lower melting points (unsaturated) are preferentially bound to position 1 of the glycerol, and those with higher melting points (saturated) are bound to position 2. It must be stressed that this unusual positioning of the fatty acids is not shared by *Acholeplasma* species, in which the unsaturated fatty acids show a much higher affinity for position 2, and the saturated acids for position 1, as is usual in nature[207,329,330,349] (Table 4). The biological significance of this interesting finding is not clear as yet.

Table 4
POSITIONAL DISTRIBUTION OF [^{3}H]PALMITATE AND [^{14}C]OLEATE IN PHOSPHATIDYLGLYCEROL PREPARATIONS FROM VARIOUS *MYCOPLASMA* AND *ACHOLEPLASMA*

Organism[a]	Distribution of label (^{3}H/^{14}C ratio)[b] Phosphatidyl-glycerol	Lysophosphati-dylglycerol	Free fatty acid	P_1/P_2 ratio of palmitate in phosphatidyl-glycerol
Mycoplasma fermentans	2.5	0.4	10.2	0.04
M. mycoides subsp. *mycoides*	1.7	0.3	10.5	0.03
M. pneumoniae	1.4	0.2	4.9	0.04
M. capricolum	1.4	0.7	3.7	0.19
M. gallinarum	1.3	0.6	3.9	0.15
M. gallisepticum	1.1	0.2	7.9	0.03
Acholeplasma laidlawii	0.9	2.8	0.6	4.76
A. granularum	0.8	2.1	0.7	3.00

[a] The organisms were grown in a medium containing 5% horse serum supplemented with [9, 10-^{3}H]palmitate and [1-^{14}C]oleate.

[b] The distribution of label was determined after phospholipase A_2 treatment of the phosphatidylglycerol preparations isolated from the lipids of the organisms. The labeling ratios in lysophosphatidylglycerol and in the free fatty acid fraction represent the labeling ratios at positions 1 (P_1) and 2 (P_2) of the phosphatidylglycerol. Data of Rottem and Markowitz.[330]

5. *Glycerol Ether Lipids*

The ability of *T. acidophilum* to grow at high temperatures and at very low pH is reflected in the unique properties of its membrane lipids. The glycolipids and phosphoglycolipids of this organism contain glycerol ether residues rather than the usual fatty acid ester-linked diglycerides. The ether linkages resist hydrolysis under the highly acidic conditions at which ester linkages are unstable. The glycerol ethers contain long, fully saturated isopranoid branched alkyl chains, primarily $C_{40}H_{82}$ and $C_{40}H_{80}$ instead of fatty acids. The long hydrocarbon chains appear relevant to maintenance of appropriate membrane fluidity at high temperatures.[179,200,345] Moreover, the thermoplasma lipids are also unique in shape, consisting of symmetrical diglycerol tetraether molecules[175] (Figure 4). Since the length of the diglycerol tetraether approximates 3.6 nm, about the thickness of a lipid bilayer, it has been suggested[175] that the lipid backbone of the *T. acidophilum* membrane consists of a monolayer of these unique molecules. Accordingly, the lipid domain of the thermoplasma membrane is formed by hydrocarbon chains which are covalently linked across the membrane rather than by intercalation of the hydrocarbon chains from two separate and opposite hydrophobic residues, as is usual in biomembranes. The diglycerol tetraethers may thus provide remarkable structural stability to the thermoplasma membrane. This concept is in accord with electron paramagnetic resonance studies.[379,451] which indicate that the thermoplasma membrane is the most rigid membrane known, and with unpublished findings by P. Ververgaert (cited in Reference 175) that *T. acidophilum* membranes cannot be freeze-fractured along their hydrophobic region.

The thermoplasma lipids are also distinguished by an extremely high percentage of carbohydrate-containing lipids. A glycerophosphoryl glycosyl diglycerol tetraether accounts for nearly 50% of the membrane lipids and another 25% are glycolipids comprised of monoglycosyl and diglycosyl diglycerol tetraether.[176,177] The preponderance of carbohydrate-containing lipids may also be a response to the high temperature, as glycolipids were shown to have a rigidifying effect on membranes of thermophiles.[244]

The finding of the peculiar diglycerol tetraether lipids in *Sulfolobus acidocaldarius* is not surprising, as this bacterium lacks peptidoglycan and inhabits acidic hot springs.[176] However, it appears that these lipids are not exclusive to the thermoacidophilic bacteria, as diphytanyl glycerol diethers are long known to occur in the extremely halophilic bacteria,[169] and very recently were shown to comprise the lipids of methanogenic bacteria as well.[414] These findings support the notion that methanogenic bacteria have a close genealogical relationship to the extremely halophilic and thermoacidophilic bacteria, and together form the group of *Archaebacteria*.[441] On the other hand, the finding of these peculiar lipids in the mesophilic methanogenic bacteria suggests that the tetraether assembly is not necessarily an adaptation to high temperature or low pH, but may reflect a more profound evolutionary development characteristic of early life in extreme environments. In any case, it may be argued that the membrane-rigidifying properties of the glycerol ether lipids are advantageous to the methanogenic and halophilic bacteria, which lack peptidoglycan in their cell envelope.

6. *Sterols*

Most mycoplasmas require cholesterol for growth, a property unique among procaryotes.[293] Large quantities of cholesterol are found in membranes of the sterol-requiring *Mycoplasma, Spiroplasma,* and *Ureaplasma* species, reaching levels comparable to those found in plasma membranes of eukaryotes (25 to 30% by weight of total membrane lipids).[20,217,245,273,285] The sterol-nonrequiring *Acholeplasma* species are also capable of incorporating cholesterol from the growth medium, but to a significantly lower degree (up to 10% by weight of total membrane lipids) (Table 2). No mycoplasma tested, including the sterol-nonrequiring species, is capable of cholesterol synthesis.[386] Moreover, the unesterified cholesterol incorporated from the medium is generally not esterified or changed in any way.[20,102,293,302] The cholesterol esters detected in mycoplasma membranes (Table 2) originate in the growth medium, as their fatty acid composition resembles that of the esterified cholesterol fraction in the serum supplement.[329,338]

a. Specificity of Sterol Requirement

In order to promote mycoplasma growth, the sterol must possess a planar steroid nucleus, a free hydroxyl group at the 3 β-position, and a hydrocarbon side chain.[293,386] These are precisely the structural features required for a sterol to exert a regulatory effect on the fluidity of both artificial membrane systems and biological membranes.[84,293] Hence, sterols capable of growth promotion include β-sitosterol, stigmasterol, ergosterol, and cholesterol,[102,333] whereas sterols such as coprostanol and epicholesterol cannot substitute for cholesterol and may even inhibit growth.[333,383] *M. mycoides* subsp. *capri* and the related *M. capricolum* differ from other *Mycoplasma* species in being able to grow with minimal amounts of cholesterol, so that their membrane cholesterol content can be lowered to less than 3% of the total membrane lipids.[15,65,183,343] Still, these *Mycoplasma* species will not grow in the total absence of cholesterol, differing from the *Acholeplasma* species.[65,307,343] Recent reports by Odriozola et al.[232] and Lala et al.[172a] raise serious problems. They found that the sterol requirement of *M. capricolum* can be met by the methylcholestane derivatives lanosterol, cycloartenol, 4,4-dimethylcholesterol, 4β-methylcholestanol, cholesteryl methyl ether, cholesteryl acetate, and 3α-methyl-cholestanol. All of these sterols are deficient in one or more of the structural attributes thought necessary for sterol function in mycoplasma and eukaryotic membranes. The surprisingly broad sterol specificity of *M. capricolum* may be associated with its ability to grow with very low cholesterol concentrations and with its marked capacity to take up large quantities of cholesteryl esters from the medium.[285] It should be noted however that Archer[15] reported that the closely related *M. mycoides* subsp. *capri* could not grow with 5β-cholestan-3-α-ol, indicating a requirement for the 3β-hydroxy group, a finding contradictory to that of Odriozola et al.[232] with *M. capricolum*.

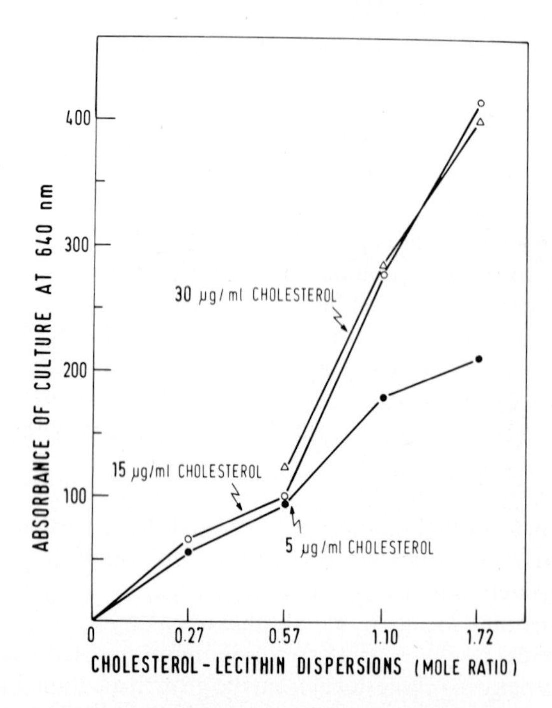

FIGURE 5. Growth response of *Mycoplasma hominis* to various concentrations of cholesterol added to the growth medium as cholesterol-phosphatidylcholine dispersions of various molar ratios. The improved growth-promoting efficiency of dispersions with higher cholesterol-to-phospholipid ratios can be seen. (From Kahane, I. and Razin, S., *Biochim. Biophys. Acta*, 471, 32, 1977. With permission.

b. Cholesterol Donors

Serum lipoproteins are the natural cholesterol donors for animal mycoplasmas *in vivo* and in conventional mycoplasma media *in vitro*. When purified human serum lipoproteins were used as the sole cholesterol source, the amounts of cholesterol incorporated into the mycoplasma membranes from low-density lipoproteins (LDL) were much higher than those taken up from the high-density lipoproteins (HDL). The very-low-density lipoproteins (VLDL) inhibited growth of some mycoplasmas, though they served as effective cholesterol donors.[375-377] These results support the notion that LDL is a far better cholesterol donor than HDL. The much higher molar ratio of unesterified cholesterol to phospholipid in LDL (about 0.8 to 1) as compared with that of HDL (about 0.15 to 1) may be responsible, at least in part, for the better performance of LDL as a cholesterol donor. Cooper et al.[68] postulated that the higher the unesterified cholesterol-to-phospholipid ratio of the cholesterol donor is, relative to that of the membrane, the more effective it is in donating cholesterol. This suggestion is supported by our observation[165] (Figure 5) that cholesterol-phosphatidylcholine vesicles with a molar ratio of 0.3 to 1 served as inefficient cholesterol donors and only permitted poor growth of *M. hominis*, whereas cholesterol-phosphatidylcholine vesicles at a molar ratio of 1:1 and above were very effective cholesterol donors and supported excellent growth of the organisms.

An interesting question is whether cholesterol is transferred to the membrane during a transient contact of the lipoprotein particle with the membrane, or as a result of a more intimate association in which the lipoprotein particle fuses and becomes part of the membrane. Experiments with lipoproteins in which the protein moieties were selectively labeled with ^{125}I favor the first alternative, at least for *Acholeplasma.* Up to 45% of the lipoprotein unesterified cholesterol was taken up by *A. laidlawii* membranes, with little or no concomitant uptake of the labeled protein. The total absence of esterified cholesterol from membranes of *Acholeplasma* species grown with serum (Table 2) also speaks against the adherence of lipoprotein particles to their membranes. Moreover, the lipoproteins exposed to the *A. laidlawii* membranes did not undergo any noticeable degradation apart from their significant depletion of unesterified cholesterol.[375,376] On the other hand, the presence of significant amounts of esterified cholesterol in *Mycoplasma* species growth with serum (Table 2) favors the possibility that serum lipoproteins do adhere to membranes of the sterol-requiring mycoplasmas. Incubation of *M. gallisepticum* cells with ^{125}I-labeled LDL in buffer resulted in the binding of about 10 μg of the LDL protein per mg of cell protein.[340] If this value represents adherent lipoprotein particles, these particles could account for only 15% of the unesterified cholesterol found to be taken up by the organisms during the incubation period, leading to the conclusion that most of the cholesterol in the membranes was transferred through transient contact of the lipoprotein particles with cells. As *M. gallisepticum* is distinguished by its low esterified cholesterol content, *M. capricolum* appears to be a better candidate for lipoprotein binding experiments, as it contains more esterified than unesterified cholesterol when grown with serum (Table 2). Definite conclusions as to the ability of *Mycoplasma* species to bind lipoprotein particles cannot, therefore, be drawn until further experimentation is done.

Serum lipoproteins, the natural cholesterol donors in mycoplasma media, can be replaced by phospholipid-cholesterol vesicles,[165,285] Tween® 80-cholesterol micelles, or by an ethanolic solution of cholesterol.[296] Lipid vesicles appear to be the best of the three artificial cholesterol donors, as the detergent Tween® 80 may damage the membrane and inhibit mycoplasma growth,[296] and when cholesterol is added as an ethanolic solution most of it crystallizes and co-sediments with the cells or membranes upon centrifugation.[292] Lipid vesicles, on the other hand, can be easily separated from cells or membranes by differential centrifugation.[165,285]

c. Control of Cholesterol Uptake

A problem of great biological significance concerns the mechanism by which cells control the amount of exogenous cholesterol incorporated into their plasma membrane. Mycoplasms offer several unique advantages for investigating this problem as their plasma membrane interacts directly with exogenous lipid donors and the cholesterol taken up is not esterified or modified in any other way. Moreover, mycoplasma lipid composition and membrane fluidity can be manipulated in a controlled manner, facilitating the study of the factors influencing cholesterol uptake.[293] In addition, the *Acholeplasma* species incorporate much less cholesterol into their membranes, compared to the sterol-requiring species (Table 2). Hence, elucidation of the factors which restrict cholesterol uptake by *Acholeplasma* may lead to the better understanding of the mechanisms controlling cholesterol uptake.

Cholesterol uptake by isolated mycoplasma membranes can be defined as a physical adsorption process with an activation energy of 6 kcal/mole.[109] We recently found that the marked difference in cholesterol binding capacity between *A. laidlawii* and *M. capricolum* cells is retained in their isolated membranes,[285] indicating that the mechanism restricting cholesterol uptake in *A. laidlawii* is a physicochemical one. This mechanism probably depends upon a specific composition and organization of membrane components rather than on active cholesterol exclusion dependent on metabolic energy as was previously suggested.[297]

Our observations that the ratio of membrane lipid to protein decreases markedly on aging of mycoplasma cultures[9,12,269,323,340] and increases on the addition of chloramphenicol[12,162,269] were used to show that the amount of cholesterol incorporated into the cell membrane of any specific mycoplasma depends on the polar lipid content of the membrane and is not influenced by variations in membrane protein content.[269] Digestion of membrane phospholipids by phospholipase A_2 decreased the cholesterol binding capacity of isolated *A. laidlawii* and *M. capricolum* membranes, roughly in proportion to the amount of phospholipid digested.[455] An increase in the phospholipid content of *M. capricolum* membranes following the uptake of exogenous phosphatidylcholine and sphingomyelin resulted in increased cholesterol uptake.[285] Our results support the principle postulated by Cooper et al.[69] that the amount of cholesterol incorporated into biomembranes depends on the amount of phospholipid available for interaction with it. If this is the case, membranes with a higher polar lipid content should be able to incorporate larger quantities of cholesterol. However, the ratio of polar lipid (including glycolipids) to membrane protein is not significantly higher in the cholesterol-rich *Mycoplasma* species than in the cholesterol-poor *Acholeplasma* species.[297] In fact, the free cholesterol-to-phospholipid ratio in *A. laidlawii* is much lower than in *M. capricolum.*[285] Moreover, the *A. laidlawii* membranes contain large amounts of glycolipids which may provide additional sites for cholesterol binding, and these are unaccounted for in the cholesterol-to-phospholipid ratio.

Glycolipids are major components in *Acholeplasma* but not in *Mycoplasma* membranes. The possibility that glycolipids have lower affinity for cholesterol than phospholipids cannot be ruled out as yet. Preliminary data[456] show that lipid dispersions prepared by sonication of cholesterol with isolated *A. laidlawii* glycolipids contain cholesterol in amounts comparable to those in dispersions made of *A. laidlawii* or *M. hominis*. Similar experiments carried out by McCabe and Green[202] with eukaryotic glycolipids (cerebrosides and gangliosides) showed that the two glycolipids incorporated smaller quantities of cholesterol into their dispersions than of choline-containing phospholipids. Our recent experiments[464] show that complete hydrolysis of the "pure" phospholipids of *A. laidlawii* membranes by phospholipase A_2 decreased but did not abolish cholesterol binding. However, the finding that the removal of the "pure" phospholipids (phosphatidylglycerol and diphosphatidylglycerol) which consist only 30% of the polar lipids caused a decrease of about 55% in cholesterol uptake, suggests that the glycolipids have a lower binding capacity for cholesterol.

Our finding that growing cells of *Mycoplasma* species take up large quantities of free and esterified cholesterol and phospholipids from serum and lipid vesicles, whereas *Acholeplasma* species take up only low amounts of free cholesterol (Table 2), may point to another explanation for the difference in cholesterol uptake between *Mycoplasma* and *Acholeplasma* species. It can be speculated that cells of *Mycoplasma* species have receptors for serum lipoproteins and lipid vesicles on their surface, which facilitate the transfer of cholesterol and phospholipids to the growing cell membrane (see Section III. A. 1.). *Acholeplasma* cells lack these receptors, so that the transient contact between the lipid donors and the cell surface suffices for cholesterol transfer only, not being close or prolonged enough for phospholipid transfer.[39] The possibility that serum lipoproteins actually adhere to the cell surface of growing mycoplasmas, particularly *M. capricolum,* cannot be ruled out (Section III.A.1.). The binding of significant quantities of cholesteryl esters to *Mycoplasma* membranes supports this interpretation, as cholesteryl esters are not usually found in biomembranes[171] and the amounts of cholesteryl esters that can be incorporated into artificial phospholipid bilayers are much lower than those found associated with *Mycoplasma* membranes.[146]

The organization and physical state of the polar lipids in the membrane may influence cholesterol uptake, as has been indicated by studies on the cholesterol-binding capacity

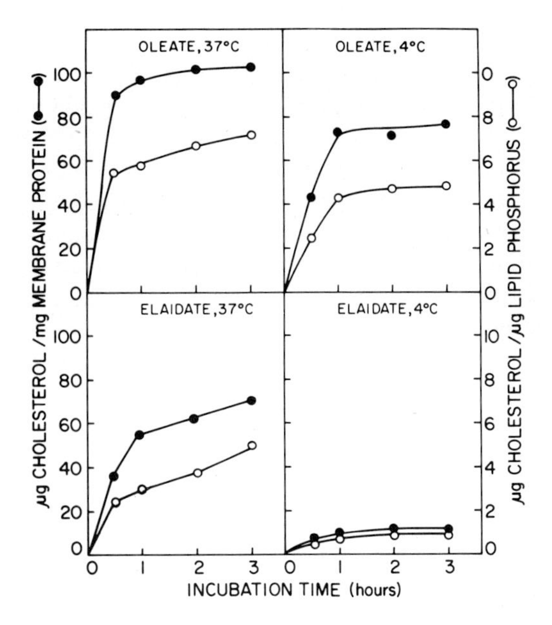

FIGURE 6. Effect of temperature on cholesterol uptake by oleate- and elaidate-enriched *Acholeplasma laidlawii* cells. Cholestorol uptake at 4°C was minimal by the elaidate-enriched cells in which membrane lipids were in the gel state, but significant in the oleate-enriched cells in which membrane lipids were still in the liquid-crystalline state. (From Razin, S., *Biochim. Biophys. Acta,* 513, 401, 1978. With permission.)

of the outer and cytoplasmic membranes of the Gram-negative bacterium *Proteus mirabilis.*[270] When the bacteria were grown with serum, the cytoplasmic membrane, though located inside the outer membrane, incorporated over four times as much cholesterol per mg of phospholipid as did the outer membrane. In this case, the two membrane types are known to contain the same phospholipid species, but their fluidity and molecular organization differ markedly.[327] It is conceivable that phospholipids which interact with membrane proteins or lipopolysaccharides (constituting the so-called "boundary" or "annular" lipids) are less available to interaction and binding of cholesterol.[69] The higher percentage of the boundary phospholipids in the outer membrane[327] may explain its lower cholesterol-binding capacity. The possibility that *Acholeplasma* membranes contain a higher percentage of boundary lipids than those of mycoplasmas was tested in our laboratory by employing the dry ether extraction procedure.[172] Dry ether supposedly extracts the unbound fraction of membrane phospholipids representing the protein-free lipid bilayer.[108] Our results failed to show any significant difference in the quantity of ether-extractable phospholipids between the *Acholeplasma* and *Mycoplasma* membranes tested.

The dependence of cholesterol uptake on membrane fluidity has been convincingly demonstrated with *A. laidlawii* cells enriched with elaidate or oleate.[272] The transfer of elaidate-enriched cells in culture from 37°C to 4°C virtually arrested cholesterol incorporation into the cell membrane. Cholesterol uptake continued, though at a slower rate, in the oleate-enriched cells undergoing a similar temperature shift-down (Figure 6). It can be concluded that the incorporation of exogenous cholesterol into the cell membrane of living mycoplasmas is rapid when the membrane lipid bilayer is in the liquid-crystalline state and very slow when the lipid bilayer is in the gel state.

d. Agents Complexing with Membrane Sterols

The ability to manipulate the amounts and types of sterols incorporated into *A. laidlawii* and *M. mycoides.* subsp. *capri* membranes has made these organisms most useful models for studies on the mode of action of agents which interact with membrane sterols. The polyenes filipin and amphotericin B lysed *A. laidlawii* cells grown with cholesterol, but had no effect on cells grown in its absence,[98,433] a finding providing perhaps the most convincing evidence for the notion that membrane cholesterol is the target of polyene action. The larger polyene nystatin behaved somewhat differently. Although large quantities of it were bound to various sterol-requiring mycoplasmas,[261] it did not inhibit their growth at concentrations far exceeding those required to inhibit the growth of yeasts.[14,173,230,261] An explanation for the lower toxicity of nystatin to mycoplasmas was given by Hsuchen and Feingold,[136] who showed that *A. laidlawii* became sensitive to nystatin when grown with ergosterol instead of cholesterol. Very similar results were obtained by Archer and Gale[16,17] with amphotericin B and *M. mycoides* subsp. *capri* grown with ergosterol instead of cholesterol. Ergosterol is the predominant sterol in yeasts, explaining their much higher sensitivity to the larger polyenes than the cholesterol-containing mycoplasmas. Filipin, on the other hand, has a higher affinity for cholesterol than for ergosterol, accounting for its high toxicity to animal cells and mycoplasmas.[16,17]

A. laidlawii grown with cholesterol served as the main test organism in the extensive studies by Norman et al.[231] and de Kruyff et al.[79,81,82] aimed toward the elucidation of the mechanism of membrane damage caused by the polyene antibiotics. Only membranes with sterols having a planar steroid nucleus, a 3β-OH group, and a hydrophobic side chain at C_{17} interacted with the polyenes. The polyenes complexed the sterol in the membrane, rendering it unavailable for interaction with membrane phospholipids. The formation of the polyene-sterol complexes was accompanied by permeability changes. Amphotericin B and nystatin caused the release of small ions and glucose from cells, probably through hydrophilic pores formed as a part of the polyene-sterol complexes. Filipin was harsher and disrupted the membrane structure, as was reflected by the release of macromolecules such as glucose-6-phosphate dehydrogenase.[81,82]

Cholesterol is apparently both the receptor and the target site for the lytic thiol-activated bacterial toxins produced by *Streptococcus pyogenes* (streptolysin O), *Clostridium perfringens* (perfringolysin O), *Clostridium tetani* (tetanolysin), and *Bacillus cereus* (cereolysin). Again, experiments with mycoplasmas were instrumental in reaching this conclusion.[70] Bernheimer and Davidson[28] were the first to demonstrate that staphylococcal and streptococcal toxins lyse sterol-requiring mycoplasmas. Later studies have established the association of the thiol-activated bacterial toxins with membrane cholesterol. Thus, streptolysin O and cereolysin were shown to produce ring- and arc-shaped structures in *A. laidlawii* membranes containing cholesterol but not in membranes of the same organism devoid of cholesterol.[71] However, cholesterol-grown *A. laidlawii* may not be a suitable model for these types of studies, as this organism is far less affected by the toxins than are the sterol-requiring mycoplasmas.[28,324] The reasons for the relative resistance of *A. laidlawii* to the toxins are not clear. The low cholesterol content in the *A. laidlawii* membrane (Table 2) and the possibility that the polyhexosamine polymer (Section II. C.), lipopolysaccharides, and other carbohydrate moieties on the membrane surface mask the cholesterol, can be cited as contributing factors for resistance. Hence, the use of *M. capricolum* or *M. mycoides* subsp. *capri* grown with high and low cholesterol concentrations may be preferable to the use of *A. laidlawii,* as was shown to be the case with tetanolysin.[324]

It has been proposed that after the toxin binds to cholesterol, the toxin-cholesterol complexes aggregate in the plane of the membrane by lateral diffusion. This process is influenced by membrane viscosity, as it is restricted at low temperatures. As with polyene antibiotics,[79] the damage to the membrane is apparently caused by the removal of cholesterol from its normal interaction with membrane phospholipids,[70,324] resulting in a more permeable and structurally fragile membrane.

7. Carotenoids

All the established *Acholeplasma* species are capable of *de novo* synthesis of carotenoids, as evidenced by the incorporation of ^{14}C-acetate and ^{14}C-mevalonate into polyterpenes.[386] Carotenoids can be visualized by the yellow color they impart to pellets of cells and isolated membranes. Boiling ethanol extracts the carotenoids from the membranes, and an estimate of their concentration in the extract can be obtained by absorbancy measurements.[279,380] *A. axanthum,* so named because it appears unpigmented, synthesizes carotenoids which are virtually colorless. This apparently reflects a lower number of conjugated double bonds in their molecules.[388] The polyterpenoid nature of the *Acholeplasma* pigments has been proven beyond doubt, but their exact structural characterization has yet to be accomplished.[386] Although the carotenoids are conspicuous because of their bright coloration, they constitute only a minor fraction of membrane lipids. The neutral lipid fraction, which contains triglycerides and free fatty acids in addition to the carotenoids, comprised only 1.1% of the total membrane lipids in *A. oculi* and 4.4% in *A. modicum.*[197] As carotenoids are synthesized *de novo,* it is possible to alter their amounts in the membrane, either by stimulating synthesis by adding acetate to the growth medium,[140,279,380] or inhibiting synthesis by adding propionate to the medium.[331,335] The directed alterations in the membrane carotenoid content have been recently used to study the effect of these compounds on membrane fluidity[140,331] (Section III.C.4.).

B. Transbilayer Distribution of Lipids

It has been proposed that the unequal distribution of the different membrane lipids in the outer and inner halves of the lipid bilayer, known as transbilayer lipid asymmetry, is a general property of biomembranes.[47,316] However, until now the erythrocyte membrane is the only biomembrane in which lipid asymmetry has been convincingly demonstrated, although the evidence for asymmetry in several viral membranes is fairly compelling.[234] Recent evidence supporting lipid asymmetry in mycoplasma membranes is, therefore, of great importance. The absence of a cell wall in mycoplasmas facilitates membrane lipid disposition studies, as these depend primarily on the use of macromolecules such as enzymes and antibodies which are unable to penetrate the bacterial cell wall barrier.

1. Phospholipids

The external location of at least some of the mycoplasma membrane phospholipids has been hinted at by electron microscopy of several *Mycoplasma* and *Acholeplasma* species treated with ferric oxide hydrosols in propanoic acid[359] or polycationic ferritin[358] which bind to negatively-charged groups on the membrane surface. The anionic sites appear to be lipid phosphate groups rather than protein carboxyls, since extraction of the membranes with lipid solvents abolished labeling, whereas pronase treatment increased labeling.[358,359] Striking membrane asymmetry was detected upon labeling of isolated membranes of *M. mycoides* subsp. *capri* with polycationic ferritin. The probe was found to bind to one membrane surface only, presumably the outer one.[358] On the other hand, *M. hominis* membranes, even when isolated from cells, failed to react with the iron-containing labels unless membrane protein was first digested by pronase. This is in accord with the findings that the major phospholipid of the *M. hominis* membrane, phosphatidylglycerol, resists hydrolysis by phospholipase C[326] and fails to interact with antibodies to it,[355] unless the membranes are first treated with pronase. Similarly, membrane phospholipids of *M. hominis* resisted the action of lipid-splitting enzymes of rabbit granulocytes, unless the membranes were first treated with deoxycholate.[246] Apparently membrane phospholipids in *M. hominis* are masked by proteins on both membrane sides. On the other hand, antibodies to phosphatidylglycerol reacted with *M. pneumoniae, M. mycoides* subsp. *capri,* and *A. laidlawii* cells, indicating that at least some of the phosphatidylglycerol molecules are exposed on the membrane surface.[85]

The chemical modification of membrane phospholipids by agents which do not penetrate through the membrane, and thus label only lipids in the outer layer, has been of limited use in mycoplasmas. The reason is that nonenzymatic modifications can be easily obtained with phospholipids containing a reactive amino group (i.e., phosphatidylethanolamine), and these are absent from mycoplasmas (Section III. A. 1.). The reagents required for chemical modification of the other polar headgroups are generally too harsh and either lead to disruption of the membane or penetrate readily into the cell.[234] The recent finding that the lactoperoxidase-mediated iodination technique, useful for membrane protein localization (Section IV.C.1.) can also be applied to lipid disposition studies is therefore of great interest.[123] Although the labeled iodine predominantly binds to membrane proteins, about 5% of the label is found in membrane lipids, probably through the iodination of the α-position of the carbonyl groups. The comparison of the labeling intensities of the various lipid species in iodinated whole *A. laidlawii* cells and isolated membranes revealed that the phospholipids and phosphoglycolipids of this organism are almost equally distributed in the outer and inner halves of the membrane, whereas the glycolipids are localized almost exclusively in the outer half of the bilayer[123] (Figure 7).

The digestion of phospholipids in the outer monolayer of a membrane by exogenous phospholipase is another widely-used means for studying the distribution of phospholipids between the two membrane layers.[234] However, the use of phospholipases suffers from several serious deficiencies. Thus, not all membrane phospholipids are susceptible to digestion. The phosphoglycolipids of *A. laidlawii* resist phospholipase digestion.[31] Even when susceptible, the phospholipids in the outer half of the bilayer may be shielded by other membrane components and protected from exogenous phospholipase, as was found with *M. hominis.*[326] The susceptibility of phosphatidylglycerol in *A. laidlawii* to digestion by phospholipase A_2 was recently shown to depend on the state of membrane energization.[29] Energization of the cells by glucose decreased the rate of phosphatidylglycerol hydrolysis and limited its extent to about 40% of the lipid. In contrast, when glucose was absent or when cells were treated with the ionophore nigericin, hydrolysis was much faster and nearly reached completion (Figure 8). The reasons for this phenomenon are unclear. A possible explanation can be based on the recent observation by Amar et al.[11] that membrane proteins are more exposed on the surface of energized *A. laidlawii* cells, possibly shielding the phospholipid from the exogenous phospholipase more effectively.

Another deficiency of phospholipases is that the hydrolysis products, lysophospholipids, free fatty acids, and diglycerides destabilize the bilayer structure and cause cell lysis.[234] In the case of *A. laidlawii* this problem is overcome by the intrinsic membrane-bound lysophospholipase,[425] which instantaneously degrades the lyso compounds produced by exogenous phospholipase A_2 treatment. The free fatty acids can be effectively removed by adding albumin to the reaction mixture.[31]

Information on the localization of phosphatidylglycerol in *A. laidlawii,* using phospholipase A_2 as a tool, was recently obtained by Bevers et al.[31] Treatment of intact cells with the enzyme at 37°C caused the degradation of all the phosphatidylglycerol,

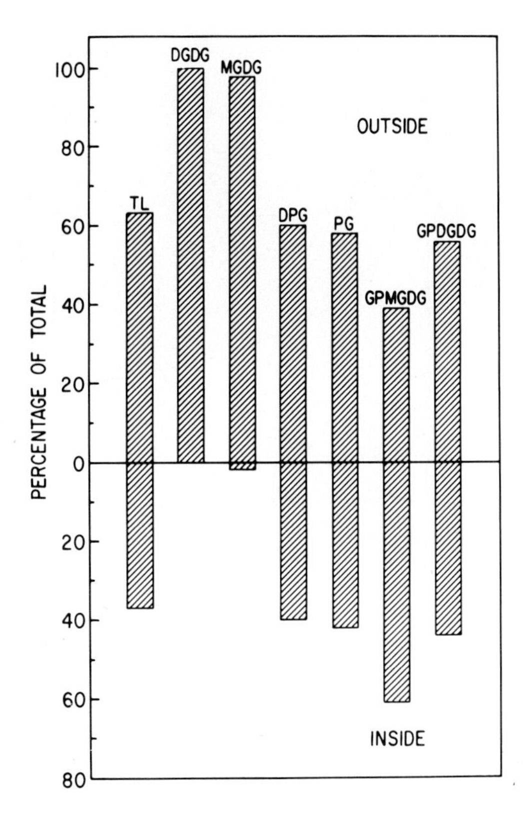

FIGURE 7. Transbilayer distribution of the major lipid species of *Acholeplasma laidlawii* membranes, as revealed by the lactoperoxidase-mediated iodination technique. From Gross, S. and Rottem, S., *Biochim, Biophys. Acta,* 555, 547, 1979. With permission.

which constitutes about 30% of the membrane lipids, without causing cell lysis or K^+ leakage. Although in itself this observation suggests that all phosphatidylglycerol is externally located, other findings tend to discredit this explanation. When linoleic acid-grown cells were kept at 5°C, a temperature at which membrane lipids are still in the liquid-crystalline state, about 50% of the phosphatidylglycerol was rapidly hydrolyzed. The residual phosphatidylglycerol could only be hydrolyzed at elevated temperatures and at much slower rates. Moreover, when membranes isolated from these cells were treated at 5°C, about 70% of the phosphatidylglycerol was hydrolyzed immediately, whereas hydrolysis of the residual phosphatidylglycerol was again strongly temperature-dependent. These results led Bevers et al.[31] to suggest the presence of three different phosphatidylglycerol pools: one, consisting of about 50% of the lipid, exposed on the external membrane surface; the second (about 20%) exposed on the inner membrane surface; and the third (about 30%) located in a region protected from the enzyme at low temperatures, probably by association with membrane protein. Experimental support for the hypothesis that the "protected" phosphatidylglycerol constitutes a boundary lipid closely associated with intrinsic membrane proteins was more recently provided by Bevers et al.[33] Accordingly, the diffusion of the boundary lipid from the protected region into the bulk lipid phase is sufficiently slow at 5°C, thus becoming the rate-limiting step of phosphatidylglycerol hydrolysis in the cold.

The susceptibility of all the phosphatidylglycerol in intact *A. laidlawii* cells to

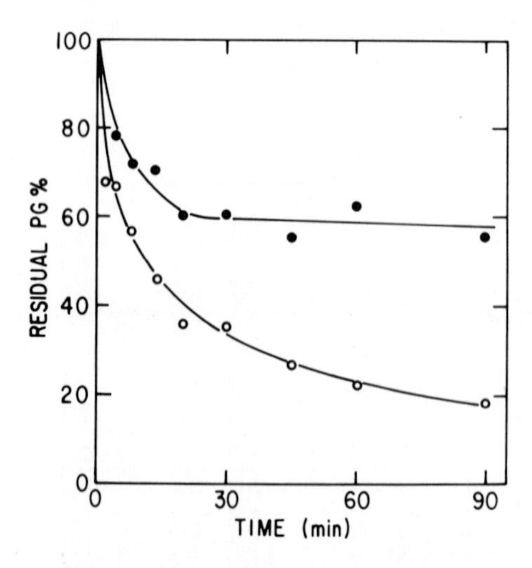

FIGURE 8. Inhibition of phosphatidylglycerol hydrolysis in energized cells of *Acholeplasma laidlawii*. Cells incubated with (●) or without (○) glucose were treated with pancreatic phospholipase A_2 and analyzed at various time intervals for the residual phosphatidylglycerol (PG) content. (From Bevers, E.M., Leblanc, G., Op den Kamp, J.A.F., and van Deenen, L.L.M., *FEBS Lett.*, 87, 49, 1978. With permission.)

hydrolysis by phospholipase A_2 at 37°C points to rapid transbilayer translocation ("flip-flop") of the lipid at this temperature. However, it appears that the transbilayer movement of the phospholipid in the treated cells is much faster than in membranes of untreated cells. The selective removal of the phospholipid at the outer membrane half disturbs the equilibrium distribution of the phospholipid over the two membrane layers, enhancing its transbilayer movement from the inner to the outer membrane half. This is a serious drawback of the phospholipase method in lipid disposition studies.[234] In fact, the radioiodination technique showed no difference in phosphatidylglycerol labeling in *A. laidlawii* cells treated at 0°C and at 37°C, suggesting a low rate of phosphatidylglycerol translocation in membranes where there was no depletion of externally located phosphatidylglycerol.[123]

2. Glycolipids and Lipopolysaccharides

One of the important common features of glycolipids in eukaryotic cells is their asymmetrical location in the outer half of the surface membrane bilayer. This characteristic qualifies glycolipids to be among the group of molecules involved in the information-controlling mechanism of cells.[449] Carbohydrate residues are also ubiquitous on the prokaryotic cell surface as constituents of cell wall polymers. Thus, the lipopolysaccharides of Gram-negative bacteria are exclusively located on the external surface of the cells[220] and play an important role in host-parasite interactions. Evidence that many mycoplasmas are covered by carbohydrate slime layers is summarized in Section II.C. Indications that membrane glycolipids are also exposed on the cell surface were first provided by studies with *M. pneumoniae* in which an antiserum to glycolipids was found to agglutinate the cells.[289] More recently, antibodies to a phosphoglucolipid of *M. mycoides* subsp. *capri* were also shown to react with intact cells. The antibodies could be absorbed by the cells as effectively as by isolated membranes, suggesting that a significant part, if not all, of the phosphoglucolipid molecules are located in the outer half of the lipid bilayer.[356]

Lectins have also been found useful for the localization of carbohydrate-containing components in mycoplasma membranes. Thus the surface carbohydrates on *T. acidophilum* cells were visualized by the aid of concanavalin A, horseradish peroxidase and diaminobenzidine staining procedure.[199] However, unlike specific antibodies, the lectins cannot distinguish between carbohydrate residues of glycolipids, lipopolysaccharides and glycoproteins. In *T. acidophilum* mannose and glucose residues, which react with concanavalin A, are constituents of phosphoglycolipids,[179] lipopolysaccharides,[200] and of a recently described glycoprotein.[450]

The almost equal binding of labeled lectins to intact cells and isolated membranes of a variety of *Mycoplasma* species[167] suggests that in mycoplasmas, as in erythrocytes and enveloped viruses,[316] all the carbohydrate-containing membrane components are exposed on the cell surface. Concanavalin-ferritin bound exclusively to one surface of isolated *M. neurolyticum* membranes, presumably that belonging to the external cell surface.[361] Nevertheless, the finding that glycolipids and phosphoglycolipids constitute more than 60% of the total membrane lipids in *A. laidlawii*[241,348,435] appears to oppose the notion that all carbohydrate-containing lipids must be externally located. In fact, the radioiodination technique revealed that only the "pure" glycolipids of *A. laidlawii* (monglucosyl diglyceride and diglucosyl diglyceride) are exclusively located in the outer half of the bilayer, whereas the phosphoglycolipids are almost equally distributed in the outer and inner halves of the membrane.[123]

3. Cholesterol

Cholesterol taken up from serum lipoproteins, or from any other exogenous source, must first be incorporated into the outer half of the lipid bilayer of the mycoplasma membrane. A question then arises: do the cholesterol molecules flip-flop and move from the outer to the inner half of the lipid bilayer, and if so, how fast is this movement? Evidence that cholesterol is distributed in both halves of the lipid bilayer was obtained by two different experimental approaches: (1) rapid kinetic measurements of filipin-cholesterol association in the membrane[37] and (2) kinetics of cholesterol exchange between cells or isolated membranes and high-density lipoproteins[340] or phosphatidylcholine vesicles.[457] The filipin-cholesterol association is easily monitored by absorbance or fluorescence intensity measurements.[36,231] To minimize membrane perturbations by filipin, which may lead to cell lysis, the interaction of the cells or membranes with filipin is done in the cold for very short reaction periods. This is accomplished by stopped-flow kinetic measurements and calculation of initial rates of filipin-cholesterol interactions.[37] Conclusions based on the comparison of initial rates of a reaction of any reagent with membrane lipids are, however, liable to some criticism.[234] A biological membrane is not a homogeneous dispersion of lipid molecules, so that the reactivity of lipids may differ greatly within one membrane layer, as well as between the inner and outer layers. Differences in reactivity of lipids may be caused by lipid-lipid or lipid-protein interactions, lipid-phase transition or phase separation, and environmental conditions such as differences in pH. Bittman and Rottem[37] based their calculations on the observation that the initial velocity of filipin association with vesicle-bound cholesterol depends on the accessibility of cholesterol for interaction at the bilayer surface.[36] Accordingly, the ratio of rate constants of filipin association with unsealed membrane vesicles relative to that obtained with intact cells can be used as a measure of the cholesterol distribution in the two membrane halves.[37] The ratios obtained indicate a symmetrical distribution of cholesterol in *M. gallisepticum* membranes, whereas in *M. capricolum* about two-thirds of the unesterified cholesterol is localized in the outer half of the lipid bilayer.[37] The finding that about twice as much filipin was bound to isolated membranes as to intact cells of *A. laidlawii*[82] suggests approximately equal distribution of cholesterol in the two halves of the bilayer of this organism as well.

The presence of about 50% of the membrane cholesterol in the inner half of the lipid bilayer of the *M. gallisepticum* membrane has also been indicated by the cholesterol-exchange technique.[340] Approximately 50% of ^{14}C-cholesterol from resting *M. gallisepticum* cells was readily exchanged with cholesterol of high-density lipoproteins, the half-time for exchange being about 4 hr at 37°C. The rate of exchange of the other half of the cholesterol was extremely slow, with a half-time of about 18 days. On the other hand, over 90% of the total cholesterol in isolated, unsealed *M. gallisepticum* membrane preparations was exchanged in a single kinetic process, reaching completion in 10 hr. These results suggest that cholesterol is present in approximately equal concentrations in the two halves of the lipid bilayer and that in resting *M. gallisepticum* cells the rate of movement of the cholesterol molecules from the inner to the outer half of the bilayer is extremely slow.

The very slow rate of cholesterol flip-flop in resting *M. gallisepticum* cells is somewhat surprising in light of the fast transbilayer movement of cholesterol in red blood cells, which can also be regarded as "resting" cells.[174] Recent experiments[457] in which labeled cholesterol was exchanged between *M. gallisepticum* cells and phosphatidylcholine-cholesterol vesicles, gave, in fact, a much faster rate for cholesterol flip-flop (a calculated half-time of about 12 hr at 37°C). The reasons for the discrepancy between the results obtained with lipid vesicles and those obtained with the high-density lipoproteins are not clear. In any case, we must assume that in growing cells the rate of cholesterol translocation from the outer to the inner membrane is fast enough to be accomplished within the 16 to 20 hr period used for growing the mycoplasmas with the exogenous cholesterol source. Supporting evidence for this statement has been obtained by Clejan et al.[65] Transfer of growing *M. capricolum* cells from a cholesterol-poor to a cholesterol-rich medium resulted in an approximately six-fold increase in the unesterified cholesterol content of the membrane within 4 hr of incubation of the growing culture at 37°C. The rate constants for the filipin-cholesterol association indicated that the transbilayer distribution of cholesterol was essentially invariant throughout the growth period, with 50% of the cholesterol located in the outer half and 50% in the inner half of the bilayer, implying a fast rate of cholesterol flip-flop in growing mycoplasmas.

Another possible way to localize cholesterol in membranes is by comparing the susceptibility of cholesterol in intact cells and isolated membranes to oxidation by cholesterol oxidase. Cowell and Bernheimer[70] have recently reported that treatment of *A. laidlawii* with cholesterol oxidase abolished the cereolysin inhibitory activity of this mycoplasma, indicating that cholesterol in the membrane is available for interaction with the enzyme. Hence, it appears that this approach may be feasible for tackling the problem of cholesterol localization in mycoplasma.

4. *Significance of Lipid Asymmetry*

The functional significance of lipid asymmetry in biomembranes is still obscure. The virtually exclusive location of carbohydrate-containing components on the cell surface points to their role in cell-cell interactions. It is conceivable that polar group asymmetry, in conjunction with variations in the fatty acid constituents among lipid classes, can result in different fluidities for the two monolayers comprising the bimolecular leaflet.[316] The greater freedom of motion of a spin-labeled fatty acid incorporated into membranes of intact *A. laidlawii* and *M. hominis* cells, as compared with its motion in isolated membranes, was taken to suggest higher fluidity in the outer half of the lipid bilayer.[317] Since the various lipid species of *A. laidlawii* show some variation in their melting temperature,[54,369] their asymmetrical transbilayer distribution may account for differences in fluidity between the two membrane halves. However, the glycolipids, which are located exclusively in the outer half of the *A. laidlawii* membrane[123] show a somewhat higher phase transition temperature than the total membrane lipids.[369] An explanation for the relative rigidity of the inner half of the lipid bilayer may thus be founded on a different basis. Most membrane proteins in *A. laidlawii* and *M. hominis* face the cytoplasm and are thus in closer contact with the inner half of the lipid bilayer.[9,10] The resulting protein-lipid interactions can considerably reduce the freedom of motion of the hydrocarbon chains of membrane lipids,[339] and in this way decrease the fluidity of the inner half of the lipid bilayer.

C. Membrane Fluidity

1. *Physical Properties of the Lipid Bilayer*

A. laidlawii membranes were the first biomembranes shown to undergo a thermal phase transition by differential-scanning calorimetry, a discovery which profoundly influenced the formation of the fluid mosaic membrane concept.[374] The pioneer studies of Steim et al.[400] provided perhaps the first strong experimental support for the hypothesis that lipids are basically organized as a bilayer in biomembranes. The numerous subsequent studies on the mode of organization and physical state of mycoplasma membrane lipids have employed a great variety of physical methods including: differential-scanning calorimetry and differential-thermal analysis,[54,203,205,208,299] X-ray diffraction,[92] electron paramagnetic resonance spectroscopy,[328,415] nuclear magnetic resonance spectroscopy,[50,77,210,233,378,401,402] light-scattering,[1] and fluorescence polarization.[321] The advantages and disadvantages of the various methods have been discussed at length in previous reviews.[271,319]

The voluminous amounts of data accumulated by the application of the physical methods have essentially confirmed the basic conclusions drawn from the early differential-scanning calorimentry experiments. The majority of mycoplasma membrane lipids associate with each other to form a bilayer, while the minority interact with the hydrophobic portions of membrane proteins, forming the so-called "boundary," "annular", or"halo" lipids. The part of the membrane lipids unbound to protein is free to undergo a thermal phase transition from the liquid-crystalline to the gel state. This event, in the case of completion, profoundly affects the structural and functional properties of the biomembrane. The transition of the membrane lipid bilayer into the gel state is accompanied by thickening of the bilayer,[92] aggregation of the intramembranous particles representing integral membrane proteins,[145,343,428,429] a decrease in membrane permeability,[78,205,321,424,427] and transport of metabolites,[298] as well as a marked decrease in the activity of membrane-bound enzymes.[83,137,321,342] It is conceivable that the lipid bilayer must be in the liquid-crystalline state, at least partially, in order for the biomembrane to function. According to McElhaney,[203] up to about one half of the membrane lipid in *A. laidlawii* may be transformed to the gel state without apparent effects on cell growth, and the existence of less than one tenth of the membrane lipid in a fluid state is sufficient to support some cell growth and replication, albeit at greatly reduced rates. Once the lipid bilayer totally crystallizes, cells stop growing and the membrane loses its elasticity, so that the cells lyse rather than swell when placed in hypotonic solutions.[427]

The early differential-scanning calorimetry studies of Steim et al.[400] have already shown that the temperature range of the *A. laidlawii* membrane lipid phase transition is extremely broad, even reaching up to 30°C.[370,371] The broadness of the transition peak obtained with membranes (compared to the sharp peak of pure phospholipids) reflects the heterogeneity of membrane lipids, which have different phase transition temperatures.[54] Two major factors may influence the phase transition temperature of a specific membrane lipid: the length and degree of unsaturation of its fatty acyl chains, and the nature of its polar headgroup. The recent finding of Silvius and McElhaney[370,371] that "fatty-acid homogeneous" *A. laidlawii* shows a very narrow lipid phase transition indicates that the broadness of the lipid phase transition peak in "normal" membranes is mainly due to fatty acid heterogeneity rather than to the diversity of the lipid headgroups.

The findings that the fatty-acid homogeneous *M. mycoides* strain Y and *A. laidlawii* grow as well as the normal organisms with "fatty-acid heterogeneous" membranes[305,306,370,371] raises doubts about the need of fatty acid heterogeneity for proper membrane function. It is clear, however, that in many membranes containing an assortment of lipids with different melting temperatures, it is possible to find neighboring regions of lipids in the gel and in the liquid-crystalline states, a phenomenon named lateral phase separation.[367] Again, *A. laidlawii* membranes were the first biomembranes in which lateral phase separation was demonstrated by freeze-fracturing electron microscopy.[428] Lateral phase separation in *A. laidlawii* membranes was also recently demonstrated by Bevers et al.,[30] making use of the inability of pancreatic phospholipase A_2 to hydrolyze phospholipids in the gel state. *A. laidlawii* grown with palmitate synthesized two phosphatidylglycerol (PG) species, dipalmitoyl PG and monopalmitoyl PG, in which only one of the two acyl chains is palmitic acid. The physical behavior of both species in the membrane, as measured by susceptibility to hydrolysis by phospholipase A_2, showed their segregation at 25°C. At this temperature there was almost complete hydrolysis of the monopalmitoyl species, whereas all the dipalmitoyl species, which were still in the highly ordered gel state, remained unhydrolyzed. The limited hydrolysis of the dipalmitoyl PG points to an important conclusion: exchange diffusion of PG between the fluid and ordered domains is restricted.

Another way to demonstrate lateral phase separation is by the method of Wieslander et al.,[434] who separated two subpopulations of membrane fragments from osmotically lysed *A. laidlawii.* Separation was achieved by countercurrent distribution of membrane fragments in a two-polymer aqueous phase system made of dextran and polyethylene glycol. This system has an electrostatic potential difference between the two phases (top phase positive). The subpopulations of membranes were almost indistinguishable in ultrastructure, buoyant density and protein composition, but differred in their polar lipid composition and enzymatic activities. The phase system used by Wieslander et al.[434] appears to separate the membrane fragments both on the basis of charge and hydrophobic/hydrophilic properties, resulting from the different polar lipid composition in the membrane subpopulations. While there is no doubt that lateral phase separations of membrane lipids do occur in membranes of growing mycoplasmas, their possible role in membrane function remains a moot point. Many studies, including the recent ones by Silvius and McElhaney,[370,371] indicate that mycoplasma cells grow very well when all of their membrane lipid is in the liquid-crystalline state (Table 3).

2. *Regulation of Membrane Fluidity by Alterations in Fatty Acid Composition*

A mechanism for regulating membrane fluidity is essential to avoid the harmful effects of membrane lipid crystallization. It appears that various mycoplasmas have developed different mechanisms for this purpose. While *A. laidlawii* regulates membrane fluidity by adjusting its fatty acid composition, the sterol-requiring mycoplasmas use cholesterol for this purpose. The fluidity of membrane lipids and the temperature range of the gel-to-liquid-crystalline phase transition are primarily

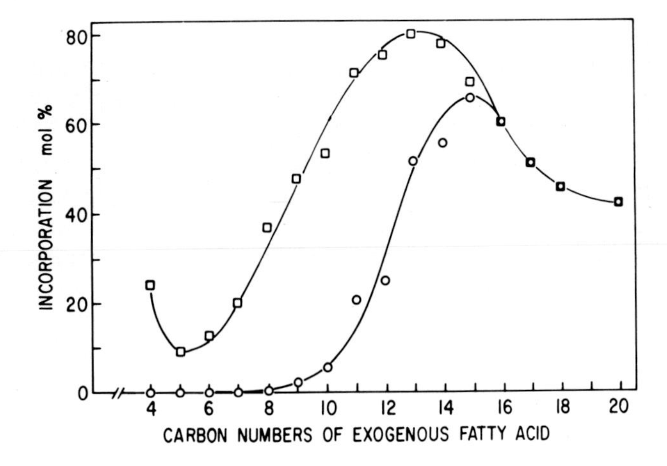

FIGURE 9. Effect of fatty acid chain length on the extent of the direct (○) and total (□) incorporation of exogenous straight-chain saturated fatty acids into membrane lipids of *Acholeplasma laidlawii*. Direct incorporation refers to fatty acids incorporated without chain elongation. Total incorporation refers to fatty acids incorporated both directly and after chain elongation. (From Saito, Y. and McElhaney, R.N., *J. Bacteriol.*, 132, 485, 1977. With permission.)

dependent on the length and structure of the hydrocarbon chains of membrane lipids.[370,371,421] The cohesive van der Waal's forces between the hydrocarbon chains are cumulative and increase with the number of methylene groups in the interacting chains. Thus, an increase in the chain length of the fatty acids augments the attractive forces between the adjacent phospholipids and results in a tightly packed lipid backbone, where the freedom of motion of the hydrocarbon chains is low, and the temperature of the gel-to-liquid-crystalline phase transition is high. On the other hand, the attractive forces between the methylene groups decrease most markedly with the increasing distance between the interacting chains. Moreover, when unsaturated bonds or methyl or cyclopropane groups are introduced into the chain, steric hindrance will drastically reduce the attractive forces, increasing membrane fluidity and consequently decreasing the temperature of the gel-to-liquid-crystalline phase transition.[271,421]

Regulation of the acyl chain length and degree of unsaturation appears to play a major role in controlling membrane fluidity in the cholesterol-nonrequiring *A. laidlawii*. The mechanism of regulation acts at various levels of *de novo* fatty acid biosynthesis, the incorporation and elongation of exogenous fatty acids, and the utilization of the fatty acids for synthesis of the complex membrane lipids. The *de novo* fatty acid biosynthesis by *A. laidlawii,* in the absence of exogenous fatty acids, was found to yield about an equimolar mixture of saturated C_{14} and C_{16} fatty acids with acetate as a primer and C_{15} and C_{17} fatty acids with propionate as a primer, a mixture which apparently provides optimal membrane fluidity at 35°C.[351] When exogenous fatty acids are provided, *A. laidlawii* incorporates most effectively those saturated fatty acids (C_{14} through C_{16}) which are most similar to the endogenous fatty acids normally produced[348] (Figure 9). The organisms can also incorporate and grow well with certain members of the *trans*-mono-unsaturated and branched-chain fatty acid series, which provide a moderate level of membrane lipid fluidity.[348]

A. laidlawii is capable of elongating exogenous fatty acids as well. Although the substrate specificity of the chain elongation system is strikingly broad, even acting on cyclopropane fatty acids,[241,352] those acids with less than 6 to 9, or more than 15 to 18 carbon atoms cannot be elongated. In this way the exogenous fatty acids are elongated to yield only acids having optimal or nearly optimal chain lengths for complex lipid synthesis, so that the fluidity of membrane lipids is maintained within the required range.[352] Moreover, exogenous medium- or long-chain fatty acids decrease the *de novo* synthesis of fatty acids in the direction of producing an intermediate membrane fluidity. Thus an exogenous unsaturated long-chain fatty acid can act to increase the average chain length of the biosynthesized saturated fatty acids by decreasing the rate of utilization of the long-chain biosynthetic products.[373] Silvius et al.[373] base their explanation for these findings on the positional distribution of the fatty acids in membrane lipids. In *A. laidlawii,* fatty acids with low melting temperatures are preferentially bound to the 2-position of the glycerol backbone, while fatty acids with high melting temperatures predominate in the 1-position.[207,314,373,330] Thus, an exogenous fatty acid with a low melting point would compete with the endogenously synthesized short-chain fatty acid for transfer to the 2-position of the glycerol. The decreased rate of acyl transfer of the short-chain biosynthetic product would, therefore, enable its elongation to reach termination.

The selective incorporation of fatty acids into the 1- and 2-positions of the glycerol backbone during glycerolipid synthesis acts also in another way to modulate membrane lipid fluidity.[349,350] It appears that the enzymes catalyzing the acylation of the 1- and 2-positions recognize the physicochemical properties of their fatty acid substrates, rather than the chemical properties of a particular electronic configuration of the fatty acid hydrocarbon chain. This mechanism restricts the production of disaturated or diunsaturated glycerolipids. Thus, intramolecular fatty acyl chain mixing is maximized, and the production of the relatively physicochemically homogeneous class of glyco- and phospholipid molecules, apparently required for optimal membrane function, is assured.[349]

Viewed as a whole, the studies of McElhaney and co-workers[348-352,369-371,373] indicate that the enzyme systems responsible for fatty acid and glycerolipid biosynthesis in *A. laidlawii* function to maintain an optimal degree of membrane fluidity. If so, one would expect this mechanism to operate during marked shifts in the growth temperature — a common situation when regulation of fluidity becomes essential.[72,235] In fact, a regulatory mechanism which senses temperature changes was demonstrated in *A. laidlawii* by Rottem et al.[328] Decreasing the growth temperature to 15°C caused a significant increase in the amount of exogenous oleic acid incorporated into membrane lipids. Membranes of *A. laidlawii* grown in the cold (15 or 28°C) were more fluid than membranes of cells grown at 37°C.[137,141,208,328] Thus, when cells were grown at 37°C, membrane lipids were fluid at this temperature, but were entirely in the gel state at 25°C, whereas when grown at 25°C membrane lipids were mostly fluid even at 25°C.[208] In *Escherichia coli* the mechanisms responsible for the temperature-induced fatty acid alterations during temperature shifts operate at the levels of both fatty acid biosynthesis (by changing the ratio of unsaturated to saturated fatty acids) and the acyltransferase-mediated incorporation of the fatty acids into the glycerol backbone.[72] The findings with *A. laidlawii* indicate, however, that changes in the growth temperature produce only minor alterations in the pattern of the saturated fatty acids derived from *de novo* synthesis and from the elongation of exogenous medium-chain fatty acids.[352] In addition, the temperature range of the thermotropic phase transition of membranes derived from cells grown in fatty acid-free media varied only slightly with growth temperaures.[203] It appears, therefore, that the mechanism responsible for temperature-induced fatty acid alterations in *A. laidlawii* does not operate at the fatty acid biosynthesis level, but rather at the level of the transacylation reaction.[323] As the growth temperature is decreased, *A. laidlawii* adjusts the fatty acid composition of its glycerolipids by incorporating exogenously supplied fatty acids with progressively lower melting points.

It should be stressed that the animal mycoplasmas, including *A. laidlawii,* are not usually exposed to significant temperature changes in their homoiothermic host, but rather may be subjected to marked variations in the composition of the fatty acids supplied by the host. The mechanisms regulating fatty acid composition described above appear quite adequate for this purpose. Furthermore, Silvius et al.[369] conclude that membrane fluidity in *A. laidlawii* need not be tightly regulated. Thus the "fatty-acid homogeneous" *A. laidlawii* grew well at 37°C with a variety of fatty acids, despite the fact that the membrane phase transition midpoint temperature varied within a broad range of temperatures, depending on the fatty acid used for growth. Accordingly, the major function of the "homeoviscous adaptation" observed in *A. laidlawii* and other prokaryotes is not to maintain an absolutely constant membrane fluidity. Rather, it adjusts the lipid phase transition temperature range, in order to maintain a proper lipid phase state (all liquid-crystalline, or a mixture of gel and liquid-crystalline lipids) for optimal membrane functioning at the growth temperature. A membrane in the gel state or in a very fluid state (such as that obtained with highly fluidizing lipid species, e.g., dilinoleyl lipids) will not enable growth, as the barrier function of the membrane will be seriously impaired.

Does the composition of the exogenously provided fatty acids or the degree of membrane fluidity influence the types of glycerolipids synthesized by *A. laidlawii?* The extensive studies of Christiansson and Wieslander[60] and Wieslander and Rilfors[435] revealed that only the synthesis of monoglucosyldiglyceride (MGDG) and diglucosyl-diglyceride (DGDG) was significantly influenced by the fatty acid composition of the growth medium. The MGDG-to-DGDG ratio increased when the medium was supplemented with a saturated or a *trans*-unsaturated fatty acid; the ratio decreased significantly in the presence of *cis*-unsaturated fatty acids. Shifting down the growth temperature from 37°C to 17°C also resulted in the preferential synthesis of MGDG accompanied by an increased incorporation of unsaturated fatty acids into this lipid. This led Christiansson and Wieslander[60] to suggest that higher membrane viscosity stimulates MGDG synthesis at the expense of DGDG. While confirming the influence of the fatty acid composition on the MGDG-to-DGDG conversion, Silvius et al.[369] do not agree with the interpretation that the changes in the MGDG-to-DGDG ratio are primarily a function of lipid fluidity, as membranes with highly different lipid phase transition midpoint temperatures may have a very similar MGDG-to-DGDG ratio.

Another intriguing problem concerns the way by which the enzymes involved in membrane lipid synthesis select the appropriate fatty acids to maintain membrane fluidity within the optimal range. The studies of McElhaney and coworkers[348-352,369-371,373] indicate that the fatty acids are selected on the basis of their physicochemical properties rather than their chemical properties or electronic configurations. Since the binding of a substrate to an enzyme reflects the molecular structure of the ligand, it is difficult to explain this selectivity by enzyme specificity. Melchior and Steim[209] therefore suggested that the temperature sensing selection mechanism is thermodynamically determined and is not dependent on the acyl-transferase specificity. According to their hypothesis, the physical state of the lipid bilayer itself determines the concentrations of the various fatty acids available to the enzymes. They provide experimental data showing that the relative affinity of phospholipid bilayers for palmitate, as compared to oleate, increases with an increase in temperature above the beginning of the phase transition. The more fluid the bilayer, the more palmitate binds to it, as compared to oleate. The acyltransferase embedded in the lipid bilayer has little or no innate ability to change its

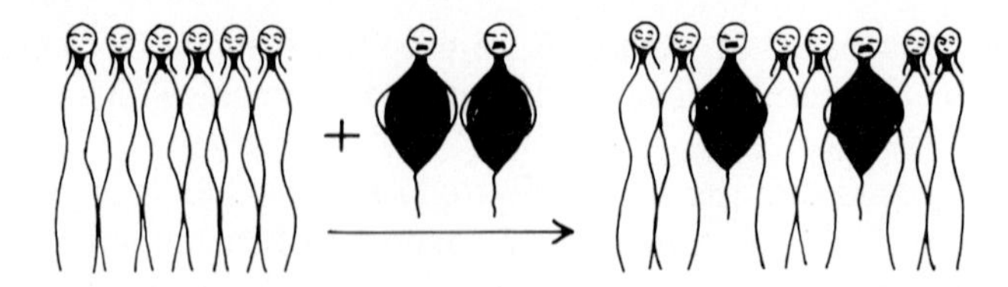

FIGURE 10. The influence of cholesterol on the packing and fluidity of a phospholipid film presented as a cartoon picture. The incorporation of cholesterol molecules with their bulky ring system (black guys) into the phospholipid film (white dolls) will produce an "intermediate fluid condition" characterized by the tighter packing of the molecules at the region closer to the polar heads, and a more loose packing at the lower parts of the hydrocarbon tails. Cholesterol prevents the gelation or crystallization of the film at temperatures below the phase transition temperature of the phospholipids in this way.

selectivity toward various fatty acids, but accepts and uses the fatty acids supplied to it by the bilayer. In this sense, lipid bilayers are capable of controlling their own physical state.

3. Regulation of Fluidity by Cholesterol

The suggestion that cholesterol functions as a regulator of the fluidity of the biomembrane lipid domain, first postulated on the basis of studies on artificial lipid membranes,[53] has gained strong experimental support from studies on mycoplasma membranes. The idea, as summarized by Rothman and Engelman,[315] is that cholesterol produces an intermediate fluid condition in the membrane lipid bilayer by virtue of its peculiar molecular shape. Cholesterol or related sterols (containing a planar ring system, a free hydroxyl group at the 3β-position, and a hydrocarbon side chain) are oriented in the membrane in such a way that the ring system is aligned parallel to the hydrocarbon chains of membrane phospholipids, while the hydroxyl group anchors the sterol molecule to the polar surface of the bilayer. The rigid ring system of the cholesterol molecule, which is about twice as large as its flexible hydrocarbon side chain, interacts strongly with the upper portions of the hydrocarbon chains of the phospholipids, exerting a condensing effect on this region. On the other hand, the bulky ring system separates the lower portions of the phospholipid hydrocarbon chains from each other, increasing the freedom of motion of the chains in this region (Figure 10). Hence, the insertion of cholesterol into a phospholipid bilayer allows a liquid-like configuration for the phospholipid hydrocarbon chains at the cholesterol tail region and a more condensed configuration at the ring region. The net result of cholesterol action is to exert a condensing effect on phospholipids at temperatures above their thermal phase transition, and to prevent the cooperative crystallization of the hydrocarbon chains at temperatures below that of the phospholipid phase transition. Complete elimination of the phase transition takes place only if the molar ratio of cholesterol to phospholipids exceeds 1 to 2.[84]

The condensing effect of cholesterol, as expressed by decreased membrane fluidity and permeability, could be observed in *A. laidlawii.*[78,80,204,205] However, the relatively low amounts of cholesterol that can be incorporated into the *A. laidlawii* membrane (Table 2) only suffice to reduce the energy content of the phase transition of membrane lipids, but not to eliminate it. Hence, the sterol-nonrequiring *A. laidlawii* is certainly not the ideal model for studying cholesterol effects. A much better model for this purpose is that developed by Rottem et al.[343] *M. mycoides* subsp. *capri,* and the closely related *M.*

capricolum, can be adapted to grow with very little cholesterol.[15,65,183,264,321,343] In this way the cholesterol content of the membranes can be reduced to less than 3% of the total membrane lipid, as compared with about 25% in membranes of the same organisms grown with optimal amounts of cholesterol.[343] The most remarkable difference between the cholesterol-poor and the cholesterol-rich membranes is that a thermotropic phase transition can be demonstrated only in the former.[321] Differential-scanning calorimetry revealed an endothermic phase transition centered at about 25°C in the cholesterol-poor membranes, whereas no transition was observed in the cholesterol-rich ones. Other techniques, such as fluorescence polarization with diphenylhexatriene as a probe, and freeze-fracturing, further confirmed these findings. Chilling of the cholesterol-poor membranes to 4°C prior to the quick freezing caused the aggregation of the intramembranous particles, leaving over two-thirds of the fracture faces particle-free.[343] Aggregation of the intramembranous particles, believed to contain integral membrane proteins, is a manifestation of the gelation of the lipid domain.[145,428] Using elaidate-enriched cholesterol-poor *M. mycoides* subsp. *capri* membranes, aggregation of the particles was observed even when membranes were quenched from 37°C before freeze-fracturing. Differential scanning calorimetry of these membranes showed a large part of the acyl chains to be in the gel state at 37°C.[183] No aggregation of particles was discernible in the cholesterol-rich membranes, even when quenched from 4°C.[343]

The experiments carried out with the cholesterol-poor *M. mycoides* subsp. *capri*[321,343] membranes provided perhaps the first clear-cut evidence with membranes of growing cells, to support the hypothesis that cholesterol regulates membrane fluidity, maintaining an "intermediate fluid condition" during changes in growth temperature, or following alterations in the fatty acid composition of membrane lipids. In accordance with this supposition, growth of the adapted cholesterol-poor strain was almost completely arrested at 25°C, the temperature at which most of the membrane lipids crystallized, whereas the cholesterol-rich cells of the same organism grew well, though at a much slower rate than at 37°C.[343] There can be little doubt that the near arrest of growth of the cholesterol-poor organisms at 25°C was associated with the steep decline in the membrane-bound ATPase and transport activities which have been shown to occur at this temperature.[184,321]

Why are mycoplasmas the only procaryotes dependent on cholesterol for growth? One would tend to associate this requirement with the lack of a cell wall — undoubtedly the single most important property distinguishing the mycoplasmas from all other prokaryotes. It was long ago suggested that cholesterol stabilizes the lipid bilayer and increases the tensile strength of the mycoplasma cell membrane, thus facilitating its survival and growth without the protection of a rigid cell wall.[265,266] Cells of the cholesterol-poor *M. mycoides* subsp. *capri* were in fact found to be quite fragile, frequently undergoing lysis even in the growth medium.[264,343] Moreover, the K^+ level was found to be much lower in these cells than in the sterol-rich organisms, reflecting a higher permeability of the sterol-poor membrane to K^+ and H^+.[183] However, clearly the cholesterol requirement cannot be attributed only to the lack of a cell wall, as the *Acholeplasma* and *Thermoplasma* species and the wall-less bacterial L-forms do not require cholesterol for growth. An explanation can thus be based on the proven ability of cholesterol to act as a regulator of membrane fluidity. As discussed in a previous section, *A. laidlawii* appears capable of regulating its membrane fluidity by changing the fatty acid composition of its membrane lipids, resembling many eubacteria and eukaryotes in this respect. The *Mycoplasma* and *Spiroplasma* species, which are totally incapable of fatty acid synthesis, obviously cannot regulate their membrane fluidity in this way. Furthermore, it is not clear whether or not the sterol-requiring mycoplasmas are capable of adjusting the incorporation of exogenous fatty acids in response to variations in temperature and growth conditions. Studies with *M. hominis*[323] and *S. citri*[102,217] showed that these organisms preferentially incorporate palmitate from a mixture of palmitate and oleate or linoleate added to the growth medium. In this case, the incorporation of large quantities of cholesterol into the membrane may be necessary to prevent the membrane from becoming overly viscous, even at 37°C.

In conclusion, the ability of mycoplasmas to incorporate large quantities of cholesterol into their membrane appears to compensate for their inability to regulate membrane fluidity by preferential fatty acid synthesis or incorporation. On the other hand, it is still not clear whether or not the regulation of membrane fluidity is the only role of cholesterol in mycoplasma growth. Efforts to obtain cholesterol-free *M. mycoides* subsp. *capri* cells have failed,[307,343] indicating that small amounts of cholesterol (less than 3% of total membrane lipids) are indispensible for growth. Since this small quantity of cholesterol cannot possibly exert a fluidizing effect on the bulk lipid bilayer, it must be assumed that cholesterol also fulfills a more specific role, such as the activation of membrane-bound enzyme systems.

4. *Effects of Carotenoids on Membrane Fluidity*

The finding of carotenoids in the sterol-nonrequiring *Acholeplasma* species led Smith[380] to propose that these polyterpenoid pigments fulfill functions analogous to those of cholesterol in the sterol-requiring mycoplasmas. In support of this hypothesis, Smith and Henrikson[387] showed that in a sterol-free medium, inhibitors of carotenoid synthesis also inhibited *A. laidlawii* growth, which in several cases was reversed by cholesterol. However, the same inhibitors also affected growth of *M. arthritidis* (strain 07) which lacks the enzymes for carotenoid synthesis. In further support of this hypothesis, Smith[381] reported growth of several sterol-requiring mycoplasmas with carotenoids or their precursors as substitutes for cholesterol. The recent indications that cholesterol acts as a regulator of membrane fluidity in sterol-requiring mycoplasmas prompted the re-evaluation of Smith's hypothesis. Huang and Haug,[140] and more recently Rottem and Markowitz,[331] showed by electron paramagnetic resonance spectroscopy that the freedom of motion of spin-labeled fatty acids in *A. laidlawii* membranes is influenced by their carotenoid content. Membrane carotenoid content was increased by adding the biosynthetic precursor, acetate, to the growth medium,[279,380] whereas the addition of propionate decreased carotenoid synthesis significantly.[335] Propionate is, however, a precursor of odd-numbered fatty acids in *A. laidlawii,* and its addition to the medium causes significant changes in the membrane fatty acid composition.[331,351] The finding that *A. laidlawii* membranes can be depleted of their carotenoids by incubation with phosphatidylcholine vesicles overcomes the difficulty associated with the use of propionate.[331] The carotenoid-deficient or depleted membranes were less viscous than the native membranes. Moreover, carotenoid incorporation into the phospholipid vesicles increased the viscosity of vesicles. These observations support the recent hypothesis of Rohmer et al.[310] that a series of polyterpenoids (C_{30} to C_{50}), acyclic or polycyclic, conformationally mobile or rendered rigid by conjugation or by polycyclization, can play a role as prokaryotic membrane stabilizers in place of sterols.

Although the results of Huang and Haug[140] and Rottem and Markowitz[331] indicate that carotenoids can influence the fluidity of the lipid bilayer, it remains a moot point whether they really play a significant role in the regulation of membrane fluidity in *A. laidlawii*. The extensive studies by McElhaney and co-workers.[348-352, 369-371] have provided solid support for a regulatory mechanism in this mycoplasma based on alterations in the fatty acyl chains of membrane phospho- and glycolipids. Moreover, Silvius et al.,[369] using differential thermal analysis, obtained direct evidence that the carotenoids in membranes of *A. laidlawii* grown under normal conditions (i.e., without supplemental acetate or propionate) have little influence on the membrane lipid phase transition. They conclude that variations in carotenoid levels in fatty-acid homogeneous membranes are unlikely to influence the membrane lipid thermotropic behavior significantly. As previously mentioned, the quantities of carotenoids in *Acholeplasma* membranes are usually low. Hence, it appears that carotenoids do not act as regulators of membrane fluidity in *Acholeplasma* under normal growth conditions. In fact, it was demonstrated long ago that *A. laidlawii* grows well in a sterol-free medium in the complete absence of carotenoid synthesis.[291] One well-established role for carotenoids in nature is the protection of membranes of organisms exposed to solar radiation against photodynamic destruction. Our studies have suggested that carotenoids may fulfill a similar role in *A. laidlawii.*[322] Appropriately enough, *Acholeplasma* species have frequently been isolated from sources other than the animal body, such as sewage, soil, and the surfaces of plants,[90,419] environments in which there is a danger of solar radiation.

IV. MEMBRANE PROTEINS

A. Molecular Properties

1. Electrophoretic Analysis

The great variety of enzyme and transport activities localized in the single membranous structure of the mycoplasma cell would lead one to predict a high protein content and a great number of different proteins in the membrane. In fact, about two thirds of the dry weight of mycoplasma membranes is protein, and in *M. gallisepticum* protein comprises about 80% of the total membrane mass.[185,271,274,341] The recent application of the two-dimensional electrophoresis technique to mycoplasmas disclosed several hundred polypeptide spots upon analysis of cell proteins of various *Mycoplasma, Acholeplasma,* and *Spiroplasma* species.[18,19,126,216,308] Of the approximately 320 polypeptide spots detected by fluorography of labeled *A. laidlawii* cell proteins, about 140 were associated with the cell membrane.[19] While this value may approach the real number of membrane proteins in *A. laidlawii,* several factors may influence the number of spots on the electrophoregram. Thus, complex membrane proteins, such as the proton-translocating ATPase, dissociate into their component subunits during solubilization and electrophoresis in the presence of sodium dodecyl sulfate, so that each of these complex proteins may be represented by more than one spot in the electrophoregram.[443] In addition, proteins from the growth medium which frequently contaminate cell or membrane preparations (Section II.B.) may appear in the electrophoretic patterns. Cultivation of the organisms with ^{14}C-amino acids and visualization of the radioactive polypeptide spots by fluorography overcomes this difficulty.[19,216]

The electrophoretic patterns of mycoplasma membrane proteins are highly reproducible and species specific.[336] Moreover, membrane protein composition is usually not affected by variations in growth conditions. Thus, radical changes in the fatty acid composition of *A. laidlawii*[12,248] or in the cholesterol content of *M. mycoides* subsp. *capri* membranes[15,343] were not accompanied by any significant changes in the electrophoretic patterns of membrane proteins of these organisms. Shifting down the growth temperature of *A. laidlawii* from 37°C to 25°C, or aging of *M. hominis* cultures caused only relatively minor changes in the protein profile of the membranes.[9,12] It is clear that the composition of the major protein components of the mycoplasma membrane is a much more stable characteristic of a given species than the composition of its membrane lipids. This realization led to the proposal of using electrophoretic patterns of membrane proteins for the identification and classification of mycoplasmas.[336]

The molecular weights of mycoplasma membrane proteins, assessed by electrophoresis in sodium dodecyl sulfate-containing polyacrylamide gels, range from about 15,000 to over 200,000 daltons,[10,19,130,214,434] well within the range reported for other biological membranes. The isoelectric focusing step of the two-dimensional electrophoretic procedure revealed that the majority of *A. laidlawii* and *M. gallisepticum* membrane proteins are acidic, having pH values between 4 and 7.[18,19]

2. Solubilization and Purification

Complete characterization of any membrane protein requires its purification. To achieve this, the protein must first be detached from the membrane. It is relatively easy to detach the so-called peripheral or extrinsic proteins which are not immersed in the lipid bilayer, and are attached to the membrane mainly by ionic bonds and salt bridges.[374] Changes in ionic strength, in the pH of the suspending medium, or the addition of EDTA detach these proteins in a water-soluble and lipid-free form. It is not known how many of the mycoplasma membrane proteins belong to this category. It appears reasonable to assume that the isolation of mycoplasma membranes by osmotic lysis of the organisms in deionized water, or in any dilute buffer, and the subsequent washings of the membranes cause the detachment of a significant portion of the peripheral proteins. Two-dimensional electrophoresis of the soluble and membrane fractions of osmotically- lysed *A. laidlawii* showed many protein species common to both fractions, suggesting that these are loosely bound peripheral membrane proteins partially detached from the membrane during cell lysis.[19] Hence, washed mycoplasma membranes are already deficient in some peripheral proteins, before the conventional EDTA treatment used to release this class of membrane proteins is applied. The values of 8 to 14% for peripheral membrane proteins released by EDTA treatment of *A. laidlawii* and *M. hominis* membranes isolated by osmotic shock[7,227] are apparently minimal values, as they do not take into account the proteins detached during osmotic lysis. Higher values for peripheral membrane proteins in *A. laidlawii* were reported by Archer et al.[19] Exhaustive washings in low ionic-strength solution combined with EDTA treatment of membranes released about one-half of the total membrane protein, including about 60 of the 140 membrane polypeptides detected by two-dimensional electrophoresis.

Although peripheral membrane proteins released from the membrane can, by definition, be purified by the usual protein purification procedures devised for water-soluble proteins, none of the mycoplasma membrane proteins belonging to this category has been purified and characterized. The finding that major enzymatic activities such as ATPase, NADH oxidase, and *p*-nitrophenylphosphatase, as well as the main protein antigens cannot be released from *A. laidlawii* membranes by procedures used to detach peripheral proteins[227,228] seems to have discouraged efforts to characterize the peripheral membrane proteins of this organism. This should not be interpreted to mean that peripheral membrane proteins in mycoplasmas do not fulfill important roles.

It appears that most proteins in mycoplasma membranes are integral or intrinsic proteins immersed to a varying depth within, or transversing the lipid bilayer. Detergents or organic solvents are the most common agents used for solubilizing these proteins.[128,267] Detergents have proven more effective than organic solvents in solubilizing mycoplasma membranes.[227,325] Although the ionic detergent sodium dodecyl sulfate has been very effective in membrane solubilization,[21,227,286,287] it causes the denaturation of membrane proteins, ruling out its use for the solubilization of membrane-bound enzymes.[227] The milder nonionic detergents (Triton X-100, Lubrol, Brij-58, Tween 20) and bile salts have, therefore, been used quite extensively for the solubilization and fractionation of mycoplasma membrane proteins.[130,154,227,228,446] However, these detergents fail to solubilize many of the membrane proteins, and their removal from the solubilized membrane material is usually difficult.[226] The finding that the mild nonionic detergents selectively solubilize some of the membrane proteins has been successfully applied to obtain membrane fractions specifically enriched with certain proteins. Thus, treatment of *A. laidlawii* with Tween 20 provided solubilized membrane material highly enriched with several membrane proteins,[130,151,152] facilitating the fractionation and purification of these proteins.[153] On the other hand, treatment of *S. citri* membranes with Tween 20[445,446] or sodium lauryl sarcosinate (Sarkosyl)[444] left an insoluble residue highly enriched in the major membrane protein spiralin, enabling its purification and characterization.[443,446]

Early attempts to fractionate solubilized membrane proteins on Sephadex® columns containing deoxycholate resulted in the separation of membrane proteins into several reproducible peaks, most of which were devoid of membrane lipids.[226,283] Some of the fractions retained enzymatic and immunologic activities. The recent application of the more advanced methodology of membrane protein fractionation to mycoplasma membranes resulted in the purification and characterization of several mycoplasma membrane proteins (Table 5). Basically, the fractionation procedure consisted of membrane solubilization by mild detergents and fractionation of the solubilized material by gel filtration, agarose-suspension electrophoresis, or preparative polyacrylamide- or dextran gel electrophoresis.[151,153,87-89,446] Another promising approach to isolate relatively large quantities of a purified membrane protein has recently been suggested by Johansson.[150] He coupled Sepharose® beads with monospecific antisera to T_2, an *A. laidlawii* membrane protein purified by immunoelectrophoresis. The coupled beads were shown to specifically adsorb protein T_2 from a mixture of *A. laidlawii* membrane proteins solubilized in Tween 20. The adsorbed protein could be eluted from the beads by decreasing the pH to 2.3. A somewhat similar approach, based on affinity chromatography, has recently been applied by us to the isolation of *M. pneumoniae* membrane proteins with high affinity for glycophorin, the sialoglycoprotein which serves as a receptor site for the mycoplasmas adhering to the human erythrocyte cell surface.[21] In this case the affinity column consisted of Sepharose® beads conjugated with glycophorin (Section VI.A.3.)

3. Amino Acid Composition

Relatively few integral membrane proteins have been subjected to amino acid sequence analysis. The data obtained for these proteins indicate that the amphipatic nature of the integral proteins depends on the presence of regions in the polypeptide chain very rich in nonpolar amino acids. In most cases, total amino acid analysis of membranes does not reveal a preponderance of nonpolar amino acids.[423] Hence, it is not surprising that the early amino acid analyses of total membrane proteins of *A. laidlawii* revealed only a slightly higher percentage of nonpolar amino acids than in water-soluble globular proteins.[58,93,214] Similarly, membranes of *T. acidophilum,* when analyzed as a whole, contained about the same percentage of nonpolar amino acids as the cytoplasmic proteins.[345] Clearly, analysis of isolated membrane proteins may yield more meaningful data. Thus, the amino acid composition of the recently isolated glycoprotein from *T. acidophilum* shows 62 mol% nonpolar residues.[450] However, the amino acid composition of the few other mycoplasma membrane proteins isolated (Table 5), though unusual in the case of the *M. pneumoniae* glycoprotein (having an extremely high content of glycine and histidine residues),[158] does not reveal any significant excess

Table 5
PROPERTIES OF PURIFIED MYCOPLASMA MEMBRANE PROTEINS

Protein	Organism	Molecular weight	Amino acid composition	Carbohydrate moiety	Localization in membrane	Ref.
T_2	*A. laidlawii*	52,000	No excess of hydrophobic amino acids; No cysteine	None	External surface	153
T_3		110,000	As for T_2	None	Inner surface	
T_{4a}		34,000	As for T_2; contains flavin	None	Inner surface	
T_{4b}		52,000	As for T_2, but contains half-cysteine	None	Inner surface	
D_{12}		140,000	As for T_2	Low content of amino sugars, galactose and glucose	Inner surface	458
Spiralin	*S. citri*	26,000	Lacks methionine, histidine and tryptophan; contains two half-cystine per molecule; possesses amphiphilic properties	None	Probably spans the membrane; tends to form oligomers in membranes	442—446
Glycoprotein	*M. pneumoniae*	60,000	Very high glycine and histidine	About 7% carbohydrate, mostly glucose, galactose and glucosamine	External surface	158
Glycoprotein	*T. acidophilum*	152,000	62 mol% hydrophobic residues, more acidic than basic amino acids	About 8 to 10% carbohydrate, mannose, glucose and galactose (molar ratio 20:2:1)	External surface	450

of nonpolar amino acids. Determination of the amino acid sequence of the polypeptide chains is apparently required to clarify this point.

A peculiar property of *T. acidophilum* membrane proteins has been reported by Smith et al.[390] The ratio of free-COOH to free-NH_2 groups on the membrane surface (4 to 1) is identical to nonthermoacidophilic mycoplasmal membranes. However, the total quantity of these charged radicals is less than half in *Thermoplasma,* indicating a far more hydrophobic cell surface. The repulsion of ionized carboxyl groups at high pH values may be responsible for membrane and cellular disruption. *Thermoplasma* has apparently taken advantage of its acidic environment by requiring H^+ ions for protonation of free-COOH groups on its cell surface to maintain cellular integrity.[176,177]

4. *Glycoproteins*

Glycoproteins appear to be far less abundant in bacterial plasma membranes than in animal cell membranes.[124,354] The fact that the procaryotic plasma membrane is covered by a carbohydrate-rich cell wall is perhaps related to the scarcity of glycoproteins in the wall-covered bacteria. Since mycoplasmas are the only prokaryotes in which the plasma membrane is exposed to the environment, it seems likely that their membranes contain glycoproteins. The available data are too meager for making generalizations, but suffice to indicate that at least some mycoplasmas are totally devoid of glycoproteins. Thus, a previous claim that a glycoprotein is present in *M. gallisepticum* membranes[116] was disproved very recently by showing the absence of sugar residues in any of the numerous membrane proteins separated by two-dimensional gel electrophoresis.[18] Nevertheless, for two mycoplasmas at least, *M. pneumoniae* and *T. acidophilum,* there is good evidence for the presence of one or more glycoproteins. Electrophoresis of *M. pneumoniae* membranes in polyacrylamide gels containing sodium dodecyl sulfate revealed a protein band (molecular weight about 60,000) which stained red with the periodic acid-Schiff (PAS) reagent.[161] This protein was partially characterized and found to consist of about 80 to 90% amino acids and 7% carbohydrate (Table 5). Lactoperoxidase-mediated iodination indicated that the glycoprotein is exposed on the external cell surface, suggesting that it possibly serves as a binding site in *M. pneumoniae* attachment to epithelial cells.[158] Although the participation of the glycoprotein in adherence has not yet received any experimental support,[21] the presence of the glycoprotein in *M. pneumoniae* has been confirmed by both one- and two-dimensional-electrophoretic analysis of the membranes.[18,21] However, the possibility that this glycoprotein is a serum contaminant, adsorbed to the cell surface during growth, cannot be ruled out so long as data showing its biosynthesis by the organism are lacking. The danger of drawing erroneous conclusions due to the contamination of membranes with serum glycoproteins has been recently illustrated by Paroz and Nicolet.[243] A major PAS-positive band (molecular weight about 60,000, like that of the *M. pneumoniae* glycoprotein), detected in electrophoretic patterns of a series of *Mycoplasma* and *Acholeplasma* species, was shown to originate from the serum component of the growth medium.

Electrophoretic analysis of *T. acidophilum* membranes revealed two PAS-positive bands with an apparent molecular weight of 180,000 and 152,000.[450] The major 152,000-dalton protein, isolated and purified by concanavalin A-affinity chromatography, accounted for 32% of the total membrane protein. The carbohydrate moiety, consisting mostly of mannose residues (Table 5), amounted to less than 10% of the glycoprotein by weight. It was found to be *N*-glycosidically linked to the polypeptide and to possess a highly branched pattern of carbohydrate chains. These properties are highly reminiscent of eukaryotic glycoproteins, and have not yet been described in prokaryotes. The notion that glycolipids act as membrane reinforcers (Section III. A.

2.) may, of course, apply to the glycoprotein of *T. acidophilum* as well. In fact, Yang and Haug[450] suggest that the highly branched carbohydrate moieties of the major glycoprotein of the thermoplasma form a coating on the cell surface, where water molecules are immoblized. The resulting water turgor and the interactions between carbohydrates may contribute to the high rigidity of the thermoplasma membrane. Since *T. acidophilum* grows in a protein-free medium, glycoproteins detected in its membrane are doubtless of endogenous origin. Yet, the marked compositional similarity of the carbohydrate moieties of the glycoprotein[450] and the lipopolysaccharide of *T. acidophilum*[176,200] (Section III.A.3.) raise the question of a possible relationship between these compounds, a point which is not treated by Yang and Haug.[450]

An indirect way to look for glycoproteins in mycoplasma membranes is to examine the binding of ^{125}I-labeled lectins or lectins conjugated with ferritin or iron-dextran to mycoplasma cells.[167,360,361] The near-equal binding of the labeled lectins to cells of a variety of *Mycoplasma* and *Acholeplasma* species and to their isolated membranes suggests that all the carbohydrate-containing components of these mycoplasmas are exposed on the cell surface.[8,167] To determine whether the carbohydrate groups interacting with the lectins are part of glycoprotein molecules, proteolytic digestion or lipid extraction were performed on membranes. With most *Mycoplasma* and *Acholeplasma* species, proteolytic digestion increased lectin binding, whereas lipid extraction abolished it,[167,357] indicating that the lectin-binding sites are glycolipids, lipopolysaccharides, or carbohydrate slime layers. Nevertheless, proteolytic digestion of *M. hominis*[167] and *S. citri*[160] membranes diminished their lectin binding capacity and their carbohydrate content, suggesting that the lectin binding sites are parts of glycoprotein molecules in these organisms. Electrophoretic analysis of these membranes failed to reveal any PAS-positive bands, however, which may be due to a low carbohydrate content of the proteins and low sensitivity of the PAS staining technique when applied to glycoproteins devoid of sialic acid.[160] Protein D_{12}, isolated from *A. laidlawii* membranes (Table 5), which appears to contain a low amount of carbohydrate, failed to stain with PAS.[458]

5. Contractile Proteins

Although actin and myosin are not usually regarded as membrane proteins, the association of contractile elements made of actin and myosin with membranes plays a paramount role in many biological processes in eukaryotic cells.[64] Prokaryotes have long been believed lacking in actin and myosin. Recent studies, however, raise some doubts about the validity of this. The initial claim by Minkoff and Damadian[212] that they isolated a crude protein fraction with actin-like characteristics from *E. coli* was later confirmed by Nakamura and Watanabe[224] who identified a myosin-like protein in addition to the actin-like protein in *E. coli.* Both proteins were localized in the membrane fraction. Typical "arrowhead" complexes formed on interaction of the actin-like rich fraction of *E. coli* with rabbit heavy meromyosin were demonstrated by electron microscopy,[187] further supporting the presence of an actin-like protein in this prokaryote. Nevertheless, the available data do not appear solid enough to enable drawing a definite conclusion on this issue. Since the *E. coli* elongation factor EF-Tu resembles eukaryotic actin in many ways, it may be confused with actin.[344] The different paracrystalline arrays, and the lack of antigenic relationship indicate clearly that EF-Tu and actin are different proteins.[447]

A very strong case can be built supporting the existence of a contractile protein system in mycoplasmas. Contractile processes can easily be observed in these plastic wall-less prokaryotes. Mycoplasmas divide by constriction of their plasma membrane[46,280] and reversible shape changes have been observed in *M. mycoides*[420] and *M.*

hominis cells.[46] *M. pulmonis, M. pneumoniae*, and *M. gallisepticum* glide on liquid-covered surfaces,[44] and spiroplasmas twitch and rotate.[74] It is therefore not surprising that mycoplasmas were among the first prokaryotes to be examined for the presence of contractile proteins. Neimark[229] identified a protein in cell extracts of *M. pneumoniae* that resembled eukaryotic actin in solubility in high-ionic salt solutions and insolubility at low ionic strength, in molecular weight, and its ability to form, in the presence of ATP and Mg^{2+} long curvilinear filaments 5 to 6 nm wide, which closely resemble eukaryotic filamentous actin. Typical arrowhead complexes were formed on the addition of vertebrate heavy meromyosin.[229] Claims for detection of actin-like proteins in *A. laidlawii, M. gallisepticum,* and *T. acidophilum* sprang up almost simultaneously.[111,166,365] However, in all cases the presumed actin-like protein was not purified. Recent work [219] revealed that fractions isolated from *A. laidlawii* cytoplasm by DEAE-cellulose chromatography inhibited pancreatic DNase I activity, a characteristic of eukaryotic actin. However, we failed to isolate the DNase I inhibitory substance by affinity chromatography on an immobilized DNase I column, as did other investigators.[309,459]

Although actin is one of the few proteins which has been well conserved during the evolution of eukaryotes, it is not likely that an identical protein is to be found in prokaryotes. In fact, two-dimensional electrophoretic analysis of *M. pneumoniae, M. mycoides, M. gallisepticum, M. hominis, S. citri,* and *A. laidlawii* cell proteins labeled with radioactive amino acids during growth failed to show any labeled protein coelectrophoresing with eucaryotic α-actin.[309] The protein missing fram a nonhelical and nonmotile *S. citri* strain could not be identified as actin either.[418] The recent demonstration of a strong positive reaction of formaldehyde-fixed *S. citri* cells with an antiserum to actin[438] is therefore most interesting. It should be noted however that the antiserum was prepared against denatured invertebrate actin, isolated by polyacrylamide gel electrophoresis in the presence of sodium dodecyl sulfate, a factor which may have radically influenced the serological activity of the protein.

Striated fibrils, 3.6 nm in diameter, with a repeat interval of approximately 9 nm along their length, have been observed in several spiroplasmas.[396,437] The spiroplasma fibrils are clearly different from actin filaments. Efforts to isolate and characterize the fibrils have only recently proved successful.[417] The fibrils isolated from *S. citri* consist of a single protein with a molecular weight of 55,000 (similar to that of tubulin), and show a marked tendency to aggregate and form hollow tubes (Figure 11). However, unlike most eucaryotic microtubules, the *S. citri* structures do not dissociate when cooled to 0°C, and are unaffected by colchicine. It has not yet been determined whether these structures are contractile.[460] The recent report by Margulis et al.[193] on microtubules, apparently composed to tubulin, in spirochetes of termites indicates that microtubules can indeed be found in prokaryotes.

In conclusion, despite the many remaining questions, the evidence available suggests that contractile proteins do exist in mycoplasmas and other prokaryotes. It is hoped that the intensive investigations currently underway will soon clear up the picture.

B. Enzymatic Activities

1. Electron-Transport Enzymes

The electron-transport system, a series of enzymes acting in concert, is usually localized in the plasma membrane of procaryotes.[353] None of the *Mycoplasma* and *Spiroplasma* species examined conform to this rule, as their electron transport activity, measured as transfer of electrons from NADH to oxygen, is located in their cytoplasmic fraction.[160,186,217,252,255,304] The finding of a truncate flavin-terminated respiratory system lacking quinones and cytochromes in these mycoplasmas[131,132] may explain its lack of

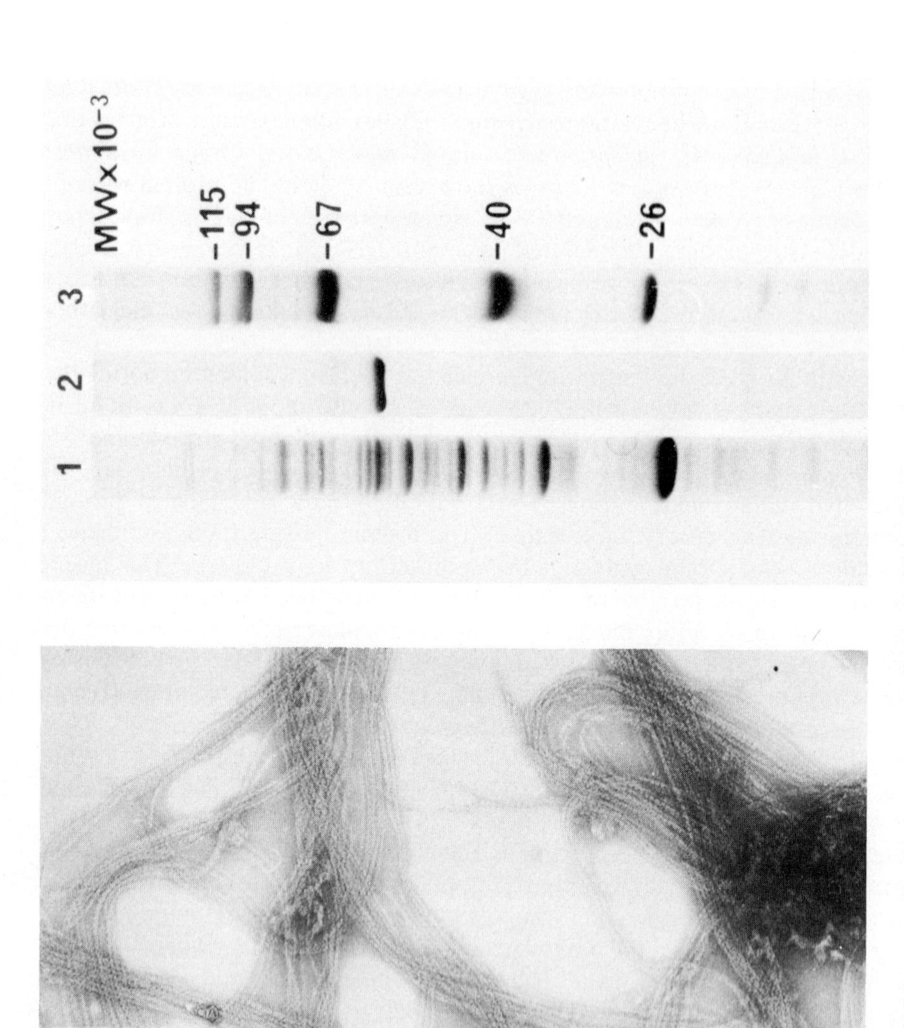

FIGURE 11. Fibrils isolated from *Spiroplasma citri*. (A) Negatively stained preparation. (Magnification × 100,000.) (B) Electrophoretic pattern in polyacrylamide gels containing SDS of: (1) *S. citri* cell proteins, (2) isolated fibrils and (3) molecular weight markers. (From Townsend R., Archer, D.B., and Plaskitt, K.A., J. Bacteriol., 142, 694, 1980. With permission.)

dependence on a membranous structure for organization. In *T. acidophilum,* the only member of the mollicutes with a complete respiratory system consisting of flavins, quinones and cytochromes, [131,132] the electron transport chain is, in fact, membrane-bound. Nevertheless, the presence of a truncated electron-transport system does not necessarily mean that it must be cytoplasmic. Thus, the NADH oxidase activity in all *Acholeplasma* species is membrane bound,[252,253] though in *A. laidlawii* it appears to consist of only two components, [147,148,180,252] like the cytoplasmic NADH oxidase system of *M. pneumoniae.*[186]

Characterization of the membrane-bound NADH oxidase system of *A. laidlawii* is still incomplete. Johansson [458] has recently isolated two proteins from *A. laidlawii* membranes, a flavoprotein (designated T_{4a}, Table 5) and an NADH dehydrogenase, both of which may constitute the NADH oxidase system in this organism. The flavoprotein is a major membrane protein, while the NADH dehydrogenase (estimated mol wt 140,000) is apparently a minor membrane component. The NADH dehydrogenase of *A. laidlawii* was also purified from an ethanol extract of the membranes by deoxycholate and gel filtration.[147] The purified enzyme (estimated to be over 90% pure) did not depend on lipids for activity. This is in accord with earlier findings showing the lack of influence of the physical state of membrane lipids on NADH oxidase activity in *A. laidlawii*[83] and the presence of this enzymatic activity in a solubilized membrane fraction devoid of lipid.[226] Moreover, the NADH oxidase activity of *A. laidlawii* has been shown to be unaffected by enzymatic hydrolysis of the membrane phosphatidylglycerol, a treatment which quelled the lipid-dependent ATPase activity.[32] The above may indicate that the electron transport enzymes in *A. laidlawii* are less integrated with membrane structure than are the complex electron transport chains in other bacteria, such as *Micrococcus lysodeikticus* and *E. coli*, where lipids were found to be required for NADH oxidase activity.[95,223] Yet, it must be stressed that NADH oxidase in *A. laidlawii* does not fit into the category of the loosely bound peripheral membrane proteins, as it cannot be released from the membranes by EDTA and low-ionic strength buffers.[227]

2. Adenosine Triphosphatase and Other Hydrolases

ATPase activity has been detected in the membrane of every mycoplasma examined, [160,194,217,255,311,334,451] indicating its key role in membrane function. Still, we know very little of its molecular properties. The greatest obstacle hampering the characterization of the mycoplasmal ATPase is its close association with the membrane lipid bilayer and the consequently great difficulties involved in the isolation of the enzyme complex in an active form. This stands in marked contrast to ATPases of the wall-covered bacteria, which resemble the mitochondrial and chloroplast proton-translocating ATPases in ease of isolation of the enzyme moiety responsible for ATP hydrolysis. The bacterial ATPase consists of a major moiety, the so-called BF_1, located outside the lipid bilayer, and a minor moiety, BF_0, which presumably forms the proton channel transversing the lipid bilayer. The BF_1 moiety, which retains the ATPase activity, can be readily detached from the bacterial membrane by washing in low-ionic strength buffers.[86] All efforts to release the ATPase from *A. laidlawii, M. mycoides* subsp. *capri*, and *S. citri* membranes by methods employed in the detachment of BF_1 have failed, [160,227,228,334] indicating that the entire mycoplasmal ATPase complex behaves like an integral membrane protein. Unfortunately, the marked sensitivity of the mycoplasmal ATPase to detergents[226-228,334] has hampered the isolation and purification of this enzyme from *A. laidlawii* membranes by techniques devised for the isolation of integral membrane proteins. The dependence of the mycoplasmal ATPase activity on membrane lipids may explain its high sensitivity to detergent action.

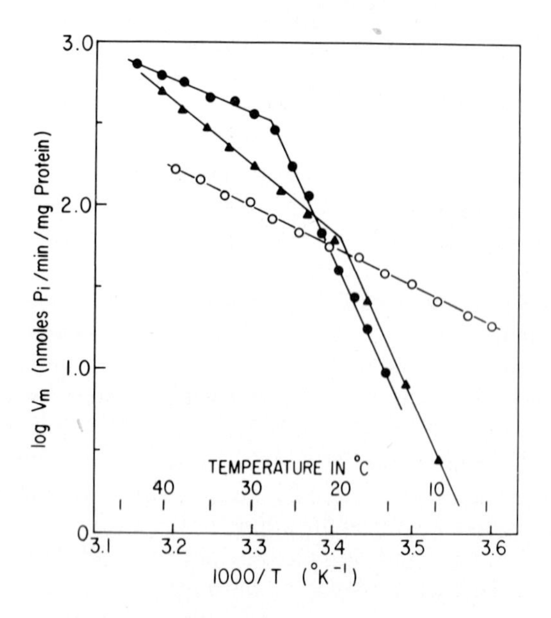

FIGURE 12. Arrhenius plots of the ATPase activity of *Acholeplasma laidlawii* membranes enriched with different fatty acids (●, 16:0, palmitic acid; ▲, $18{:}1_t$, elaidic acid; ○, $18{:}1_c$, oleic acid). (From Silvius, J.R., Jinks, D.C., and McElhaney, R.N., in *Microbiology-1979,* Schlesinger, D., Ed., American Society for Microbiology Washington, D.C., 1979, 10. With permission.)

There is now solid evidence to show that the *A. laidlawii* and *M. mycoides* subsp. *capri* ATPases are dependent on lipids. Arrhenius plots of the ATPase activity of these mycoplasmas show distinct breaks at the temperature of the membrane lipid phase transition.[83,137,321] The temperatures at which the breaks occur depend on the fatty acid composition and on the membrane cholesterol content. The break temperature in *A. laidlawii* falls between the gel and liquid-crystalline phase transition midpoint, at which about two-thirds of the membrane lipid are already in the gel state (Figure 12).[368] Silvius and McElhaney[372] suggest that the ATPase is active only in association with liquid-crystalline lipids forming an annulus around the enzyme. Accordingly, the bulk lipid fatty acyl chain structure does not affect the temperature dependence of the ATPase activity, so long as the lipids surrounding the enzyme remain in the liquid-crystalline state. The nature of the lipid comprising the annulus surrounding the *A. laidlawii* ATPase has been elucidated by Bevers et al.,[32] who showed that hydrolysis of over 90% of the membrane phosphatidylglycerol by phospholipase A_2 abolished ATPase activity. Activity could be restored by phosphatidylglycerol liposomes, but not by liposomes of any of the other A. *laidlawii* membrane lipids. The fatty acid composition of phosphatidlyglycerol determined the activation energy of the enzyme and the temperature at which the break in the Arrhenius plot occurred. It is significant that only about 10% of the membrane phosphatidylglycerol (i.e., less than 3% of the total membrane lipids) is required for ATPase activity, supporting the idea that this lipid forms an annulus of molecules closely associated with the enzyme protein.[32]

Because the mode of attachment to the membrane of the mycoplasmal ATPases is different from that of other bacterial ATPases, the crucial question arises whether the mycoplasmal ATPase differs from the bacterial ATPases in function as well. Most, if not all, of the bacterial ATPases characterized thus far resemble the proton-translocating ATPases of mitochondria and chloroplasts in composition and properties.[86] As no information is available on the composition of the mycoplasmal ATPases, comparisons cannot be made on this basis. Elucidation of the ionic requirements and susceptibility of the mycoplasmal ATPases to various inhibitors may lead to better understanding of the role of this enzyme in mycoplasma biology. Generalizations should be avoided however, since almost all the available information is restricted to the *A. laidlawii* and *M. mycoides* subsp. *capri* ATPases. The mycoplasmal ATPase requires Mg^{2+}, resembling in this respect the proton-translocating ATPases of other bacteria. Calcium cannot replace Mg^{2+} and may even inhibit ATPase activity.[149,182,334,451] The mycoplasmal ATPase is not stimulated by K^+ and Na^+, and is not inhibited by ouabain, in contrast to the (Na^+-K^+)- activated ATPases of eukaryotic cell membranes.[56,182,334,451] However, Jinks et al.[149] recently reported a three- to four-fold stimulation of the *A. laidlawii* ATPase by Na^+, though not by K^+. This observation led them to propose that the ATPase in *A. laidlawii* functions as an ion pump, extruding cations (probably Na^+) to counterbalance the osmotic stress resulting from intracellular accumulation of ions under a Donnan potential, as the (Na^+-K^+) does in eukaryotic cells. The findings that *A. laidlawii* cells suspended in isosmotic solutions of NaCl or KCl slowly lyse in the absence of an energy source to activate the presumed ion pump support this hypothesis.[149] Lysis was enhanced by thiol group reagents known to inactivate the ATPase. On the other hand, cells suspended in isotonic solutions of sucrose or $MgSO_4$ were quite resistant to lysis induced by energy depletion or thiol group reagents.

Obviously, the crucial question is whether the mycoplasmal ATPases can translocate protons. As described in detail elsewhere, there is evidence supporting the ability of the *A. laidlawii* and *M. mycoides* subsp. *capri* ATPases to serve as proton channels.[182,407] Thus, the generation of ΔpH (alkaline inside) in *M. mycoides* subsp. *capri* during cell energization was completely inhibited by $5 \times 10^{-5} M$ *N,N'*-dicyclohexylcarbodiimide (DCCD), a specific inhibitor of proton translocating ATPases.[27,182] Yet, DCCD inhibited only up to 30 to 40% of the activity of the *M. mycoides* subsp. *capri* enzyme as expressed by ATP hydrolysis. The relatively low sensitivity of the mycoplasmal ATPase activity to DCCD as compared to that of other bacteria has also been noted for *A. laidlawii*.[149] The low sensitivity of the mycoplasmal ATPases to DCCD may reflect structural differences between them and the bacterial BF_0-F_1 type ATPases,[149] or may be due to the different mode of association of the mycoplasmal enzyme with the membrane. Another possibility is that DCCD-resistant ATP hydrolyzing enzymes exist in mycoplasma membranes in addition to the DCCD-sensitive proton translocating ATPase.[182] None of these possibilities can be confirmed or refuted until more information on the structural features of the mycoplasmal ATPases becomes available.

The question of whether the ATPases and *p*-nitrophenylphosphatase activities of *A. laidlawii* are the expression of a single enzyme[228] can now be answered negatively. The different sensitivities of the two enzymatic activities to detergents,[227,228] organic solvents and pronase digestion,[228] the lack of dependence of the *p*-nitrophenylphosphatase on membrane lipids for activity[32] and its resistance to DCCD[149] all lead to the conclusion that the *p*-nitrophenylphosphatase is an enzyme entity different from the ATPase.

In addition to the ATPase and *p*-nitrophenylphosphatase, the *A. laidlawii* membrane contains a variety of hydrolytic enzymes, which are so far characterized only poorly. They include phosphatases acting on nucleotides,[334] ribonuclease and deoxyribonuclease,[228,255] peptidases,[59,247] and a lysophospholipase.[425] Degradation of macromolecules in the growth medium into small molecules capable of permeating the cells may be one function of the membrane-bound hydrolases. Thus, the nucleases on the membrane surface of *A. laidlawii*[228] enable it to utilize RNA and DNA as a source of purine and pyrimidine bases it requires for growth.[284]

3. Membrane Biosynthesis

The enzymes responsible for membrane lipid biosynthesis in mycoplasmas are mostly membrane bound. Thus, the enzymes participating in the synthesis of phospholipids, glucolipids, and phosphoglucolipids in *A. laidlawii* have been localized in the membrane of this mycoplasma.[283] The acyl-CoA:α-glycerophosphate transacylase which synthesizes phosphatidic acid, the precursor of membrane phospholipids, is associated with the membrane of *A. laidlawii* and *M. hominis,* but the activation of fatty acids by acyl-CoA synthetase has been localized in the cytoplasmic fraction of these mycoplasmas.[323] A thioesterase acting specifically on long-chain fatty acyl-CoA has been detected in membranes of several mycoplasmas. Since its activity is highest in *Acholeplasma* species, and very low in most of the *Mycoplasma* species tested, it may have a regulatory role in fatty acid biosynthesis.[342]

The rate of membrane lipid biosynthesis can differ markedly from that of membrane protein synthesis under different growth conditions, resulting in membranes with different lipid-to-protein ratios. Thus, membranes of *A. laidlawii* cells grown at high pH have a much lower lipid-to protein ratio than membranes of cells grown at a low pH value,[164] while aging of mycoplasma cultures is accompanied by a marked decrease in the membrane lipid-to-protein ratio.[9,12,217,269,323,435] The most striking results are obtained by the addition of chloramphenicol to actively growing mycoplasma cultures. Lipid synthesis continues for several hours after protein synthesis is halted, resulting in membranes much richer in lipid.[12,162,269] Hence, membrane lipid synthesis in mycoplasmas is not strictly synchronized with membrane protein synthesis, so that a stoichiometric relationship between protein and lipid is not essential for membrane function, a conclusion also supported by the findings of Mindich[211] on plasma membrane biosynthesis in *Bacillus subtilis.*

Mycoplasmas have distinct advantages over wall-covered bacteria in studies on membrane protein biosynthesis, secretion of proteins through the membrane, and the influence of the composition and physical state of the lipid bilayer on these processes. The fact that nothing has been reported on this "hot" subject so far apparently reflects the unjustified fears of many investigators of the difficulties in working with mycoplasmas.

C. Disposition of Membrane Proteins

1. Transbilayer Distribution

The asymmetrical transbilayer distribution of membrane proteins is a major feature of biological membranes. The distribution of proteins in plasma membranes shows absolute asymmetry in the sense that the proteins facing the cytoplasm are different from those exposed on the cell surface. In the case of proteins spanning the membrane, their moieties facing the cytoplasm differ from those facing the cell exterior.[316] Most of the techniques developed for studying membrane protein disposition are based on the same principle: specific labeling or proteolytic digestion of membrane proteins in intact cells and isolated membranes, assuming that the labeling agent or proteolytic enzyme has access to proteins exposed on both membrane surfaces when isolated membranes are used, but only to proteins on the outer membrane surface when intact cells are used. The osmotic sensitivity of mycoplasmas poses a serious difficulty in disposition studies. Since labeling or proteolytic digestion are carried out on washed cells suspended in buffer, it is essential to ensure that the cells do not lyse during treatment. The tendency of washed mycoplasmas to lose intracellular components and even lyse on incubation in buffer at 37°C is well known.[51,393] The recent finding of Jinks et al.[149] that *A. laidlawii* cells lyse in iso-osmotic solutions of NaCl or KCl unless an energy source (glucose) is present provides a possible explanation for the phenomenon. (Section IV.B.2.) The use of sucrose (about 0.3 to 0.4 *M*) as an osmotic stabilizer is thus recommended. Another difficulty which may hamper disposition studies is the tendency of many isolated biomembranes to reseal spontaneously, preventing the access of the labeling agent to proteins on one of the membrane surfaces. As will be discussed in detail later mycoplasma membranes do not usually reseal upon their isolation.

The lactoperoxidase-mediated iodination technique has been the method most extensively used to localize proteins in mycoplasma membranes.[7, 9-12,19,139,161] The high-molecular weight lactoperoxidase cannot penetrate mycoplasma cells, so that only the tyrosine residues of proteins exposed on the outer membrane surface are labeled by ^{125}I when intact cells are treated. A minor fraction of the label (5 to 10%) is detected in cytoplasmic proteins and membrane lipids.[10] Labeling of membrane lipids by this technique has been found useful for determination of their transbilayer distribution[123] (Section III.B.1). The main conclusion that can be drawn from the iodination experiments is that more proteins are exposed on the cytoplasmic than on the external membrane surface in mycoplasmas,[7,9,10,12,126] as is probably true for plasma membranes in general.[316] Two-dimensional electrophoresis showed that of about 140 proteins detected in *A. laidlawii* membranes, 90 to 100 proteins were iodinated on treatment of isolated membranes, and a maximum of 40 were labeled after iodination of whole cells.[19] The significant number of proteins which remained unlabeled on iodination of isolated membranes suggests that these proteins are either totally buried in the lipid bilayer or lack the iodine-binding tyrosine residues on their polypeptide moieties exposed to the aqueous surroundings.

Trypsin is preferred over pronase for protein disposition studies because it is a milder, more selective and better defined proteolytic enzyme.[9,10] The results obtained with this enzyme corroborated the iodination data by showing that many fewer membrane proteins are affected on treatment of whole mycoplasma cells than on treatment of isolated membranes.[9,10,18] The finding that most of the membrane proteins exposed on the external cell surface are of high molecular weight[9,10] may imply that these proteins span the membrane, but experimental proof is lacking. Trypsin was also found useful for the localization of several membrane-bound enzymatic activites in mycoplasmas. The only slight decrease in the ATPase, NADH oxidase, and acyl-CoA thioesterase activities upon trypsin treatment of whole *A. laidlawii* cells, compared with the essentially complete inactivation of these enzymatic activities when isolated membranes were trypsinized, indicates that these enzymes are located on the cytoplasmic face of the membrane.[228,342]

The crossed immunoelectrophoretic technique, so useful in the localization of membrane proteins and enzymes in plasma membranes of wall-covered bacteria[236,237,354] was, in fact, first applied to mycoplasma membranes by Johansson and Hjertén.[152] A major advantage to using antibodies for localization of membrane components is that they cannot penetrate the cells, and unlike the techniques employing proteolytic enzymes do not modify membrane components, thus minimizing membrane perturbation. In addition, this technique enables assessment of the relative quantities of the various membrane proteins and identification of membrane-bound enzymes by specific zymogram staining procedures.[237] Crossed immunoelectrophoresis has not been extensively used with mycoplasmas despite its merits. The early studies of Johansson and Hjertén[152] showed that only one out of four purified *A. laidlawii* membrane proteins is exposed on the external cell surface. Similar studies with *M. arginini* revealed one major antigen on the outside, and two others totally immersed in the lipid bilayer, since antibodies to them could only be adsorbed by Triton-solubilized membrane material.[4]

All mycoplasma membranes examined by freeze-fracturing show the presence of particles, 50 to 100 Å in diameter, on the two fracture faces. The interpretation of these particles as the morphologic manifestation of proteins or proteins with tightly bound lipids intercalated into the membrane is widely accepted.[429] The protein nature of the intramembranous particles in mycoplasmas has been supported by the finding of markedly fewer particles in membranes of *A. laidlawii* in which protein synthesis was halted by puromycin or by depletion of the amino acid supply in the growth medium.[416] As with other plasma membranes, the distribution of particles on the two fractures surfaces of mycoplasma membranes is asymmetrical; more particles are observed on the convex fracture face, corresponding to the inner half of the membrane lipid bilayer.[121,190,281,343,413,415] The larger number of particles in the inner part of the bilayer supports findings by other techniques that more proteins face the cytoplasm than the external cell surface.

2. Lateral Mobility

The lateral diffusion of membrane lipids and proteins within the plane of the membrane is a well-established phenomenon. The rate of diffusion depends on the nature and size of the molecule, the physical state of the lipid bilayer, association of the molecule with other membrane components, etc.[316] The relatively rapid lateral mobility of the membrane components would predict their homogeneous distribution in the plane of the membrane. This is not always the case. One of the most common reasons for nonhomogeneous distribution of membrane components is the lateral phase separation of membrane lipids which may take place within the temperature range permitting normal growth of mycoplasmas (Section III.C.1.) The differential crystallization of lipids in the bilayer influences the distribution of proteins partially or wholly immersed in it, as evidenced by freeze-fracture electron microscopy. The progressive crystallization of lipids in the bilayer causes the aggregation of the intramembranous particles into patches of the more fluid lipids. Since phase transition of lipids is averted in cholesterol-rich membranes, aggregation of intramembranous particles could be demonstrated only in *A. laidlawii* [145,428,431] or in the cholesterol-poor *M. mycoides* subsp. *capri* membranes chilled to 4°C before freeze-fracturing.[343] The distribution of particles on the fracture faces of the cholesterol-rich mycoplasma membranes, such as those of *M. gallisepticum,*[190] *M. mycoides* subsp. *capri* and subsp. *mycoides,*[343] *M. meleagridis,*[121] and *S. citri*[281] was largely homogeneous.

A nonhomogeneous lateral distribution of proteins in the membrane is illustrated by *M. gallisepticum.* A small bleb bounded by the plasma membrane can be seen at one or both ends of this cocobacillary organism. A membrane fraction rich in blebs contains most of the ATPase activity of the cells,[192] confirming the histochemical localization of this membrane-bound enzyme at the bleb area.[221] In addition, freeze-fracturing shows fewer intramembranous particles on the membrane fracture faces around the bleb.[111] The bleb may thus serve to distinguish one area of the membrane from the other both morphologically and biochemically.

The planar distribution of proteins exposed on the membrane surface can be studied by an avidin-ferritin label.[431,432] The membranes are treated with a biotin-avidin-ferritin complex which specifically links to free amino groups of the exposed membrane proteins. By visualizing the distribution of the ferritin-containing complex on the surface of *A. laidlawii* membranes exposed to different temperatures, Wallace et al.[431,432] concluded that the labeled sites were dispersed relatively homogeneously above or below the lipid phase transition, whereas in membranes labeled at the midtransition, low- and high-density patches of label were observed. As in the case of freeze-fracturing, these results signify that lateral mobility of membrane proteins does exist and is influenced by the physical state of membrane lipids. However, in contrast to the results of freeze-fracturing where the intramembranous particles remained aggregated below the phase transition, the labeling data indicate that lateral mobility of membrane proteins also takes place when membrane lipid is entirely in the gel state. Evidently, freeze-fracturing and the avidin-ferritin labeling focus on two classes of membrane proteins which are differently influenced in their distribution by the physical state of membrane lipids. According to Wallace and Engleman[431] the proteins represented by the intramembranous particles are those deeply immersed in the lipid bilayer, so that their mobility is influenced by the lipids on both the cytoplasmic and the exoplasmic halves of the bilayer. Surface proteins, on the other hand, are superficially immersed in the lipid bilayer, and thus are influenced by lipids in only one half of the bilayer. As the transbilayer distribution of lipids is very frequently asymmetric (Section III.B.1.), the physical properties of the two leaflets of the bilayer may be different, causing the differences in the distribution of the surface and deeply immersed proteins.

3. Vertical Displacement

The dependence of the lateral mobility and distribution of membrane proteins on the physical state of membrane lipids leads to the intriguing question of whether the physical state of the lipid bilayer also influences the vertical disposition of the proteins immersed in it. Borochov and Shinitzky[42] have postulated that the position of the amphipathic proteins in the membrane reflects an equilibrium state between the interactions of their hydrophobic parts with membrane lipids and their hydrophilic parts with the aqueous surroundings. Accordingly, with decreasing lipid fluidity, the interaction of the hydrophobic parts of the proteins with the lipids diminishes, resulting in the squeezing-out of the protein which will then occupy a new equilibrium position. Obviously, the opposite vectorial displacement will occur on increasing lipid fluidity. The ease of introducing controlled alterations in *A. laidlawii* membrane fluidity renders this organism an excellent tool for testing this hypothesis.[12] Changes in *A. laidlawii* membrane fludity were brought about in several ways: altering the fatty acid composition of membrane lipids, changing the growth temperature, aging of cultures, and inducing changes in the membrane lipid-to-protein ratio by treatment with chloramphenicol. No consistent correlation could be found between altered membrane fluidity and degree of exposure of membrane proteins to lactoperoxidase-mediated iodination.[12] Nevertheless, iodination values of intact cells regularly decreased upon exposure of the organisms to suboptimal growth conditions employed to introduce changes in membrane fluidity. The iodination data of isolated membranes, on the other hand, remained unchanged regardless of growth conditions.[12]

A possible explanation for the data of Amar et al.[12] is that the variations in the exposure of proteins on the cell surface are caused by changes in the energized state of the membrane of the intact cell, rather than from alternations in membrane fluidity. Supportive evidence for this hypothesis has been provided by Amar et al.[11] Exposure of *A. laidlawii* cells to the ionophore valinomycin, which dissipates the K^+ gradient, or to carbonylcyanide *m*-chlorophenylhydrazone (CCCP), which causes collapse of the proton gradient, resulted in a rapid drop in the iodination values of proteins exposed on the *A. laidlawii* cell surface (Table 6). It has been proposed that the decreased availability of iodine-binding sites on the mycoplasma cell surface may reflect conformational changes in membrane proteins triggered by short-circuiting the membrane potential and proton gradient.

The effect of membrane potential on the mode of organization of membrance components has so far received little attention. However, there are some studies indicating that changes in the electrochemcial ion gradient across cell membranes cause conformational changes in a variety of membrane proteins.[362] The increased availability of the *A. laidlawii* membrane phosphatidylglycerol to digestion by phospholipase A_2,

Table 6
EFFECTS OF VALINOMYCIN AND CARBONYLCYANIDE *m*-CHLOROPHENYLHYDRAZONE (CCCP) ON THE LACTOPEROXIDASE-MEDIATED IODINATION OF INTACT CELLS AND ISOLATED MEMBRANES OF *A. LAIDLAWII*

Inhibitor[a]	^{125}I label (10^5 cpm/mg membrane protein) Isolated membranes	Intact cells	Labeling ratio (membranes/cells)
No inhibitor	30.9	7.5	4.2
Valinomycin ($10^{-5}M$)	30.5	5.8	5.3
CCCP ($10^{-5}M$)	30.4	5.3	5.7
Valinomycin ($10^{-5}M$) +CCCP ($10^{-5}M$)	30.8	2.7	11.2

[a] Cells were treated with inhibitors for 15 min at 37°C. Part of the cells were osmotically lysed and the isolated membranes as well as the intact cells were subjected to lactoperoxidase-mediated iodination. Data of Amar et al.[11]

accompanying decrease in the energized state of the cells,[29] appears relevant to our discussion. Our finding that membrane energization influences the degree of exposure of proteins on the cell surface may have important implications for understanding interactions between cells. Many cell to cell interactions depend on the availability of protein receptors on the cell surfaces. If membrane energization indeed influences the degree of exposure of the proteins on the cell surface, it will consequently influence cell to cell interaction. Testing of this assumption is now underway (Section VI.A.4.).

V. TRANSPORT MECHANISMS

Mycoplasmas are attractive subjects for transport studies because the cells are bounded by a single membrane whose lipid composition can be easily altered in a controlled manner. Hydrophilic solutes are transported into mycoplasma cells by different mechanisms, including simple diffusion (small polyhydric alcohols), active transport (K^+, amino acids, sugars), and group translocation (sugars). Hence, mycoplasmas can be extremely useful in studies on the effects of membrane lipid composition and physical state on a variety of transport phenomena.[62] Recent research has also focused on the role played by chemiosmotic coupling in generating the energy needed for transport in mycoplasmas.

A. Passive Diffusion

1. Unmediated Transport of Polyhydric Alcohols

Transport of small polyhydric alcohols (glycol, glycerol, erythritol) through *A. laidlawii* membranes occurs by simple diffusion, without mediation of a carrier.[76] *A. laidlawii* cells and liposomes prepared from their extracted lipids offered a simple system for studying the effects of alterations in the composition and physical state of membrane lipids on the permeability of small noncharged hydrophilic molecules through the hydrophobic membrane barrier. The permeability of *A. laidlawii* cells and liposomes to the polyhydric alcohols was measured by determining the initial swelling rate of the cells and liposomes at different temperatures.[76,78,204,205,314] The fatty acid composition and cholesterol content of membrane lipids were altered by growing the cells in a lipid-poor medium supplemented with palmitate and either stearate, oleate, elaidate, or linoleate, with and without cholesterol. The intact cells were slightly more permeable than the lipsomes, but the rates for both increased in this order: linoleic acid > oleic acid > elaidic acid > stearic acid. The rate of transport was decreased by the inclusion of cholesterol in the membranes.

Several conclusions can be drawn from these experiments:

1. The remarkable similarity of the results obtained with cells and liposomes strongly indicates that unmediated transport is determined primarily by membrane lipids.
2. The rate of uptake is correlated with membrane fluidity. The more fluid the membrane, the faster the rate of transport.
3. The energy of activation, ΔH^*, for the permeation of the alcohols depends on the solute, rather than on the nature of membrane lipid. De Gier et al.[76] identified the ΔH^* with the energy required for the dehydration of the permeant molecules (i.e., breaking the hydrogen bonds between the alcohol groups and water). Using a value of 5 kcal/mole needed to break one hydrogen bond, and allowing for intramolecular hydrogen bonds to be formed in the dehydrated molecules, the ΔH^* for dehydration of glycol, glycerol, and erythritol would be 15, 20, and 25 kcal/mole, respectively. The values calculated by de Gier et al.[76] from the experimental data were 14.3, 19.4, and 20.8 kcal/mole, respectively, close enough to the aforementioned theoretical values.

B. Active Transport

1. Sugar Transport in A. laidlawii

Early studies of sugar transport in *A. laidlawii* were hampered by the impermeability of the cells to nonmetabolizable sugars such as α-methyl-glucoside and 2-deoxyglucose.[409] Thus metabolizable sugars, such as glucose, maltose and fructose, had to be used, hindering the dissociation of transport from assimilation. Despite this difficulty, these studies[298,409] sufficed to indicate that glucose permeation into *A. laidlawii* cells occurs via a carrier-mediated process. Definite evidence that this process falls within the category of active transport rather than facilitated diffusion came only after membrane vesicles were introduced for the study of sugar transport in *A. laidlawii.*[96,97] In addition, the nonmetabolizable analog, 3-O-methyl-D-glucose (3-O-MG) was found to be taken up by *A. laidlawii* cells against a concentration gradient, [405,406] enabling the dissociation of transport from assimilation.

The evidence that glucose transport in *A. laidlawii* is carrier-mediated and active is as follows:

1. The 100-fold greater rate and the much lower enthalphy of activation for glucose uptake by cells than by liposomes[298] supports the carrier-mediated nature of the process in cells.
2. The strict substrate specificity of transport, capable of distinguishing between D and L-glucose[298] and the competitive inhibition of 3-O-MG uptake by glucose and 6-deoxyglucose[406] also indicate participation of a specific carrier.
3. Saturation kinetics with a Km of 4.6 μ*M* for 3-O-MG uptake by cells[406] and 21.2 μ*M* for glucose uptake by vesicles[239] suggest that there are a limited number of carriers in the cell membrane with a very high affinity for the sugar.
4. Heat inactivation of the glucose transport system between 40° and 45°C[298,406] and inhibition by thiol- and amino-blocking reagents[97,298,406,409] indicate the protein nature of the glucose carrier.
5. Glucose uptake by vesicles and 3-O-MG uptake by cells resulted in the accumulation of the unmodified sugars against significant concentration gradients.[97,406]

6. Accumulation of the sugar is inhibited by compounds known to dissipate a proton motive force, such as dinitrophenol and carbonyl cyanide m-chlorophenyl hydrazone(CCCP).[97,405,406] The last findings are the most crucial for identifying the glucose transport in *A. laidlawii* as an active transport process.

a. The Role of the Proton Motive Force

Uncouplers such as dinitrophenol and CCCP inhibited 3-O-MG accumulation by *A. laidlawii* cells and of glucose by vesicles, suggesting that the active transport of sugars in *A. laidlawii* is energized by a proton motive force.[97,405,406] Accordingly, the flux of neutral sugar molecules is mediated by H^+ symporters in response to the total proton motive force[127] which consists of a proton concentration difference (ΔpH) and a membrane potential ($\Delta\psi$). In fact, Tarshis and Kapitanov[407] demonstrated the concomitant uptake of H^+ with 3-O-MG in *A. laidlawii*, as was indicated by the alkalinization of the medium. The uncoupler CCCP prevented the 3-O-MG-induced pH shifts. This raises the question of how the proton motive force is generated in *A. laidlawii*. In wall-covered bacteria a proton motive force can be generated by either electron transport through the respiratory chain or by ATP hydrolysis by the membrane-bound, magnesium-dependent ATPase complex.[127] As mentioned in previous sections, mycoplasmas possess the proton-translocating ATPase complex. This complex can act in a reversible manner, producing ATP from ADP at the expense of the proton motive force generated by the proton pump of respiration. In this case an influx of protons from the outside of the cell, via the proton channel of the ATPase, is responsible for ATP formation. The reverse happens upon hydrolysis of ATP to ADP by the ATPase. In this case an efflux of protons occurs through the ATPase proton channel, resulting in the formation of a proton motive force. DCCD inhibits these activities by blocking the proton channel of the ATPase complex. Tarshis and Kapitanov[407] provide evidence supporting the reversible activity of the *A laidlawii* ATPase by showing that induced changes in the ΔpH or $\Delta\psi$ across the cell membrane result in a temporary increase in the intracellular concentration of ATP in starved *A. laidlawii* cells. Changes in the ΔpH were produced by the addition of HCl to a starved cell suspension, while changes in $\Delta\psi$ were obtained by adding valinomycin to K^+-loaded cells suspended in a K^+-free medium. Similarly, Benyoucef et al.[27] demonstrated the formation of ΔpH (alkaline inside) in *M. mycoides* subsp. *capri* cells when glucose was added to the cell suspension. The generation of the ΔpH during cell energization was completely inhibited by 5×10^{-5} *M* DCCD, while the uncoupler FCCP, which acts as a proton conductor, disrupted the ΔpH. The conclusion is that the cell energization in *M. mycoides* subsp. *capri* partly consists of the generation of a transmembrane H^+concentration gradient linked to the Mg^{2+}-ATPase activity.

The data obtained by Tarshis et al.[405-407] indicate that in *A. laidlawii* the proton motive force driving transport arise primarily, if not exclusively, from the hydrolysis of ATP generated during glycolysis. *A. laidlawii* is incapable of oxidative phosphorylation, and its truncated respiratory chain, lacking quinones and cytochromes,[254] apparently does not contribute to the formation of a proton motive force. Arsenate, an inhibitor of glycolysis, inhibited 3-O-MG transport, whereas anaerobic conditions had no effect on transport of the sugar.[405,406] Uncouplers and DCCD, while not affecting *A. laidlawii* respiration, inhibited 3-O-MG transport by dissipating the electrochemcial ion gradient or by blocking the ATPase proton channel, respectively.[405,406]

b. Membrane Vesicles as Tools

Membrane vesicles prepared from wall-covered bacteria have been most useful in transport studies in that they enable the dissociation of the transport process *per se* from assimilation.[157] The lack of a cell wall and the marked osmotic sensitivity of mycoplasmas would be expected to facilitate the preparation of membrane vesicles from mycoplasmas. Contrary to this expectation, however, repeated attempts to prepare sealed membrane vesicles from *A. laidlawii* have failed in several laboratories, including ours. Electron microscopy indicated that the regular osmotic lysis procedure results in the formation of very large tears in the membrane, which apparently cannot reseal spontaneously.[341] Moreover, the membrane preparations obtained by osmotic lysis exhibited the full activity of ATPase and NADH oxidase, enzymes localized on the inner surface of the *A. laidlawii* membrane.[228] These enzymatic activities should be masked in sealed right-side out vesicles because the substrates, ATP and NADH, cannot penetrate through the intact membrane.[33,228] The absence of sealed vesicles in the *A. laidlawii* membrane preparations was also indicated by the failure to demonstrate the formation of an electrochemical-ion gradient in the membrane preparations following the addition of glucose.[33] Recent attempts by Shinar and Rottem[457] to produce sealed vesicles from *A. laidlawii* cells lysed in dionized water or in various concentrations of Tris-maleate buffer at several pH values have also failed. Various modifications of the method devised by Kaback[157] for the preparation of membrane vesicles from wall-covered bacteria were also tried with *A. laidlawii* and invariably gave negative results.[96,461]

The reasons for the difficulties encountered in obtaining sealed right-side out vesicles from osmotically lysed *A. laidlawii* cells are not clear. The only report on the successful preparation of sealed vesicles from *A. laidlawii* comes from the group of Tarshis,[96,97,239] and is therefore of special interest. The cells were washed in dilute buffer at 37°C, which caused the cells to swell. Protectors of thiol groups were included in the wash fluid to prevent oxidation of SH-containing proteins. The swollen cells were then lysed by alternate freezing and thawing. The composition of the resulting membrane preparations resembled that of *A. laidlawii* membranes isolated by osmotic lysis. However, they differed in consisting of sealed vesicles, as was indicated by their shrinking in the presence of nonpenetrable solutes such as KCl, and the reversal of this effect by valinomycin. The right-side out conformation of the vesicles was indicated by their low NADH oxidase and ATPase activities, and by their ability to accumulate D-glucose, maltose, D-fructose, sucrose, and lactate.[96] The vesicles retained over 60% of the sugar transport activity of intact cells.[239] Howver, in contrast to the results with whole *A. laidlawii* cells,[405,406] glucose transport in the membrane vesicles was not inhibited by arsenate or by DCCD. This rules out ATP hydrolysis as the mechanism generating the proton motive force energizing transport in the vesicles. As exogenous electron donors such as lactate and pyruvate also failed to stimulate glucose uptake by the vesicles, Fedotov et al.[97] proposed that the proton motive force is generated by the oxidation of unidentified endogenous electron donors. Hence, the results obtained with vesicles show an unexplained inconsistency with the data obtained with whole cells, in which hydrolysis of ATP generated by glycolysis provides the proton motive force energizing transport.[405,406] In order to solve this inconsistency, comparative experiments with whole cells and membrane vesicles are clearly required.

c. *Effects of Membrane Lipid Fluidity on Sugar Transport*

The rate of glucose permeation into *A. laidlawii* cells showed a direct dependence on the fluidity of membrane lipids (Figure 13), similar to that observed for the passive diffusion of the polyhydric alcohols into this organism. The rate of glucose uptake increased as the gel-to-liquid crystalline phase transition temperature of membrane lipids decreased.[298] Arrhenius plots of the initial rate of glucose uptake by membrane vesicles of *A. laidlawii* showed a break at 15°C, probably corresponding to the transition temperature of membrane lipids. Membrane vesicles isolated from cells grown with

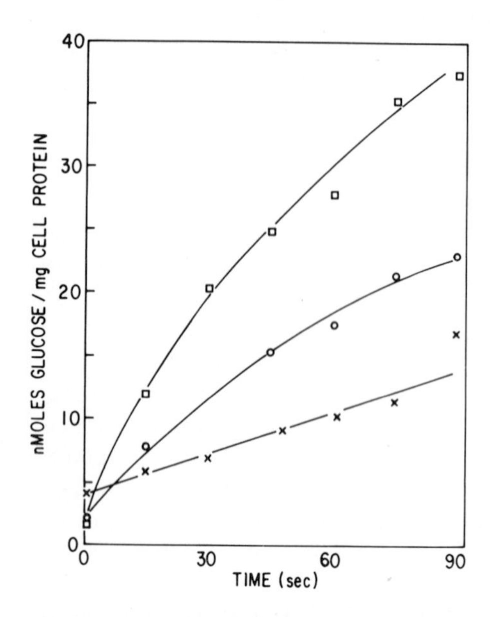

FIGURE 13. Uptake of glucose by *Acholeplasma laidlawii* cells grown with palmitate and: linoleate, □; oleate, ○; or elaidate, x. (From Reads, B.D. and McElhaney, R.N., *J. Bacteriol.*, 123, 47, 1975. With permission.)

oleic acid showed the break at 10°C.[239] Hence, as was found for many active transport systems in prokaryotes,[235] sugar transport in *A. laidlawii* is markedly affected by the physical state of membrane lipids.

2. K^+ Transport

The ability of mycoplasmas to accumulate K^+ against a concentration gradient by an energy-dependent process was first demonstrated in *A. laidlawii.*[56,334] More recently, *M. mycoides* subsp. *capri* and *M. gallisepticum* were also shown to share this property.[183,364] In light of the importance of K^+ in cell metabolism,[118] transport systems for this cation are presumably common to all mycoplasmas. The level of K^+ accumulated in *M. mycoides* subsp. *capri* cells during growth may reach 300 m*M*, i.e., about 25 times higher than in the growth medium. This is approximately equivalent to the level of K^+ accumulated in *E. coli.*[182,363]

Glucose serves as an adequate energy source for K^+ transport in the fermentative mycoplasmas.[181,182,334] Extensive studies by LeGrimellec and Leblanc[181-184] on *M. mycoides* subsp. *capri* have led to the conclusion that the membrane-bound Mg^{2+}-ATPase drives K^+ transport by generating a proton motive force through ATP hydrolysis. Inhibition of K^+ influx by DCCD supports this conclusion (Figure 14). As expected, valinomycin, the K^+ ionophore, induced a rapid loss of K^+ in nonmetabolizing cells, or in metabolizing cells when added with DCCD, or with uncouplers such as dinitrophenol or FCCP. Valinomycin or the uncouplers alone were ineffective.[182] Measurements of the electrical membrane potential by the flurescent dye merocyanine corroborated the above results by showing a marked increase in the membrane potential $\Delta\psi$ (reaching -140 mV) when glucose was added. This potential was abolished by the combined treatment of cells with valinomycin and FCCP.[182] The finding that uncoupling agents in the absence of valinomycin failed to modify K^+ distribution and $\Delta\psi$ is unexpected, as it contradicts Mitchell's hypothesis.

The identity of the K^+ carrier in the membrane raises an interesting question. The

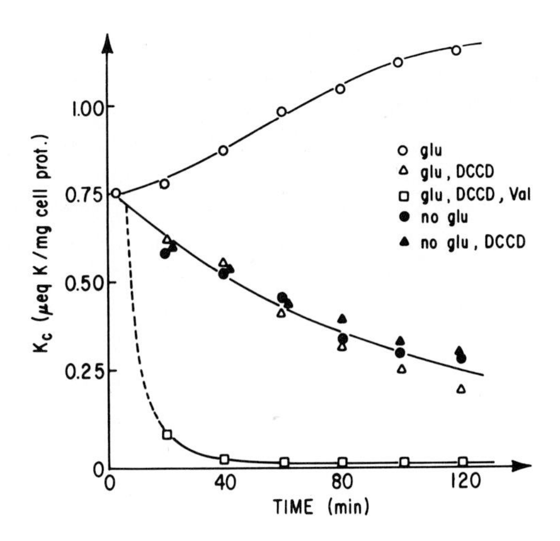

FIGURE 14. Effect of dicyclohexylcarbodiimide (DCCD) on the K^+ level of *Mycoplasma mycoides* subsp. *capri*. Washed organisms were resuspended in 1 m*M* potassium phosphate buffer at 37°C. When present, glucose (20 m*M*) was added at the beginning of incubation; DCCD (5×10^{-5}*M*) and valinomycin (10^{-7}*M*) were added 5 and 10 min later. ●, no glucose; ○, glucose; △, glucose + DCCD; ▲, DCCD alone; □, glucose + DCCD + valinomycin. (From Leblanc, G. and LeGrimellec, C., *Biochim. Biophys. Acta*, 554, 168, 1979. With permission.)

possibility that a membrane-bound ATPase serves as the K^+ carrier is attractive, particulary in light of recent reports that a K^+ stimulated ATPase serves as a K^+ carrier in *E. coli.*[94] LeGrimellec and Leblanc[184] examined this possibility by comparing the effects of temperature on K^+ influx, Mg-ATPase activity and transmembrane potential ($\Delta\psi$) in *M. mycoides* subsp. *capri* in which membrane lipid composition had been altered. Although the K^+ influx was affected by changes in membrane lipid composition, comparison of the Arrhenius plots of K^+ uptake with those of the ATPase activity did not show close correspondence. Moreover, the temperature-dependent K^+ influx curves did not parallel the curves of $\Delta\psi$ values obtained by merocyanine fluorescence. These results led LeGrimellec and Leblanc[184] to conclude that the K^+ carrier cannot be identified with the Mg^{2+}-ATPase, though it is clear that this enzyme complex is instrumental in energizing K^+ transport.

Although both the ATPase and the presumed K^+ carrier appear to be directly affected by alterations in membrane lipid composition,[32,83,184,321] these alterations may also influence K^+ uptake indirectly by affecting membrane permeability. This was demonstrated by experiments with *M. mycoides* subsp. *capri,* in which the membrane cholesterol content was drastically reduced from 25% to less than 2% of membrane lipids.[183] The intracellular K^+ level in the sterol-poor cells was much lower than in the sterol-rich cells. The cause for this was traced to a higher permeability of the sterol-poor membranes to K^+ and H^+. The increased proton permeability is liable to affect the proton motive force which energizes K^+ uptake.

C. Group Translocation

1. The PEP: Sugar Phosphotransferase System (PTS)

This highly efficient sugar transport system is unique to prokaryotes. As phosphorylation of the sugar is an obligatory step in its transport by the PTS, this process is distinguished from facilitated diffusion and active transport, where the transported molecules are not chemically altered.[347] Information on the distribution of PTS among prokaryotes has led to the generalization that this transport system is confined to anaerobes and facultative anaerobes metabolizing sugars via anaerobic glycolysis. A principal end product in glycolysis is phosphoenol pyruvate (PEP), the phosphorylating agent of the PTS. Those prokaryotes that normally metabolize sugars aerobically via the Entner-Doudoroff pathway lack the PTS and accumulate sugars by active transport systems energized by ATP and a proton motive force.[347] The distribution of the PTS among the Mollicutes appears to conform to this rule. Thus, all the glycolytic *Mycoplasma* species tested and *S. citri* possess the PTS,[63,143,144,422] whereas *T. acidophilum,* an aerobe with a complete respiratory chain, lacks it.[62] The *Acholeplasma* species, however, appear to be an exception to this rule. Despite their resemblance to the fermentative *Mycoplasma* and *Spiroplasma* species in metabolizing sugars by glycolysis, the acholeplasmas lack PTS and accumulate sugars by active transport (see previous section).

The PTS system of *M. capricolum* has been characterized by Jaffor Ullah and Cirillo.[143,144] The mycoplasmal PTS resembled that of other procaryotes in consisting of nonsugar-specific cytoplasmic proteins, enzyme I and HPr, and membrane-bound sugar-specific proteins (enzymes II). The purified *M. capricolum* HPr resembles that of *E. coli* and *Staphylococcus aureus* in molecular weight (about 9000 daltons), heat stability, and in having a single histidine residue. However, it differs from the other HPrs in having two cysteine residues. Moreover, the mycoplasmal HPr does not cross-react with antibodies to the *E. coli* HPr.[143] The purified enzyme I from *M. capricolum,* which catalyzes the transfer of the phosphoryl moiety from PEP to HPr, differs from that of *E. coli, Salmonella typhimurium* and *S. aureus* in molecular properties.[144] The mycoplasmal enzyme I has a molecular weight of about 220,000 daltons and appears to be comprised of four subunits (two of 44,500 daltons, one of 62,000, and one of 64,000), whereas the enzymes of the eubacteria mentioned above consist of a single polypeptide having a molecular weight between 70,000 and 90,000 daltons. The enzymes II specific for glucose, fructose, and mannose are constitutive in *M. mycoides* subsp. *capri,* whereas in *M. capricolum* the enzyme II for fructose is inducible.[218] None of these membrane-bound enzymes has been purified from any prokaryote. Thus we have no information on their molecular properties, nor do we understand the mechanism by which the enzymes II mediate transmembrane sugar translocation.[347] Despite the significant molecular differences between the purified PTS components of *M. capricolum* and those of *E. coli,* and despite the wide phylogenetic gap between these organisms, complementation experiments showed that they can replace each other quite effectively. Thus the phosphorylated mycoplasmal HPr functioned nearly as well as the phosphorylated *E. coli* HPr with *E. coli* membranes,[143] and the mycoplasmal enzyme I phosphorylated the *E. coli* Hpr at about 25% of the rate observed for the mycoplasmal HPr.[144] It is not surprising, therefore, that the PTS components from different *Mycoplasma* and *Spiroplasma* species completely complemented each other.[143]

The complex PTS system appears to fulfill important regulatory functions apart from its primary function in sugar transport. These include the regulation of the intracellular level of cyclic 3′,5′-adenosine monophosphate (cyclic AMP) and the regulation of the uptake of carbohydrates transportable by systems different from the PTS. Data showing that the intracellular level of cyclic AMP in *M. capricolum* can be regulated by sugars transportable by the organism PTS have recently been obtained by Mugharbil and Cirillo.[218] In the wild type of *M. capricolum* both glucose and fructose reduced the intracellular level of the nucleotide. In wall-covered bacteria possessing the PTS, transportable sugars inhibit the induction of enzymes required for the utilization of alternate energy sources. Inhibition of induction is caused by a combination of a reduction in the intracellular level of cyclic AMP and inhibition of the inducer uptake. The data of Mugharbil and Cirillo[218] show that the PTS substrates can fulfill a similar regulatory role in mycoplasms as well.

VI. MEMBRANE COMPONENTS AS PATHOGENICITY FACTORS

A. Adherence to Host Cells

All the mycoplasmas, apart from *T. acidophilum*, are parasites of animals, insects, or plants. In animals they usually adhere to and colonize the epithelial linings of the respiratory and urogential tracts, only rarely invading the tissues and blood stream.[397] Hence, the mycoplasmas may be regarded as surface parasites. Their attachment is firm enough to prevent elimination by the action of the ciliated epithelium or by the urine. Moreover, the intimate association of mycoplasmas with their host cell surface provides a nutritional advantage to the parasites. The attached mycoplasmas not only enjoy a higher concentration of nutrients adsorbed onto their host cell membrane,[113] but may also utilize fatty acids and cholesterol of the host membrane itself. Moreover, the intimate association between adhering mycoplasmas and their host cells gives rise to a situation in which local concentrations of toxic metabolites can build up and cause cell damage. Thus, the H_2O_2 excreted by the attached mycoplasmas may attack the cell membrane without being rapidly destroyed by catalase or peroxidase present in the extracellular body fluids.[67] Hydrolytic enzymes produced by the mycoplasmas may damage the host cell membrane as well. It has also been suggested,[273] though not experimentally proven, that adhering mycoplasmas may take up cholesterol from the host cell membrane, thus depleting it of an essential component.

Mycoplasma adherence to host cells differs in one important aspect from that of wall-covered bacteria: there is no wall to separate the plasma membrane of the parasite from that of its host. This may theoretically enable the fusion of the parasite's cell membrane with that of its host, or at least allow the exchange of components between the two membranes. For the latter some experimental evidence is already available, as will be discussed below.

1. *Experimental Models*

Mycloplasmas have been shown to adhere to erythrocytes,[22,49,75,99,188,394] HeLa cells,[189] fibroblasts in monolayer cultures,[48,106] spermatozoa,[411] macrophages,[155,156] lymphocytes,[399,440] tracheal epithelial cells,[394] tracheal organ cultures,[67,104] and inert surfaces, such as glass and plastic.[100,117,159,395,410] Thus, there is a great variety of experimental systems for studying mycoplasma adherence. For reasons of simplicity and convenience, the adherence of mycoplasmas to erythrocytes has been studied most extensively, though in recent years the tendency has been to use host cells to which the mycoplasmas attach *in vivo* ("target cells"), such as tracheal epithelial cells. The term "receptor site" has become accepted to denote the sites on the host cell surface participating in adherence, whereas the term "binding site" is used to specify the sites on the mycoplasma cell membrane responsible for attachment of the parasite.

2. *Receptor Sites*

Attachment of the human and avian respiratory pathogens *M. pneumoniae, M. gallisepticum,* and *M. synoviae* is usually affected, and in some cases nearly abolished, by pretreatment of the host cells with neuraminidase.[3,22,99,110,189] Hence, it is widely believed that these mycoplasmas attach to sialic acid moieties on the cell surface (Figure

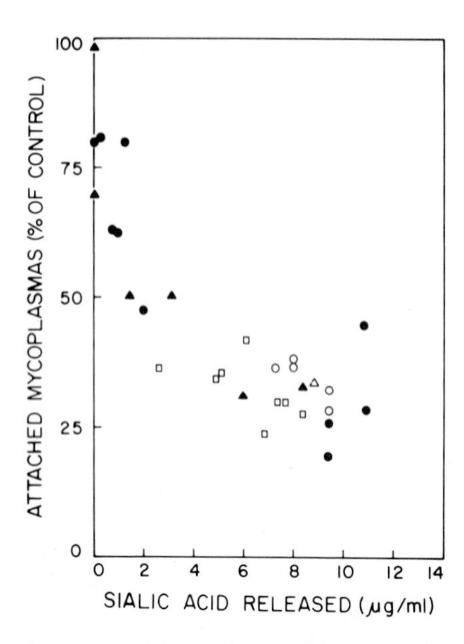

FIGURE 15. Effects of sialic acid removal from human red blood cells by neuraminidase treatment on their ability to attach *Mycoplasma gallisepticum.* (From Banai, M., Kahane, I., Razin, S., and Bredt, W., Infec. Immun., 21, 365, 1978. With permission.)

15). Our finding[22] that glycophorin, the erythrocyte membrane protein carrying almost all the sialic acid moieties of the erythrocyte, inhibits attachment of *M. gallisepticum* to human erythrocytes, supports the above conclusion. Nevertheless, it appears that sialic acid moieties are not the only receptors for these *Mycoplasma* species. Thus, treatment of tracheal epithelial cells with neuraminidase only decreased attachment of *M. pneumoniae* by about 50 to 65%.[91,259] The residual background attachment apparently involves receptors other than sialic acid. Recent studies indicate that various eukaryotic cells may differ in the nature of their receptor sites for the same mycoplasma. Thus, extensive neuraminidase treatment of sheep erythrocytes decreased their ability to bind *M. pneumoniae* by about 80 to 90%, but had little effect on the binding capacity of human erythocytes and none on that of rabbit erythrocytes.[99] Similarly, neuraminidase treatment of lung fibroblasts decreased *M. pneumoniae* attachment by 83 to 96%, as compared to a decrease of about 50% in neuraminidase-treated tracheal epithelial cells.[105,106] It thus appears that some cells have sialic acid containing receptors in addition to other still unidentified receptors.

In conclusion, there are strong indications that glycophorin serves as the major receptor site for *M. gallisepticum* in the human erythrocyte membrane[22] while *M. pneumoniae* binds to other receptors on this membrane as well.[99] In lung fibroblasts, the major receptor sites for *M. pneumoniae* appear to be sialoglycoproteins rather than sialoglycolipids.[105]

No information is available on the chemical nature of the receptors for mycoplasmas other than *M. pneumoniae, M. gallisepticum,* and *M. synoviae.* Receptors on HeLa cells for *M. hominis* and *M. salivarium,* though resistant to neuraminidase, were inactivated by proteolytic enzymes,[189] suggesting their protein nature, while the erythrocyte receptor sites for *M. dispar* resisted proteolytic treatment as well as treatment with neuraminidase or periodate.[135] Similarly, the attachment of *M. pulmonis* to mouse macrophages was unaffected by pretreatment of the macrophages with neuraminidase, trypsin, chymotrypsin, and glutaraldehyde,[156] suggesting that the receptor sites are neither sialic acid residues nor membrane proteins.

3. Binding Sites

Adherence to surfaces of some wall-covered bacteria is mediated through slime layers covering the bacterial cells. Thus, the dextran produced by *Streptococcus mutans* from sucrose supports its adherence to dental plaques.[113] Slime layers are common on many mycoplasmas, but their role in adherence is uncertain. *M. mycoides* subsp. *mycoides* and *A. laidlawii,* the mycoplasmas with the best-characterized slime layers, do not hemadsorb.[188] Hence, it seems that the mycoplasmal binding sites are constituents of the membrane *per se.*

Satisfactory, though largely indirect, evidence is available for the protein nature of the binding sites in *M. pneumoniae* and *M. gallisepticum.* Pretreatment of these mycoplasmas by heat, merthiolate, glutaraldehyde, or formalin inhibits their attachment to cells or inert surfaces partially or completely.[75,100,110,117,394] More relevant are the findings that mild trypsin treatment of *M. pneumoniae* cells nullifies their ability to adhere to erythrocytes[21,99,117] and tracheal cells.[139,394] The ability of *M. pneumoniae* to adhere, lost by trypsin treatment, could be restored by incubating the treated cells in growth medium for 4 hr at 37°C, but not at 4°C. Chloramphenicol, mitomycin C, or ultraviolet radiation inhibited restoration.[99,117] Polyacrylamide-gel electrophoresis of *M. pneumoniae* cells briefly treated with trypsin showed the disappearance of a major protein band designated P_1 (molecular weight about 100,000) exposed on the cell surface. Hu et al.[139] proposed P_1 to be the *M. pneumoniae* binding site, as the failure of the mycoplasmas to attach correlated with its absence. Moreover, regeneration of P_1 by transfer of the trypsinized organisms to fresh growth medium for 6 hr renewed attachment capacity. Erythromycin inhibited the resynthesis of the protein and regeneration of attachment capacity.[139] However, further studies were somewhat disappointing in showing that an avirulent *M. pneumoniae* strain, with a reduced adherence capacity, did possess the P_1 protein.[126] By two-dimensional electrophoresis Hansen et al.[126] detected three proteins synthesized by the virulent strain, but not by the homologous avirulent one. One of these proteins was located on the cell surface by the lactoperoxidase iodination method, and was removable by trypsin. Nevertheless, this did not prove to be P_1. Hansen et al.[126] suggest three possible explanations as to why the existence of P_1 in the avirulent strain does not discount its role in adherence: (1) the P_1 protein has been altered to a nonfunctional state by spontaneous mutation; (2) more than one protein is required for adherence; (3) a previously undetected protein on the *M. pneumoniae* surface is responsible for the trypsin-sensitive attachment mechanism. The picture became even more complicated when *M. pneumoniae* mutants, defective in hemadsorption capacity, were isolated and tested electrophoretically.[125] Generally, the cell protein patterns in slab gels were very similar, but some of the mutants lacked a high-molecular weight protein (~ 190,000) found in the wild type. Nevertheless, the finding that the protein patterns of some hemadsorption-negative mutants were indistinguishable from that of the hemadsoprtion-positive wild type led them to propose that mycoplasma membrane surface components other than proteins may be involved in the hemadsorption process.[125]

A more direct approach to the characterization of the *M. pneumoniae* and *M. gallisepticum* binding sites has recently been tried in our laboratory.[21] The principle is to use glycophorin as a ligand in affinity chromatography for the isolation of binding sites specific to sialic acid receptors from the membranes of both mycoplasmas. Treatment of *M. pneumoniae* membranes with 0.5% deoxycholate solubilized over 45% of the total membrane protein, leaving the binding site(s) in the insoluble residue. This finding indicates the tight association of the binding sites with other membrane components. Since the residue resisted further solubilization by nonionic detergents, we had to apply the strong and highly denaturative detergent sodium dodecyl sulfate (SDS). The SDS-solubilized material was subjected to affinity chromatography on a glycophorin-Sephadex© column, and a fraction was obtained containing several proteins in the medium- and low-molecular weight ranges. Due to the inevitable presence of SDS in this fraction, glutaraldehyde-fixed human erythrocytes had to be used to prevent hemolysis. The binding capacity of the fraction to the fixed erythrocytes was no higher (based on protein) than that of native membranes, but was specific in the sense that glycophorin inhibited binding of the fraction to the erythrocytes. Unfortunately, the use of SDS has serious deficiencies, as it causes protein denaturation which may lead to decreased biological activity of the presumed binding sites. We are therefore now searching for milder agents suitable for solubilization of the binding sites.

4. Relationship of Adherence to Metabolic Activity

The association of metabolic activity with the adherence capacity of mycoplasmas has been stressed Hu et al.[138] and Powell et al.,[259] who go so far as to conclude that only metabolically active *M. pneumoniae* cells are capable of attachment to respiratory ciliated epithelium. This conclusion is substantiated by findings that adherence capacity of mycoplasmas decreases, or is even nullified by, energy metabolism inhibitors,[117] aging of cultures,[21,188] killing of the mycoplasmas,[100,259] and at low temperatures.[22,100,117,259] Heating *M. pneumoniae* cells to 56°C for 30 min decreased by about 40% their ability to bind fibroblast membrane glycoproteins, presumably the receptor sites for this mycoplasma.[105] Most relevant to this discussion is our recent finding[100] that addition of a metabolizable sugar (glucose) to *M. pneumoniae* cells suspended in buffer enhanced their attachment to glass most significantly. We have proposed a working hypothesis to explain these findings.[282] According to this hypothesis, the degree of exposure of binding sites on mycoplasma membranes depends upon membrane energization. Conditions which may decrease the electrochemical ion gradients across the cell membrane such as aging, low temperature, growth inhibitors, and ionophores have been shown to decrease the degree of exposure of membrane proteins on the mycoplasma cell surface[11] (Section IV.C.3). Experiments are now underway to test the effects of ionophores on adherence of mycoplasmas to cells and inert surfaces.

Since isolated mycoplasma membranes are obviously not energized, it would be predicted, according to the above hypothesis, that they would exhibit a much lower binding capacity than metabolically active cells. This has in fact been shown in our laboratory with *M. gallisepticum* membranes.[462] Our results confirm those of Gabridge et al.[104] that isolated *M. pneumoniae* membranes can attach to tracheal epithelial cells, though less effectively than viable cells, and differ from those of Hu et al.[138] who failed to show any attachment of isolated *M. pneumoniae* membranes to respiratory epithelium. In our opinion membrane isolation is expected to decrease the degree of exposure of binding sites on the membrane surface, but does not eliminate them.

5. Distribution of Binding Sites

An interesting question is whether the binding sites are homogeneously distributed over the mycoplasma cell surface. The two most studied adhering mycoplasmas, *M. pneumoniae* and *M. gallisepticum,* possess specialized tip structures. Phase-contrast microscopy[45] and scanning-electron microscopy[34,222] show the filamentous *M. pneumoniae* to have a bulb-like neck and a tapered tip (Figure 16). In thin sections the tip is seen to consist of a dense central rodlike core surrounded by a lucent space and enveloped by the cell membrane.[439] A bleb at the tip of the worm-like cells of *M.*

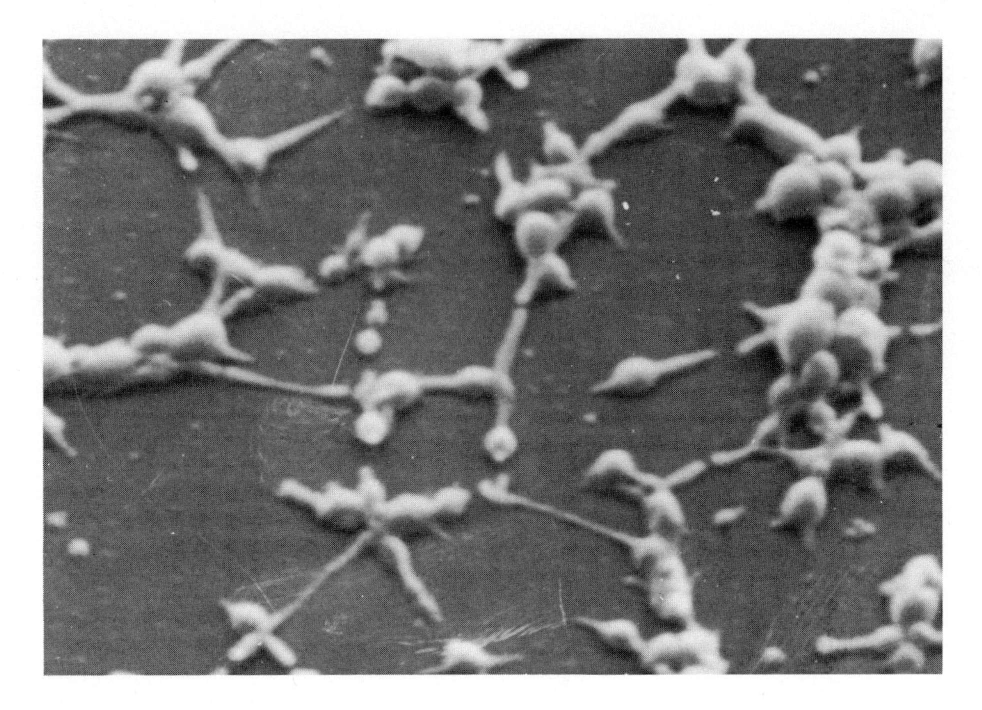

FIGURE 16. Scanning electron micrograph of *Mycoplasma pneumoniae* cells showing the characteristic bulb-like neck with a tapered tip shape. (Magnification × 20,000.) (From Razin, S., Banai, M., Gamliel, H., Pollack, A., Bredt, W., and Kahane, I., *Infect. Immun.*, 30, 538, 1980. With permission.)

gallisepticum has also been observed by a variety of electron microscopical techniques[107,191] (Figure 17). In thin sections, the bleb appears as a differentiated ellipsoid structure bounded by the plasma membrane. Since the terminal structures have not been isolated, we have no information on their chemical composition. Indirect evidence obtained with specific stains suggests that the terminal structure of *M. pneumoniae* is composed of basic proteins and is covered by a mucoprotein layer heavier than that of the rest of the cell surface.[439]

The conception that terminal structures are associated with adherence originated from the excellent electron micrographs of Zucker-Franklin et al.,[452] which showed *M. gallisepticum* cells clustered around leukocytes like iron filings on a magnet or flukes attached to their host, the terminal bleb being the site of contact more frequently than chance could account for. Later studies have shown that *M. pneumoniae* also attaches to the respiratory epithelium of tracheal organ cultures through its terminal structure.[67] Both *M. pneumoniae* and *M. gallisepticum* have also been shown adhering to inert surfaces, such as glass and plastic, through the tip structures and crawling on these surfaces with the tip leading the way.[44] Clearly, the polarity exhibited by *M. pneumoniae* and *M. gallisepticum* adhering to cells or inert surfaces through their specialized tip structures would lead one to suggest a high concentration of binding sites on the surface of these structures. A possible approach to test this hypothesis may be based on interaction of the cells with ferritin-labeled antibodies to the binding sites, or with labeled receptors such as glycophorin, and localization of these reagents on the mycoplasma cell surface by electron microscopy. Such tests have not yet been performed, but labeling of *M. pneumoniae* and *M. gallisepticum* cells with the less specific reagents polycationized ferritin and lectins failed to show any concentration of groups reactive with these reagents in the tip areas of the organisms.[49]

Evidence suggesting that binding sites are not restricted only to the tip area is

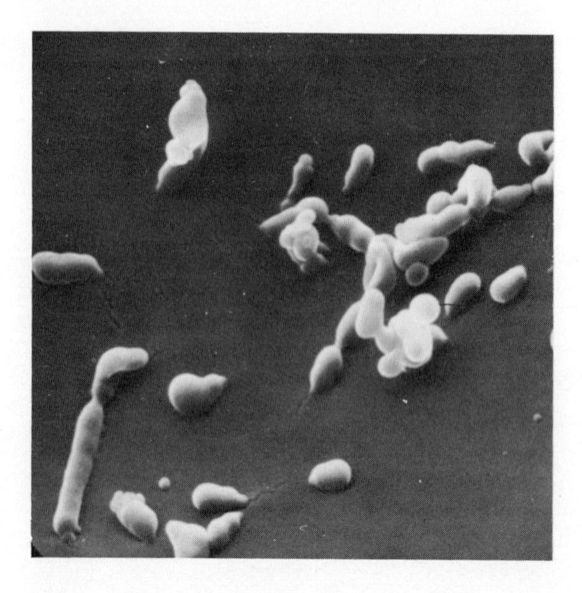

FIGURE 17. Scanning electron micrograph of *Mycoplasma gallisepticum* cells with characteristic blebs. (Magnification × 11,000).

accumulating. Thus, Brunner et al.[49] showed by scanning electron microscopy that *M. pneumoniae* adheres to erythrocytes with membrane sites other than the tip structure. Organisms of spherical, pear-shaped, and irregular forms having no visible tip structure were found to adhere well to the erythrocyte surface. It has been speculated that the perpendicular position of *M. pneumoniae* filaments attaching by their tips to ciliated respiratory epithelium is due to the motility of the organisms. Thus, the tip may be the first part of the organisms, making their way along the cilia, to reach and adhere to the epithelial cell surface. The densely packed cilia may permit the thin and flexible filamentous organisms to move forward, but obstruct spherical organisms from reaching the epithelial cell surface.[49] Accordingly, the perpendicular orientation of the mycoplasmas in tracheal organ cultures is supported by the closely packed cilia.[40] When support is absent, as with lung fibroblasts[106] or on a glass surface,[222] the mycoplasma filaments attach throughout their entire length.

B. Fusion, Capping, and Mitogenicity

Since there is no barrier, i.e., a cell wall, separating the plasma membrane of the mycoplasma from that of its host cell, the plausibility of fusion of the two membranes has attracted the attention of many investigators. Obviously, if fusion does occur, a wide variety of potentially cytotoxic proteins (enzymes) and lipids can be introduced directly into the host cells. The new "patch" on the eucaryotic cell membrane may comprise a relatively weak area, able to leak ions or essential metabolites.[104] Although the fusion hypothesis is appealing, the only available supporting evidence is based on electron microscopy showing *M. gallisepticum* fusing to erythrocytes.[13] Even in this case, only a small proportion of the mycoplasmas fused, and the factors influencing this process were not defined. The claims by Grant and McConnell[120] and Teuber and Bader[412] of fusion of phosphatidylcholine vesicles with membranes of growing *A. laidlawii* cells could not be confirmed by us.[165,285] Almost all published electron micrographs of mycoplasmas adhering to eukaryotic cells show a distance of about 10 nm separating the mycoplasma membrane from the host membrane,[40,135,155,439,440] a finding

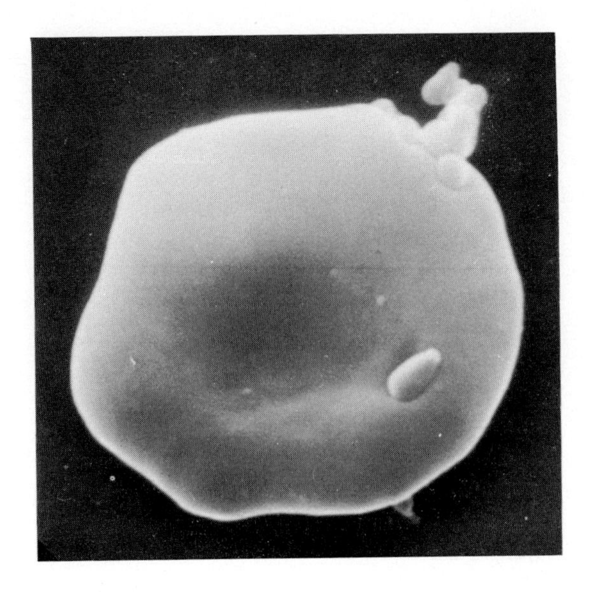

FIGURE 18. Scanning electron micrograph of *Mycoplasma gallisepticum* adhering to a human red blood cell. The tight association of the parasite with the eucaryotic cell surface can be seen. (Magnification × 13,000.) (From Razin, S., Banai, M., Gamliel, H., Pollack, A., Bredt, W. and Kahane, I., *Infect. Immun.*, 30, 538, 1980. With permission.)

which does not support fusion. However, fine threads of material can be seen bridging the gap between the two membranes,[40,135,439] indicating the tight association of the parasite with its host. This is illustrated nicely in the scanning-electron micrographs of mycoplasmas adhering to erythrocytes (Figures 18 and 19).

The intimate association between mycoplasmas and the host cell membrane is elaborated by several most intriguing findings reported recently by Stanbridge and Weiss[399] and by Wise et al.[440] *M. hyorhinis* cells infecting the surface of mouse lymphocytes were shown to cap by a process resembling capping by concanavalin A, albeit at a slower rate.[399] Towards the latter stages of the 24 hr infection period, the capped mycoplasmas were shed from the surface of the lymphocytes as large aggregates containing membranous vesicles of presumed host origin. This is the first report of an infectious organism behaving similarly to soluble multivalent ligands and capping on the lymphocyte surfaces in the absence of a specific antibody. Certain viruses are distributed in patches on the lymphocyte surfaces in the absence of antibody, but addition of specific antibody is required to induce capping.[399]

The phenomenon of capping by *M. hyorhinis* could be associated with the well-established mitogenic effect of mycoplasma cells[35] and membranes.[225] About 40% of mouse splenic lymphocytes showed evidence of infection by *M. hyorhinus* as measured by specific fluorescence. Although relatively few blast cells were visible, virtually all were capped. According to Stanbridge and Weiss,[399] this indicates patching and capping to be closely related to induction of blast transformation by mycoplasmas. In this case the mitogenic activity of the mycoplasmas was even higher than that of bacterial lipopolysaccharides. The biological significance of these observations is clear: if mycoplasmas are capable of inducing B-cell activation *in vivo* as *in vitro,* forbidden clones may possibly be activated to produce autoantibodies, which are rather common in mycoplasma infections.[101] Alternatively, mycoplasma-host cell interaction may lead

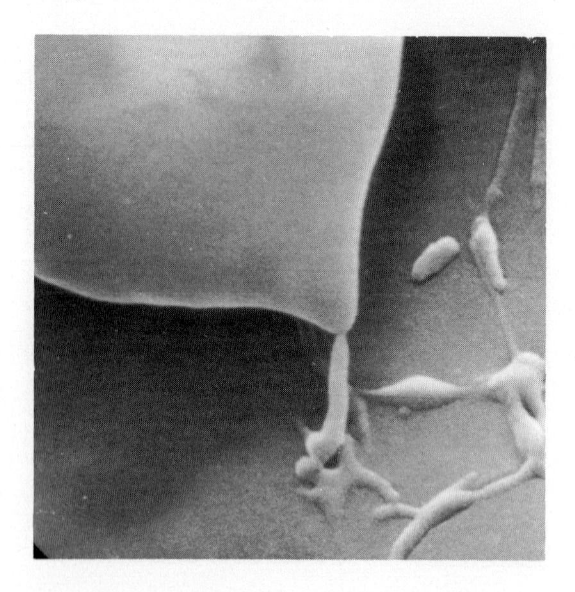

FIGURE 19. Scanning electron micrograph of a *Mycoplasma pneumoniae* cell adhering through its tip to a human red blood cell. The stretching of the red blood cell at the point of attachment with the parasite demonstrates the tight association between the two. (Magnification × 20,000.) (From Razin, S., Banai, M., Gamliel, H., Pollack, A., Bredt, W., and Kahane, I., *Infect Immun.*, 30, 538, 1980. With permission.)

to autoimmune reactivity by antigenic modulation. Stanbridge and Weiss[399] mention preliminary evidence of depletion of host cell surface antigens, probably resulting from capping and shedding during mycoplasma infection. In this way the antigenic architecture of the infected cell surface can be modified, a factor which may induce an autoimmune response.

The notion of antigenic modulation of cell membranes resulting from interaction of mycoplasmas with eukaryotic cells is strongly supported by the work of Wise et al.[440] They could detect several antigens of the host cell membrane in *M. hyorhinis* infecting a murine T-lymphoblastoid cell line. The mycoplasmas acquired only certain host cell surface antigens, accumulating them selectively. This finding may represent an important mechanism by which mycoplasmas avoid or alter the immunological response of the host. Moreover, the intimate association of mycoplasmas with host membrane appears to alter the immunological response of the host cells as well. Thus, an *M. hyorhinis* antiserum mediated specific complement-dependent cytolysis of infected lymphoblastoid cells.[440] These experiments did not reveal whether lysis resulted from recognition of *M. hyorhinis* antigens integrally inserted into the membrane of host cells, or from innocent bystander damage arising from complement activation by antigen-antibody complexes formed on organisms residing on the host cell membrane. The phenomenon does, however, suggest a mechanism by which tissue damage may result from an immunological reaction to surface-associated mycoplasmas *in vivo*.[440]

The findings of Stanbridge and Weiss[399] and of Wise et al.[440] demonstrate most vividly the dynamic nature of the interaction between the cell membranes of the host and its parasite, e.g., the mycoplasma. The association between the two membranes appears tight enough to enable exchange of integral membrane components, an event which may trigger immunological responses of serious consequences to the host cell. Whether or not membrane fusion is involved in this process remains unknown.

VII. CONCLUSIONS AND PROSPECTS

The mycoplasmas are unique in being the only self-replicating organisms with a single membranous structure—the plasma membrane. This can be regarded as an extremely useful property for membrane studies, for it facilitates the isolation of pure plasma membranes uncontaminated by other membrane types. Moreover, lack of cell walls of the mycoplasmas enables the application of gentle and simple techniques for membrane isolation.

The restricted genetic information in the small mycoplasma genome dictates limited biosynthetic abilities. Thus, mycoplasmas are partially or totally incapable of fatty acid synthesis, depending on the growth medium for their supply. This dependence on exogenous fatty acids has been exploited most effectively to introduce controlled alterations in the fatty acid composition of mycoplasma membranes and to study the consequent effects on the physical and biochemical properties of the membrane. Such studies provided perhaps the strongest evidence supporting the bilayer configuration of lipids in biomembranes, and brought the importance of membrane lipid composition and physical state on transport processes and membrane-bound enzymatic activities into focus.

The major lipid species synthesized by mycoplasmas are typical prokaryotic glycerophospholipids and glycosyl diglycerides. In some mycoplasmas the carbohydrate moiety of the glycosyl diglycerides is markedly extended, conferring physical properties on the molecules which resemble those of lipopolysaccharides of Gram-negative bacteria, though differing significantly in chemical structure.

The wall-less *Thermoplasma acidophilum* differs from the parasitic mycoplasmas in many biochemical and physiological properties, including synthesis of symmetrical diglycerol tetraether phospho- and glycolipids containing long saturated isopranoid branched alkyl chains, rather than fatty acids. The recent finding of similar lipids in methanogenic bacteria supports the opinion that *T. acidophilum* rightfully should be classified among the *Archaebacteria.*

The requirement of mycoplasmas for cholesterol, unique among procaryotes, has been utilized to prove that it functions in regulating membrane fluidity during changes in growth temperature or alterations in the fatty acid composition of membrane lipids. Acholeplasmas, which do not require cholesterol for growth, regulate membrane fluidity by adjusting the chain length of saturated fatty acids which they synthesize and by selective elongation and incorporation of exogenous fatty acids.

It remains to be clarified whether regulation of membrane fluidity is the sole function of cholesterol in mycoplasma membranes. Growth of *M. mycoides* subsp. *capri* with very little exogenous cholesterol, insufficient to exert a fluidizing effect on the bulk of the lipid bilayer, as well as growth promotion of the related *M. capricolum* by a variety of sterols having no effect on membrane fluidity suggests the existence of an additional, more specific function for sterols, at least in these mycoplasmas.

The marked difference in the cholesterol uptake capacity of the *Mycoplasma* and *Acholeplasma* species has been utilized to study the factors controlling the amount of exogenous cholesterol incorporated into the membranes. The recent finding that growing *Mycoplasma* cells take up considerable quantities of phospholipids in addition to free and esterified cholesterol, whereas *Acholeplasma* species take up only low amounts of free cholesterol, led to the still-unproven proposal that *Mycoplasma* species possess receptors for serum lipoproteins, but *Acholeplasma* species do not.

The ease of manipulation of cholesterol quantities in mycoplasma membranes makes them useful models for studying the mechanism of action of toxic agents which damage cells by complexing with membrane cholesterol. Experiments with *A. laidlawii, M mycoides* subsp. *capri,* and *M. capricolum,* grown with varying amounts of different sterols, provided perhaps the most convincing evidence for the notion that membrane sterol is the target of polyenes and thiol-activated bacterial toxins.

The fact that all of the mycoplasma membrane cholesterol is of exogenous origin has been utilized to show that cholesterol taken up by growing cells flip-flops rapidly from the outer to the inner half of the lipid bilayer. Whether this transbilayer movement depends on growth or occurs also in "resting" cells remains to be elucidated.

The asymmetrical transbilayer distribution of phospholipids and glycolipids and the presence of "boundary" lipids closely associated with proteins in mycoplasma membranes has been indicated in several recent studies. Again, the absence of a cell wall in mycoplasmas facilitates disposition studies, as these depend primarily on use of macromolecular labeling agents, such as enzymes and antibodies, which may be unable to penetrate the bacterial wall barrier.

The first strides towards the molecular characterization of mycoplasma membrane proteins have already been made following development of methods for their solubilization and fractionation. Amino acid analysis of only a few of these proteins revealed a preponderance of hydrophobic amino acids, while in other proteins the amino acid composition resembled that of hydrophilic ones. Hence, amino acid sequencing is clearly required to elucidate the amphiphilic nature of these proteins. Finding carbohydrates associated with some mycoplasma membrane proteins is of particular interest, considering the rarity of glycoproteins in prokaryotes.

A very strong case can be built supporting the existence of a contratile protein system associated with the mycoplasma membrane, as a variety of contractile processes are observable in these plastic wall-less prokaryotes. Several reports suggesting the presence of actin-like proteins in mycoplasmas are already available, and fibrils comprised of a protein resembling tubulin have recently been observed in *Spiroplasma citri.* Electron microscopical evidence for a fibrilar "cytoskeleton" in *M. pneumoniae* is also available. Nevertheless, other reports claim that mycoplasmas lack any protein identical to eukaryotic actin, so that the picture has yet to be clarified.

Of the many membrane-bound enzymes in mycoplasmas, none has yet been characterized. The membrane-bound NADH oxidase of *A. laidlawii,* representing a truncated electron transport chain, appears to consist of three proteins. One of these contains flavin.[463] The NADH oxidase complex of *Mycoplasma* and *Spiroplasma* species is soluble, but the possibility that it is loosely bound to the membrane *in situ* cannot be ruled out. The mycoplasmal ATPase differs from that of other prokaryotes in being tightly associated with the membrane and in dependence on membrane lipids for activity. There are indications that the mycoplasmal ATPase acts in translocation of protons and sodium ions, but the complete characterization of the mycoplasmal enzyme complex must await its successful isolation in an active form.

As in other biomembranes, the disposition of proteins in mycoplasma membranes is asymmetrical. More proteins face the cytoplasm than the cell exterior. The finding that the degree of exposure of proteins on the *A. laidlawii* cell surface is influenced by cell energization is intriguing, and should be investigated further. Results of these studies may be relevant to the understanding of adherence of mycoplasmas to host cells, which is mediated through specific protein binding sites on the mycoplasma cells surface.

Being dependent on many nutrients supplied by the host or growth medium, mycoplasmas have a variety of active transport systems, which basically resemble those of other prokaryotes. Sugar transport in *A. laidlawii* is energized by the proton motive force generated through the activity of the membrane-bound ATPase, while fermentative *Mycoplasma* and *Spiroplasma* species utilize the highly efficient phosphoenolpyruvate-dependent sugar phosphotransferase system for sugar transport. Although one would expect the preparation of sealed membrane vesicles active in transport from the wall-less mycoplasmas to be easy, results have usually been disappointing. The reasons for the problems of preparing mycoplasma membrane vesicles and the way to overcome these difficulties are yet to be discovered.

Animal mycoplasmas are surface parasites firmly adhering to and colonizing the epithelial lining of the respiratory and genital tracts of infected animals. In the case of these parasites, no wall separates their plasma membrane from that of the host, so fusion of the two membranes is possible. Although evidence supporting fusion is insufficient, recent studies suggest exchange of antigenic components between the parasite and the host membranes, which may trigger immunological responses of serious consequences to the host.

Sialic acid moieties on the host cell surface serve as specific, though not exclusive, receptors for the pathogens *M. pneumoniae, M. gallisepticum,* and *M. synoviae.* There are sound indications that the binding sites in *M. pneumoniae* and *M. gallisepticum* membranes are protein in nature, but these proteins have not yet been isolated and characterized, nor is it clear whether these binding sites are concentrated on the surface of the terminal tips and blebs characteristic of these mycoplasmas. The isolation and identification of the mycoplasma surface components responsible for adherence may be rewarding, for hopefully it will lead to development of highly specific vaccines.

VIII. ACKNOWLEDGMENTS

The preparation of this review was supported, in part, by a grant from the United States-Israel Binational Science Foundation (BSF), Jerusalem, Israel, in continuation of the collaborative Health Research Communications Program between the *Israel Journal of Medical Sciences,* Jerusalem, and the National Library of Medicine, Bethesda, Maryland. Investigations originating from this laboratory were supported by grants from the Deutsche Forschungsgemeinschaft and the United States-Israel BSF.

I thank my colleagues at the Department of Membrane and Ultrastructure Research for their contributions incorporated into this review, and the authors who kindly provided me with copies of their papers prior to publication. My sincere thanks are also due to Yael Segal for her expert help with the manuscript.

REFERENCES

1. **Abramson, M. B. and Pisetsky, D.,** Thermal-turbidimetric studies of membranes from *Acholeplasma laidlawii, Biochim. Biophys. Acta,* 282, 80, 1972.
2. **Ajufo, J. C. and Whithear, K. G.,** Evidence for a ruthenium red-staining extracellular layer as the haemagglutinin of the WVU 1853 strain of *Mycoplasma synoviae, Aust. Vet. J.,* 54, 502, 1978.
3. **Aldridge, K. E.,** Growth and cytopathology of *Mycoplasma synoviae* in chicken embryo cell cultures, *Infect. Immun.,* 12, 198, 1975.
4. **Alexander, A. G. and Kenny, G. E.,** Characterization of membrane and cytoplasmic antigens of *Mycoplasma arginini* by two-dimensional (crossed) immunoelectrophoresis, *Infect. Immun.,* 15, 313, 1977.
5. **Allan, E. M. and Pirie, H. M.,** Electronmicroscopical observations on mycoplasmas in pneumonic calves, *J. Med. Microbiol.,* 10, 469, 1977.
6. **Al-Shammari, A. J. N. and Smith, P. F.,** Lipid and lipopolysaccharide composition of *Acholeplasma oculi, J. Bacteriol.,* 139, 356, 1979.
7. **Amar, A.,** Protein Disposition in the Mycoplasma Membrane, Ph.D. thesis, The Hebrew University, Jerusalem, 1977.

8. **Amar, A., Kahane, I., Rottem, S., and Razin, S.,** Binding of lectins to membranes of mycoplasmas from aging cultures, *Microbios,* 24, 93, 1979.
9. **Amar, A., Rottem, S., Kahane, I., and Razin, S.,** Characterization of the mycoplasma membrane proteins. VI. Composition and disposition of proteins in membranes from aging *Mycoplasma hominis* cultures, *Biochim. Biophys. Acta,* 426, 258, 1976.
10. **Amar, A., Rottem, S., and Razin, S.,** Characterization of the mycoplasma membrane proteins. IV. Disposition of proteins in the membrane, *Biochim. Biophys. Acta,* 352, 228, 1974.
11. **Amar, A., Rottem, S., and Razin, S.,** Disposition of membrane proteins as affected by changes in the electrochemical gradient across mycoplasma membranes, *Biochem. Biophys. Res. Commun.,* 84, 306, 1978.
12. **Amar, A., Rottem, S., and Razin, S.,** Is the vertical disposition of mycoplasma membrane proteins affected by membrane fluidity?, *Biochim. Biophys. Acta,* 552, 457, 1979.
13. **Apostolov, K. and Windsor, G. D.,** The interaction of *Mycoplasma gallisepticum* with erythrocytes. 1. Morphology, *Microbios,* 13, 205, 1975.
14. **Arai, S., Yoshida, K., Izawa, A., Kumagai, K., and Ishida, N.,** Effect of antibiotics on growth of *Mycoplasma pneumoniae* MAC, *J. Antibiot. Tokyo,* 19, 118, 1966.
15. **Archer, D. B.,** Modification of the membrane composition of *Mycoplasma mycoides* subsp. *capri* by the growth medium, *J. Gen. Microbiol.,* 88, 329, 1975.
16. **Archer, D. B.,** Effect of the lipid composition of *Mycoplasma mycoides* subspecies *capri* and phosphatidylcholine vesicles upon the action of polyene antibiotics, *Biochim. Biophys. Acta,* 436, 68, 1976.
17. **Archer, D. B. and Gale, E. F.,** Antagonism by sterols of the action of amphotericin and filipin on the release of potassium ions from *Candida albicans* and *Mycoplasma mycoides* subsp. *capri, J. Gen. Microbiol.,* 90, 187, 1975.
18. **Archer, D. B. and Rodwell, A. W.,** personal communication.
19. **Archer, D. B., Rodwell, A. W., and Rodwell, E. S.,** The nature and location of *Acholeplasma laidlawii* membrane proteins investigated by two-dimensional gel electrophoresis, *Biochim. Biophys. Acta,* 513, 268, 1978.
20. **Argaman, M. and Razin, S.,** Cholesterol and cholesterol esters in mycoplasma, *J. Gen. Microbiol.,* 38, 153, 1965.
21. **Banai, M., Razin, S., Bredt, W., and Kahane, I.,** Isolation of binding sites to glycophorin from *Mycoplasma pneumoniae* membranes, Infect. Immun., 30, 628, 1980.
22. **Banai, M., Kahane, I., Razin, S., and Bredt, W.,** Adherence of *Mycoplasma gallisepticum* to human erythrocytes, *Infect. Immun.,* 21, 365, 1978.
23. **Barile, M. F., Razin, S., Tully, J. G., and Whitcomb, R. F., Eds.,** *The Mycoplasmas,* Vol. I to III, Academic Press, New York, 1979.
24. **Basu, S. K., Goldstein, J. L., and Brown, M. S.,** Characterization of the low density lipoprotein receptor in membranes prepared from human fibroblasts, *J. Biol. Chem.,* 253, 3852, 1978.
25. **Bebear, Ch., Latrille, J., Fleck, J., Roy, B., and Bové, J. M.,** *Spiroplasma citri:* un mollicute, in *Les Mycoplasmes,* Bové, J. M., and Duplan, J. F., Eds., INSERM, Paris, 1974, 35.
26. **Beckman, B. L. and Kenny, G. E.,** Immunochemical analysis of serologically active lipids of *Mycoplasma pneumoniae, J. Bacteriol.,* 96, 1171, 1968.
27. **Benyoucef, M., Rigaud, J. L., and Leblanc, G.,** The electrochemical potential gradient for ^{+}H ($\Delta\mu H$) in *Mycoplasma mycoides* var. *capri:* estimate of the chemical concentration component ΔpH, *Zbl. Bakt. Parasitenkd.-1, Abt.-A,* 241, 184, 1978.
28. **Bernheimer, A. W. and Davidson, A.,** Lysis of pleuropneumonia-like organisms by staphylococcal and streptococcal toxins, *Science,* 148, 1229, 1965.
29. **Bevers, E. M., Leblanc, G., Op den Kamp, J. A. F., and van Deenen, L. L. M.,** Disposition of phosphatidylglycerol in metabolizing cells of *Acholeplasma laidlawii, FEBS Lett.,* 87, 49, 1978.
30. **Bevers, E. M., Op den Kamp, J. A. F., and van Deenen, L. L. M.,** The distribution of molecular classes of phosphatidylglycerol in the membrane of *Acholeplasma laidlawii, Biochim. Biophys. Acta,* 511, 509, 1978.
31. **Bevers, E. M., Singal, S. A., Op den Kamp, J. A. F. and van Deenen, L. L. M.,** Recognition of different pools of phosphatidylglycerol in intact cells and isolated membranes of *Acholeplasma laidlawii* by phospholipase A_2, *Biochemistry,* 16, 1290, 1977.
32. **Bevers, E. ., Snoek, G. T., Op den Kamp, J. A. F., and van Deenen, L. L. M.,** Phospholipid requirement of the membrane-bound Mg^{2+}-dependent adenosine-triphosphatase in *Acholeplasma laidlawii, Biochim. Biophys. Acta,* 467, 346, 1977.
33. **Bevers, E. M., Wang, H. H., Op den Kamp, J. A. F., and van Deenen, L. L. M.,** On the interaction between intrinsic proteins and phosphatidylglycerol in the membrane of *Acholeplasma laidlawii, Arch. Biochem. Biophys.,* 193, 502, 1979.

34. **Biberfeld, G. and Biberfeld, P.,** Ultrastructural features of *Mycoplasma pneumoniae, J. Bacteriol.,* 102, 855, 1970.
35. **Biberfeld, G. and Gronowicz, E.,** *Mycoplasma pneumoniae* is a polyclonal B-cell activator, *Nature* (London), 261, 238, 1976.
36. **Bittman, R., Chen, W. C., and Anderson, O. R.,** Interaction of filipin III and amphotericin B with lecithin-sterol vesicles and cellular membranes. Spectral and electron microscope studies, *Biochemistry,* 13, 1364, 1974.
37. **Bittman, R. and Rottem, S.,** Distribution of cholesterol between the outer and inner halves of the lipid bilayer of mycoplasma cell membranes, *Biochem. Biophys. Res. Commun.,* 71, 318, 1976.
38. **Black, F. T., Birch-Andersen, A., and Freundt, E. A.,** Morphology and ultrastructure of human T-mycoplasmas, *J. Bacteriol.,* 111, 254, 1972.
39. **Bloj, B. and Zilversmit, D. B.,** Complete exchangeability of cholesterol in phosphatidylcholine/cholesterol vesicles of different degrees of unsaturation, *Biochemistry,* 16, 3943, 1977.
40. **Boatman, E., Cartwright, F., and Kenny, G.,** Morphology, morphometry and electron microscopy of HeLa cells infected with bovine *Mycoplasma, Cell Tissue Res.,* 170, 1, 1976.
41. **Boatman, E. S. and Kenny, G. E.,** Three-dimensional morphology, ultrastructure and replication of *Mycoplasma felis, J. Bacteriol.,* 101, 262, 1970.
42. **Borochov, H. and Shinitzky, M.,** Vertical displacement of membrane proteins mediated by changes in microviscosity, *Proc. Nat. Acad. Sci. U.S.A.,* 73, 4526, 1976.
43. **Bradbury, J. M. and Jordan, F. T. W.,** Studies on the adsorption of certain medium proteins to *Mycoplasma gallisepticum* and their influence on agglutination and hemagglutination reactions, *J. Hyg.,* 70, 267, 1972.
44. **Bredt, W.,** Motility, in *The Mycoplasmas,* Vol. I, Barile, M. F., and Razin, S., Eds., Academic Press, New York, 1979, 141.
45. **Bredt, W. and Bierther, M. F. W.,** Light and electron microscopy of *Mycoplasma pneumoniae* cells, *Zbl. Bakt. Parasitenk.-1 Abt.-A,* 229, 249, 1974.
46. **Bredt, W., Heunert, H. H., Höfling, K. H., and Milthaler, B.,** Microcinematographic studies of *Mycoplasma hominis* cells, *J. Bacteriol.,* 113, 1223, 1973.
47. **Bretscher, M. S.,** Asymmetrical lipid bilayer structure for biological membranes, *Nature* (London), 236, 11, 1972.
48. **Brown, S., Teplitz, M., and Revel, J. P.,** Interaction of mycoplasmas with cell cultures, as visualized by electron microscopy, *Proc. Nat. Acad. Sci. U.S.A.,* 71, 464, 1974.
49. **Brunner, H., Krauss, H., Schaar, H., and Schiefer, H.-G.,** Electron microscopic studies on the attachment of *Mycoplasma pneumoniae* to guinea pig erythrocytes, *Infect. Immun.,* 24, 906, 1979.
50. **Butler, K. W., Johnson, K. G., and Smith, I.C.P.,** *Acholeplasma laidlawii* membranes: an electron spin resonance study of the influence on molecular order of fatty acid composition and cholesterol, *Arch. Biochem. Biophys.,* 191, 289, 1978.
51. **Butler, M. and Knight, B. C. J. G.,** The survival of washed suspensions of mycoplasma, *J. Gen. Microbiol.,* 22, 470, 1960.
52. **Buttery, S. H., Lloyd, L. C., and Titchen, D. A.,** Acute respiratory, circulatory and pathological changes in the calf after intravenous injections of the galactan from *Mycoplasma mycoides* subsp. *mycoides, J. Med. Microbiol.,* 9, 379, 1976.
53. **Chapman, D.,** Liquid crystals and cell membranes, *Ann. N. Y. Acad. Sci.,* 137, 745, 1966.
54. **Chapman, D. and Urbina, J.,** Phase transitions and bilayer structure of *Mycoplasma laidlawii* B, *FEBS Lett.,* 12, 169, 1971.
55. **Chelton, E. T. J., Jones, A. S., and Walker, R. T.,** The chemical composition of the nucleic acids and the proteins of some mycoplasma strains, *J. Gen. Microbiol.,* 50, 305, 1968.
56. **Cho, H. W. and Morowitz, H. J.,** Characterization of the plasma membrane of *Mycoplasma laidlawii.* VI. Potassium transport, *Biochim. Biophys. Acta,* 183, 295, 1969.
57. **Cho, H. W. and Morowitz, H. J.,** Characterization of the plasma membrane of *Mycoplasma laidlawii.* VIII. Effect of temperature shift and antimetabolites on K^+ transport, *Biochim. Biophys. Acta,* 274, 105, 1972.
58. **Choules, G. L., and Bjorklund, R. F.,** Evidence of β structure in mycoplasma membranes. Circular dichroism, optical rotary dispersion, and infrared studies, *Biochemistry,* 9, 4759, 1970.
59. **Choules, G. L. and Gray, W. R.,** Peptidase activity in the membranes of *Mycoplasma laidlawii, Biochem. Biophys. Res. Commun.,* 45, 849, 1971.
60. **Christiansson, A., and Wieslander, Å.,** Membrane lipid metabolism in *Acholeplasma laidlawii* A EF 22. Influence of cholesterol and temperature shift-down on incorporation of fatty acids and synthesis of membrane lipid species, *Eur. J. Biochem.,* 85, 65, 1978.
61. **Chu, H. P. and Horne, R. W.,** Electron microscopy of *Mycoplasma gallisepticum* and *Mycoplasma mycoides* using the negative staining technique and their comparison with Myxovirus, *Ann. N. Y. Acad. Sci.,* 143, 190, 1967.

62. **Cirillo, V. P.,** Transport systems, in *The Mycoplasmas,* Vol. I, Barile, M. F. and Razin, S., Eds., Academic Press, New York, 1979, 323.
63. **Cirillo, V. P., and Razin, S.,** Distribution of a phosphoenolpyruvate-dependent sugar phosphotransferase system in mycoplasmas, *J. Bacteriol.,* 113, 212, 1973.
64. **Clarke, M. and Spudich, J. A.,** Nonmuscle contractile proteins: the role of actin and myosin in cell motility and shape determination, *Annu. Rev. Biochem.,* 46, 797, 1977.
65. **Clejan, S., Bittman, R., and Rottem, S.,** Uptake, transbilayer distribution, and movement of cholesterol in growing *Mycoplasma capricolum* cells, *Biochemistry U.S.A.,* 17, 4579, 1978.
66. **Cole, R. M., Tully, J. G., and Popkin, T. J.,** Ultrastructure of the agent of citrus "stubborn" disease, *Ann. N. Y. Acad. Sci.,* 225, 471, 1973.
67. **Collier, A. M.,** Mycoplasmas in organ culture, in *The Mycoplasmas,* Vol. II, Tully, J. G. and Whitcomb, R. F., Eds., Academic Press, New York, 1979, 475.
68. **Cooper, R. A., Arner, E. C., Wiley, J. S., and Shattil, S. J.,** Modification of red cell membrane structure by cholesterol-rich lipid dispersions: a model for the primary spur cell defect, *J. Clin. Invest.,* 55, 115, 1975.
69. **Cooper, R. A., Leslie, M. H., Fischkoff, S., Shinitzky, M., and Shattil, S. J.,** Factors influencing the lipid composition and fluidity of red cell membranes *in vitro:* production of red cells possessing more than two cholesterols per phospholipid, *Biochemistry,* 17, 327, 1978.
70. **Cowell, J. L. and Bernheimer, A. W.,** Role of cholesterol in the action of cereolysin on membranes, *Arch. Biochem. Biophys.,* 190, 603, 1978.
71. **Cowell, J. L., Kim, K.-S., and Bernheimer, A. W.,** Alteration by cereolysin of the structure of cholesterol-containing membranes, *Biochim. Biophys. Acta,* 507, 230, 1978.
72. **Cronan, J. E. and Gelmann, E. P.,** Physical properties of membrane lipids: biological relevance and regulation, *Bacteriol. Rev.,* 39, 232, 1975.
73. **Dahl, J. S., Hellewell, S. B., and Levine, R. P.,** A mycoplasma mutant resistant to lysis by C: variations in membrane composition and altered response to the terminal C complex, *J. Immunol.,* 119, 1419, 1977.
74. **Davis, R. E. and Worley, J. F.,** Spiroplasma: motile, helical microorganism associated with corn stunt disease, *Phytopathology,* 63, 403, 1973.
75. **Deas, J. E., Janney, L. T., and Howe, C.,** Immune electron microscopy of cross-reactions between *Mycoplasma pneumoniae* and human erythrocytes, *Infect. Immun.,* 24, 211, 1979.
76. **de Gier, J., Mandersloot, J. G., Hupkes, J. V., McElhaney, R. N., and van Beek, W. P.,** On the mechanism of non-electrolyte permeation through lipid bilayers and through biomembranes, *Biochim. Biophys. Acta,* 233, 610, 1971.
77. **de Kruijff, B., Cullis, P. R., Radda, G. K., and Richards, R. E.,** Phosphorus nuclear magnetic resonance of *Acholeplasma laidlawii* cell membranes and derived liposomes, *Biochim. Biophys. Acta,* 419, 411, 1976.
78. **de Kruyff, B., de Greef, W. J., van Eyk, R. V. W., Demel, R. A., and van Deenen, L. L. M.,** The effect of different fatty acid and sterol composition on the erythritol flux through the cell membrane of *Acholeplasma laidlawii, Biochim. Biophys. Acta,* 298, 479, 1973.
79. **de Kruijff, B. and Demel, R. A.,** Polyene antibiotic-sterol interactions in membranes of *Acholeplasma laidlawii* cells and lecithin liposomes. III. Molecular structure of the polyene antibiotic-cholesterol complexes, *Biochim. Biophys. Acta,* 339, 57, 1974.
80. **de Kruyff, B., Demel, R. A., and van Deenen, L. L. M.,** The effect of cholesterol and epicholesterol incorporation on the permeability and on the phase transition of intact *Acholeplasma laidlawii* cell membranes and derived liposomes, *Biochim. Biophys. Acta,* 255, 331, 1972.
81. **de Kruijff, B., Gerritsen, W. J., Oerlemans, A., Demel, R. A., and van Deenen, L. L. M.,** Polyene antibiotic-sterol interactions in membranes of *Acholeplasma laidlawii* cells and lecithin liposomes. I. Specificity of the membrane permeability changes induced by the polyene antibiotics, *Biochim. Biophys. Acta,* 339, 30, 1974.
82. **de Kruijff, B., Gerritsen, W. J., Oerlemans, A., van Dijck, P. W. M., Demel, R. A., and van Deenen, L. L. M.,** Polyene-antibiotic-sterol interactions in membranes of *Acholeplasma laidlawii* cells and lecithin liposomes. II. Temperature dependence of the polyene antibiotic-sterol complex formation, *Biochim. Biophys. Acta,* 339, 44, 1974.
83. **de Kruyff, B., van Dijck, P. W. M., Goldbach, R. W., Demel, R. A., and van Deenen, L. L. M.,** Influence of fatty acid and sterol composition on the lipid phase transition and activity of membrane-bound enzymes in *Acholeplasma laidlawii, Biochim. Biophys. Acta,* 330, 269, 1973.
84. **Demel, R. A. and de Kruyff, B.,** The function of sterols in membranes, *Biochim. Biophys. Acta,* 457, 109, 1976.
85. **Dorner, I., Brunner, H., Schiefer, H.-G., and Wellensiek, H.-J.,** Complement-mediated killing of *Acholeplasma laidlawii* by antibodies to various membrane components, *Infec. Immun.,* 13, 1663, 1976.

86. **Downie, J. A., Gibson, F., and Cox, G. B.,** Membrane adenosine triphosphatases of prokaryotic cells, *Annu. Rev. Biochem.,* 48, 103, 1979.
87. **Dresdner, G.,** Efficiency and specificity of the sequential extraction of membrane proteins of *Acholeplasma laidlawii* with the neutral detergent Tween 20, *FEBS Lett.,* 89, 69, 1978.
88. **Dresdner, G. and Cid-Dresdner, H.,** Tween 20-soluble membrane proteins of *Acholeplasma laidlawii.* Fractionation in the presence of a Tween 20 concentration slightly above its critical micelle concentration and in the absence of detergent by means of agarose-suspension electrophoresis, *FEBS Lett.,* 72, 243, 1976.
89. **Dresdner, G. and Cid-Dresdner, H.,** The removal of the neutral detergent Tween 20 from solubilized membrane proteins of *Acholeplasma laidlawii* and other proteins by agarose suspension electrophoresis, *Anal. Biochem.,* 78, 171, 1977.
90. **Eden-Green, S. J., and Tully, J. G.,** Isolation of *Acholeplasma* spp. from coconut palms affected by lethal yellowing disease in Jamaica, *Curr. Microbiol.,* 2, 311, 1979.
91. **Engelhardt, J. A. and Gabridge, M. G.,** Effect of squamous metaplasia on infection of hamster trachea organ cultures with *Mycoplasma pneumoniae, Infect. Immun.,* 15, 647, 1977.
92. **Engelman, D. M.,** Lipid bilayer structure in the membrane of *Mycoplasma laidlawii, J. Mol. Biol.,* 58, 153, 1971.
93. **Engelman, D. M. and Morowitz, H. J.,** Characterization of the plasma membrane of *Mycoplasma laidlawii.* IV. Structure and composition of membrane and aggregated components. *Biochim. Biophys. Acta,* 150, 385, 1968.
94. **Epstein, W., Whitelaw, U., and Hesse, J.,** A K^+ transport ATPase in *Escherichia coli, J. Biol. Chem.,* 253, 6666, 1978.
95. **Esfahani, M., Rudkin, B. B., Cutler, C. J., and Waldron, P. E.,** Lipid-protein interactions in membranes. Interaction of phospholipids with respiratory enzymes in *Escherichia coli* membranes, *J. Biol. Chem.,* 252, 3194, 1977.
96. **Fedotov, N. S., Panchenko, L. F., Logachev, A. P., Bekkouzhin, A. G., and Tarshis, M. A.,** Transport properties of membrane vesicles from *Acholeplasma laidlawii.* I. Isolation and general characteristics, *Folia Microbiol. (Prague),* 20, 470, 1975.
97. **Fedotov, N. S., Panchenko, L. F., and Tarshis, M. A.,** Transport properties of membrane vesicles from *Acholeplasma laidlawii.* III. Evidence of active nature of glucose transport, *Folia Microbiol. (Prague),* 20, 488, 1975.
98. **Feingold, D. S.,** The action of amphotericin B on *Mycoplasma laidlawii, Biochem. Biophys. Res. Commun.,* 19, 261, 1965.
99. **Feldner, J., Bredt, W., and Kahane, I.,** Adherence of erythrocytes to *Mycoplasma pneumoniae, Infect. Immun.,* 25, 60, 1979.
100. **Feldner, J., Bredt, W., and Razin, S.,** Adherence of *Mycoplasma pneumoniae* to glass surfaces, *Infect. Immun.,* 26, 70, 1979.
101. **Fernald, G. W.,** Humoral and cellular immune responses to mycoplasmas, in *The Mycoplasmas,* Tully, J. G. and Whitcomb, R. F., Eds., Academic Press, New York, 1979, 399.
102. **Freeman, B. A., Sissenstein, R., McManus, T. T., Woodward, J. E., Lee, I. M., and Mudd, J. B.,** Lipid composition and lipid metabolism of *Spiroplasma citri, J. Bacteriol.,* 125, 946, 1976.
103. **Freundt, E. A.,** *The Mycoplasmataceae,* Munksgaard, Copenhagen, 1958.
104. **Gabridge, M. G., Barden-Stahl, Y. D., Polisky, R. B., and Engelhardt, J. A.,** Differences in the attachment of *Mycoplasma pneumoniae* cells and membranes to tracheal epithelium, *Infect. Immun.,* 16, 766, 1977.
105. **Gabridge, M. G. and Taylor-Robinson, D.,** Interaction of *Mycoplasma pneumoniae* with human lung fibroblasts: role of receptor sites, *Infect. Immun.,* 25, 455, 1979.
106. **Gabridge, M. G., Taylor-Robinson, D., Davies, H. A., and Dourmashkin, R. R.,** Interaction of *Mycoplasma pneumoniae* with human lung fibroblasts: characterization of the *in vitro* model, *Infect. Immun.,* 25, 446, 1979.
107. **Gallagher, J. E., and Rhoades, K. R.,** Simplified preparation of mycoplasmas, an acholeplasma, and a spiroplasma for scanning electron microscopy, *J. Bacteriol.,* 137, 972, 1979.
108. **Gazitt, Y., Ohad, I., and Loyter, A.,** Increase in the phospholipid phase extractable by dry ether following ATP-depletion of erythrocytes, *Biochem. Biophys. Res. Commun.,* 72, 1359, 1976.
109. **Gershfeld, N. L., Wormser, M., and Razin, S.,** Cholesterol in mycoplasma membranes. I. Kinetics and equilibrium studies of cholesterol uptake by the cell membrane of *Acholeplasma laidlawii, Biochim. Biophys. Acta,* 352, 371, 1974.
110. **Gesner, B. and Thomas, L.,** Sialic acid binding sites: role in hemagglutination by *Mycoplasma gallisepticum, Science,* 151, 590, 1966.
111. **Ghosh, A., Maniloff, J., and Gerling, D. A.,** Inhibition of mycoplasma cell division by cytochalasin B, *Cell,* 13, 57, 1978.

112. **Gibbons, N. E. and Murray, R. G. E.,** Proposals concerning the higher taxa of bacteria, *Int. J. Syst. Bacteriol.,* 28, 1, 1978.
113. **Gibbons, R. J. and van Houte, J.,** Bacterial adherence in oral microbial ecology, *Annu. Rev. Microbiol.,* 29, 19, 1975.
114. **Gilliam, J. M. and Morowitz, H. J.,** Characterization of the plasma membrane of *Mycoplasma laidlawii.* IX. Isolation and characterization of the membrane polyhexosamine, *Biochim. Biophys. Acta,* 274, 353, 1972.
115. **Goel, M. C.,** New method for the isolation of membranes from *Mycoplasma gallisepticum, J. Bacteriol.,* 116, 994, 1973.
116. **Goel, M. C. and Lemcke, R. M.,** Dissociation of *Mycoplasma gallisepticum* membranes with lithium diiodosalicylate and isolation of a glycoprotein, *Ann. Microbiol. Inst. Pasteur,* 126 B, 299, 1975.
117. **Gorski, F. and Bredt, W.,** Studies on the adherence mechanism of *Mycoplasma pneumoniae, FEMS Lett.,* 1, 265, 1977.
118. **Gottschalk, G.,** *Bacterial Metabolism,* Springer-Verlag, New York, 1979.
119. **Gourlay, R. N. and Thrower, K. J.,** Morphology of *Mycoplasma mycoides* thread-phase growth, *J. Gen. Microbiol.,* 54, 155, 1968.
120. **Grant, C. W. M. and McConnell, H. M.,** Fusion of phospholipid vesicles with viable *Acholeplasma laidlawii, Proc. Nat. Acad. Sci. U.S.A.,* 70, 1238, 1973.
121. **Green, F. and Hanson, R. P.,** Ultrastructure and capsule of *Mycoplasma meleagridis, J. Bacteriol.,* 116, 1011, 1973.
122. **Greenstein, S.,** Characterization of the *Spiroplasma citri* membrane, M. Sc. Thesis, The Hebrew University, Jerusalem, 1975.
123. **Gross, Z. and Rottem, S.,** Lipid distribution in *Acholeplasma laidlawii* membrane. A study using the lactoperoxidase-mediated iodination, *Biochim. Biophys. Acta,* 555, 547, 1979.
124. **Guerrero, A., Muñoz, E., and Andreu, J. M.,** Glycoproteins in bacterial membranes. *In vivo* labeling of the sugar portion of energy-transducing ATPase and a low-molecular-weight fraction from *Micrococcus lysodeikticus* membranes, *Curr. Microbiol.,* 1, 129, 1978.
125. **Hansen, E. J., Wilson, R. M., and Baseman, J. B.,** Isolation of mutants of *Mycoplasma pneumoniae* defective in hemadsorption, *Infect. Immun.,* 23, 903, 1979.
126. **Hansen, E. J., Wilson, R. M., and Baseman, J. B.,** Two-dimensional gel electrophoretic comparison of proteins from virulent and avirulent strains of *Mycoplasma pneumoniae, Infect. Immun.,* 24, 468, 1979.
127. **Harold, F. M.,** Ion currents and physiological functions in microorganisms, *Annu. Rev. Microbiol.* 31, 181, 1977.
128. **Helenius, A. and Simons, K.,** Solubilization of membranes by detergents, *Biochim. Biophys. Acta,* 415, 29, 1975.
129. **Herring, P. K., and Pollack, J. D.,** Utilization of [1-^{14}C]acetate in the synthesis of lipids by acholeplasmas, *Int. J. Syst. Bacteriol.,* 24, 73, 1974.
130. **Hjertén, S., and Johansson, K.-E.,** Selective solubilization with Tween 20 of membrane proteins from *Acholeplasma laidlawii, Biochim. Biophys. Acta,* 288, 312, 1972.
131. **Holländer, R.,** The cytochromes of *Thermoplasma acidophilum, J. Gen. Microbiol.,* 108, 165, 1978.
132. **Holländer, R., Wolf, G., and Mannheim, W.,** Lipoquinones of some bacteria and mycoplasmas, with considerations on their functional significance, *J. Microbiol. Serol.,* 43, 177, 1977.
133. **Horne, R. W.,** The ultrastructure of mycoplasma and mycoplasma-like organisms, *Micron,* 2, 19, 1970.
134. **Howard, C. J. and Gourlay, R. N.,** An electron-microscope examination of certain bovine mycoplasmas stained with ruthenium red and the demonstration of a capsule on *Mycoplasma dispar, J. Gen. Microbiol.,* 83, 393, 1974.
135. **Howard, C. J., Gourlay, R. N., and Collins, J.,** Serological comparison and haemagglutinating activity of *Mycoplasma dispar, J. Hyg. Camb.,* 73, 457, 1974.
136. **Hsuchen, C.-C. and Feingold, D. S.,** Selective membrane toxicity of the polyene antibiotics: studies on natural membranes, *Antimicrob. Agents Chemother.,* 4, 316, 1973.
137. **Hsung, J.-C. Huang, L., Hoy, D. J., and Haug, A.,** Lipid and temperature dependence of membrane-bound ATPase activity of *Acholeplasma laidlawii, Can. J. Biochem.,* 52, 974, 1974.
138. **Hu, P. C., Collier, A. M. and Baseman, J. B.,** Interaction of virulent *Mycoplasma pneumoniae* with hamster tracheal organ cultures, *Infect. Immun.,* 14, 217, 1976.
139. **Hu, P. C., Collier, A. M., and Baseman, J. B.,** Surface parasitism by *Mycoplasma pneumoniae* of respiratory epithelium, *J. Exp. Med.,* 145, 1328, 1977.
140. **Huang, L. and Haug, A.,** Regulation of membrane lipid fluidity in *Acholeplasma laidlawii:* effect of carotenoid pigment content, *Biochim. Biophys. Acta,* 352, 361, 1974.
141. **Huang, L., Lorch, S. K., Smith, G. G., and Haug, A.,** Control of membrane lipid fluidity in *Acholeplasma laidlawii, FEBS Lett.,* 43, 1, 1974.

142. **Hummeler, K., Tomassini, N., and Hayflick, L.,** Ultrastructure of a mycoplasma (Negroni) isolated from human leukemia, *J. Bacteriol.,* 90, 1965.
143. **Jaffor Ullah, A. H. and Cirillo, V. P.,** *Mycoplasma* phosphoenolpyruvate-dependent sugar phosphotransferase system: purification and characterization of the phosphocarrier protein, *J. Bacteriol.,* 127, 1298, 1976.
144. **Jaffor Ullah, A. H. and Cirillo, V. P.,** Mycoplasma phosphoenolpyruvate-dependent sugar phosphotransferase system: purification and characterization of enzyme I, *J. Bacteriol.,* 131, 988, 1977.
145. **James, R. and Branton, D.,** Lipid- and temperature-dependent structural changes in *Acholeplasma laidlawii* cell membranes, *Biochim. Biophys. Acta,* 323, 378, 1973.
146. **Janiak, M. J., Loomis, C. R., Shipley, G. G., and Small, D. M.,** The ternary phase diagram of lecithin, cholesteryl linoleate and water: phase behavior and structure, *J. Mol. Biol.,* 86, 325, 1974.
147. **Jinks, D. C. and Matz, L. L.,** The reduced nicotinamide adenine dinucleotide "oxidase" of *Acholeplasma laidlawii* membranes, *Biochim. Biophys. Acta,* 430, 71, 1976.
148. **Jinks, D. C. and Matz, L. L.,** Purification of the reduced nicotinamide adenine dinucleotide dehydrogenase from membranes of *Acholeplasma laidlawii, Biochim. Biophys. Acta,* 452, 30, 1976.
149. **Jinks, D. C., Silvius, J. R., and McElhaney, R. N.,** Physiological role and membrane lipid modulation of the membrane-bound (Mg^{2+}, Na^+)-adenosine triphosphatase activity in *Acholeplasma laidlawii, J. Bacteriol.,* 136, 1027, 1978.
150. **Johansson, K.-E.,** Production and utilization of monospecific antisera against membrane proteins from *Acholeplasma laidlawii, Zbl. Bakt. Parasitenk.-1, Abt.-A,* 241, 199, 1978.
151. **Johansson, K.-E., Blomqvist, I., and Hjertén, S.,** Purification of membrane proteins from *Acholeplasma laidlawii* by agarose suspension electrophoresis in Tween 20 and polyacrylamide and dextran gel electrophoresis in detergent-free media, *J. Biol. Chem.,* 250, 2463, 1975.
152. **Johansson, K.-E. and Hjertén, S.,** Localization of the Tween 20-soluble membrane proteins of *Acholeplasma laidlawii* by crossed immunoelectrophoresis, *J. Mol. Biol.,* 86, 341, 1974.
153. **Johansson, K.-E., Pertoft, H., and Hjertén, S.,** Characterization of the Tween 20-soluble membrane proteins of *Acholeplasma laidlawii, Int. J. Biol. Macromolecules,* 1, 111, 1979.
154. **Johansson, K.-E. and Wròblewski, H.,** Crossed immunoelectrophoresis, in the presence of Tween 20 or sodium deoxycholate, of purified membrane proteins from *Acholeplasma laidlawii, J. Bacteriol.,* 136, 324, 1978.
155. **Jones, T. C. and Hirsch, J. G.,** The interaction *in vitro* of *Mycoplasma pulmonis* with mouse peritoneal macrophages and L-cells, *J. Exp. Med.,* 133, 231, 1971.
156. **Jones, T. C., Yeh, S., and Hirsch, J. G.,** Studies on attachment and ingestion phases of phagocytosis of *Mycoplasma pulmonis* by mouse peritoneal macrophages, *Proc. Soc. Exp. Biol. Med.,* 139, 464, 1972.
157. **Kaback, H. R.,** Transport studies in bacterial membrane vesicles, *Science,* 186, 882, 1974.
158. **Kahane, I. and Brunner, H.,** Isolation of a glycoprotein from *Mycoplasma pneumoniae* membranes, *Infect. Immun.,* 18, 273, 1977.
159. **Kahane, I., Gat, O., Banai, M., Bredt, W., and Razin, S.,** Adherence of *Mycoplasma gallisepticum* to glass, *J. Gen. Microbiol.,* 111, 217, 1979.
160. **Kahane, I., Greenstein, S., and Razin, S.,** Carbohydrate content and enzymic activities in the membrane of *Spiroplasma citri, J. Gen. Microbiol.,* 101, 173, 1977.
161. **Kahane, I. and Marchesi, V. T.,** Studies on the orientation of proteins in mycoplasma and erythrocyte membranes, *Ann. N. Y. Acad. Sci.,* 225, 38, 1973.
162. **Kahane, I. and Razin, S.,** Synthesis and turnover of membrane protein and lipid in *Mycoplasma laidlawii, Biochim. Biophys. Acta,* 183, 79, 1969.
163. **Kahane, I. and Razin, S.,** Immunological analysis of mycoplasma membranes, *J. Bacteriol.,* 100, 187, 1969.
164. **Kahane, I. and Razin, S.,** pH-Dependent changes in density of plasma membranes of growing *Mycoplasma laidlawi* cells, *FEBS Lett.,* 10, 261, 1970.
165. **Kahane, I. and Razin, S.,** Cholesterol-phosphatidylcholine dispersions as donors of cholesterol to mycoplasma membranes, *Biochim. Biophys. Acta,* 471, 32, 1977.
166. **Kahane, I., Razin, S., and Muhlrad, A.,** Contractile proteins in mycoplasmas, *Zbl. Bakt. Parasitenk.-1 Abt.-A,* 241, 200, 1978.
167. **Kahane, I. and Tully, J. G.,** Binding of plant lectins to mycoplasma cells and membranes, *J. Bacteriol.,* 128, 1, 1976.
168. **Kandler, O. and Zehender, C.,** Über das vorkommen von α-ϵ-diaminopimelinsaüre bei verscheidenen L-Phasentypen von *Proteus vulgaris* und bei den pleuropneumonie-ähnlichen organismen, *Z. Naturforsch. Sect. B,* 725, 1957.
169. **Kates, M.,** Ether-linked lipids in extremely halophilic bacteria, in *Ether Lipids: Chemistry and Biology,* Snyder, F., Ed., Academic Press, New York, 1972, 351.

170. **Kenny, G. E. and Newton, R. M.,** Close serological relationship between glycolipids of *Mycoplasma pneumoniae* and glycolipids of spinach, *Ann. N. Y. Acad. Sci.,* 225, 54, 1973.
171. **Kleinig, H.,** Nuclear membranes from mammalian liver. II. Lipid composition, *J. Cell Biol.,* 46, 396, 1970.
172. **Kutner, S.,** Uptake of Exogenous Phospholipids by Mycoplasmas, M.Sc. thesis, The Hebrew University, Jerusalem, 1979.

172a. **Lala, A. K., Buttke, T. M., and Bloch, K.,** On the sterol hydroxyl group in membranes, *J. Biol. Chem.,* 254, 10582, 1979.

173. **Lampen, J. O., Gill, J. W., Arnow, P. M., and Magana-Plaza, I.,** Inhibition of the pleuropneumonia-like organism *Mycoplasma gallisepticum* by certain polyene antifungal antibiotics, *J. Bacteriol.,* 86, 945, 1963.
174. **Lange, Y., Cohen, C. M., and Poznansky, M. J.,** Transmembrane movement of cholesterol in human erythrocytes, *Proc. Nat. Acad. Sci. U.S.A.,* 74, 1538, 1977.
175. **Langworthy, T. A.,** Long-chain diglycerol tetraethers from *Thermoplasma acidophilum, Biochim. Biophys. Acta,* 487, 37, 1977.
176. **Langworthy, T. A.,** Membrane structure of thermoacidophilic bacteria, in *Strategies of Microbial Life in Extreme Environments,* Shilo, M., Ed., Dahlem Konferenzen, Berlin, Germany, 1979, 417.
177. **Langworthy, T. A.,** Special features of thermoplasmas, in *The Mycoplasmas,* Vol. I., Barile, M. F. and Razin, S., Eds., Academic Press, New York, 1979, 495.
178. **Langworthy, T. A., Mayberry, W. R., Smith, P. F., and Robinson, I. M.,** Plasmalogen composition of *Anaeroplasma, J. Bacteriol.,* 122, 785, 1975.
179. **Langworthy, T. A., Smith, P. F., and Mayberry, W. R.,** Lipids of *Thermoplasma acidophilum, J. Bacteriol.,* 112, 1193, 1972.
180. **Larraga, V. and Razin, S.,** Reduced nicotinamide adenine dinucleotide oxidase activity in membranes and cytoplasm of *Acholeplasma laidlawii* and *Mycoplasma mycoides* subsp. *capri. J. Bacteriol.,* 128, 827, 1976.
181. **Leblanc, G. and LeGrimellec, C.,** Active K^+ transport in *Mycoplasma mycoides* var. *capri.* Net and unidirectional K^+ movements, *Biochim. Biophys. Acta,* 554, 156, 1979.
182. **Leblanc, G. and LeGrimellec, C.,** Active K^+ transport in *Mycoplasma mycoides* var. *capri.* Relationships between K^+ and distribution, electrical potential and ATPase activity, *Biochim. Biophys. Acta,* 554, 168, 1979.
183. **LeGrimellec, C. and Leblanc, G.,** Effect of membrane cholesterol on potassium transport in *Mycoplasma mycoides* var. *capri* (PG3), *Biochim. Biophys. Acta,* 514, 152, 1978.
184. **LeGrimellec, C. and Leblanc, G.,** Temperature-dependent relationship between K^+ influx, Mg-ATPase activity, $\Delta\Psi$ and membrane lipid composition in mycoplasma, *Biochim. Biophys. Acta,* 599, 639, 1980.
185. **Levisohn, S. and Razin, S.,** Isolation, ultrastructure and antigenicity of *Mycoplasma gallisepticum* membranes, *J. Hyg. Camb.,* 71, 725, 1973.
186. **Low, I. E. and Zimkus, S. M.,** Reduced nicotinamide adenine dinucleotide oxidase activity and H_2O_2 formation of *Mycoplasma pneumoniae, J. Bacteriol.,* 116, 346, 1973.
187. **Malott, D.W. and McCurdy, H. D.,** Heavy meromyosin decoration of filaments from *Escherichia coli, Curr. Microbiol.,* 1, 201, 1978.
188. **Manchee, R. J. and Taylor-Robinson, D.,** Haemadsorption and haemagglutination by mycoplasmas, *J. Gen. Microbiol.,* 50, 465, 1968.
189. **Manchee, R. J. and Taylor-Robinson, D.,** Studies on the nature of receptors involved in attachment of tissue culture cells to mycoplasmas, *Brit. J. Exp. Pathol.,* 50, 66, 1969.
190. **Maniloff, J. and Morowitz, H. J.,** Cell biology of the mycoplasmas, *Bacteriol. Rev.,* 36, 263, 1972.
191. **Maniloff, J., Morowitz, H. J., and Barrnett, R. J.,** Ultrastructure and ribosomes of *Mycoplasma gallisepticum, J. Bacteriol.,* 90, 193, 1965.
192. **Maniloff, J. and Quinlen, D. C.,** Partial purification of a membrane-associated deoxyribonucleic acid complex from *Mycoplasma gallisepticum, J. Bacteriol.,* 120, 495, 1974.
193. **Margulis, L., To, L., and Chase, D.,** Microtubules in prokaryotes, *Science,* 200, 1118, 1978.
194. **Masover, G. K., Razin, S., and Hayflick, L.,** Localization of enzymes in *Ureaplasma urealyticum* (T-strain mycoplasma), *J. Bacteriol.,* 130, 297, 1977.
195. **Masover, G. K., Sawyer, J. E., and Hayflick, L.,** Urea-hydrolyzing activity of a T-strain mycoplasma: *Ureaplasma urealyticum, J. Bacteriol.,* 125, 581, 1976.
196. **Mayberry, W. R., Langworthy, T. A., and Smith, P. F.,** Structure of the mannoheptose-containing pentaglycosyldiacylglycerol from *Acholeplasma modicum, Biochim. Biophys. Acta,* 441, 115, 1976.
197. **Mayberry, W. R., Smith, P. F., and Langworthy, T. A.,** Heptose-containing pentaglycosyl diglyceride among the lipids of *Acholeplasma modicum, J. Bacteriol.,* 118, 898, 1974.

198. **Mayberry, W. R., Smith, P. F., Langworthy, T. A., and Plackett, P.,** Identification of the amide-linked fatty acids of *Acholeplasma axanthum* S743 as D(−)3-hydroxyhexadecanoate and its homologues, *J. Bacteriol.,* 116, 1091, 1973.
199. **Mayberry-Carson, K. J., Jewell, M. J., and Smith, P. F.,** Ultrastructural localization of *Thermoplasma acidophilum* surface carbohydrate by using concanavalin A, *J. Bacteriol.,* 133, 1510, 1978.
200. **Mayberry-Carson, K. J., Langworthy, T. A., Mayberry, W. R., and Smith, P. F.,** A new class of lipopolysaccharide from *Thermoplasma acidophilum, Biochim. Biophys. Acta,* 360, 217, 1974.
201. **Mayberry-Carson, K. J., Roth, I. L., and Smith, P. F.,** Ultrastructure of lipopolysaccharide isolated from *Thermoplasma acidophilum, J. Bacteriol.,* 121, 700, 1975.
202. **McCabe, P. J. and Green, C.,** The dispersion of cholesterol with phospholipids and glycolipids, *Chem. Phys. Lipids,* 20, 319, 1977.
203. **McElhaney, R. N.,** The effect of alterations in the physical state of the membrane lipids on the ability of *Acholeplasma laidlawii* B to grow at various temperatures, *J. Mol. Biol.,* 84, 145, 1974.
204. **McElhaney, R. N., de Gier, J., and van Deenen, L. L. M.,** The effect of alterations in fatty acid composition and cholesterol content on the permeability of *Mycoplasma laidlawii* B cells and derived liposomes, *Biochim. Biophys. Acta,* 219, 245, 1970.
205. **McElhaney, R. N., de Gier, J., and van der Neut-Kok, E. C. M.,** The effect of alterations in fatty acid composition and cholesterol content on the nonelectrolyte permeability of *Acholeplasma laidlawii* B cells and derived liposomes, *Biochim. Biophys. Acta,* 298, 500, 1973.
206. **McElhaney, R. N. and Tourtellotte, M. E.,** Metabolic turnover of the polar lipids of *Mycoplasma laidlawii* strain B, *J. Bacteriol.,* 101, 72, 1970.
207. **McElhaney, R. N. and Tourtellotte, M. E.,** The relationship between fatty acid structure and the positional distribution of esterified fatty acids in phosphatidyl glycerol from *Mycoplasma laidlawii* B, *Biochim. Biophys. Acta,* 202, 120, 1970.
208. **Melchior, D. L., Morowitz, H. J., Sturtevant, J. M., and Tsong, T. Y.,** Characterization of the plasma membrane of *Mycoplasma laidlawii.* VII. Phase transition of membrane lipids, *Biochim. Biophys. Acta,* 219, 114, 1970.
209. **Melchior, D. L. and Steim, J. M.,** Control of fatty acid composition of *Acholeplasma laidlawii* membranes, *Biochim. Biophys. Acta,* 466, 148, 1977.
210. **Metcalfe, J. C., Birdsall, N. J. M., and Lee, A. G.,** ^{13}C NMR spectra of acholeplasma membranes containing ^{13}C labelled phospholipids, *FEBS Lett.,* 21, 335, 1972.
211. **Mindich, L.,** Membrane synthesis in *Bacillus subtilis.* II. Integration of membrane proteins in the absence of lipid synthesis, *J. Mol. Biol.,* 49, 433, 1970.
212. **Minkoff, L., and Damadian, R.,** Actin-like proteins from *Escherichia coli:* concept of cytotonus as a missing link between cell metabolism and the biological ion exchange resin, *J. Bacteriol.,* 125, 353, 1976.
213. **Miura, Y., Imaeda, N., Shinoda, M., Tamura, H., and Ueta, N.,** An analysis of the fatty acid composition of total lipids from mycoplasmas, *Jpn. J. Exp. Med.,* 48, 525, 1978.
214. **Morowitz, H. J. and Terry, T. M.,** Characterization of the plasma membrane of *Mycoplasma laidlawii.* V. Effects of selective removal of protein and lipid, *Biochim. Biophys. Acta,* 183, 276, 1969.
215. **Morrison, T. H. and Weibull, C.,** The occurrence of cell wall constituents in stable Proteus L-forms, *Acta Pathol. Microbiol. Scand.,* 55, 475, 1962.
216. **Mouches, C. Vignault, J. C., Tully, J. G., Whitcomb, R. F., and Bové, J.-M.,** Characterization of spiroplasmas by one- and two-dimensional protein analysis on polyacrylamide slab gels, *Curr. Microbiol.,* 2, 69, 1979.
217. **Mudd, J. B., Ittig, M., Roy, B., Latrille, J., and Bové, J.-M.,** Composition and enzyme activities of *Spiroplasma citri* membranes, *J. Bacteriol.,* 129, 1250, 1977.
218. **Mugharbil, U. and Cirillo, V. P.,** Mycoplasma phosphoenolpyruvate-dependent sugar phosphotransferase system: glucose-negative mutant and regulation of intracellular cyclic AMP, *J. Bacteriol.,* 133, 203, 1978.
219. **Muhlrad, A., Peleg, I., Razin, S., and Kahane, I.,** Actin-like DNase I inhibitor protein in mycoplasma, in *Proc. Annu. Meet. Israel Biochem. Soc.,* 1979, 4.
220. **Mühlradt, P. F. and Golecki, J. R.,** Asymmetrical distribution and artifactual reorientation of lipopolysaccharide in the outer membrane bilayer of *Salmonella typhimurium, Eur. J. Biochem.,* 51, 343, 1975.
221. **Munkres, M. and Wachtel, A.,** Histochemical localization of phosphatases in *Mycoplasma gallisepticum, J. Bacteriol.,* 93, 1096, 1967.
222. **Muse, K. E., Powell, D. A., and Collier, A. M.,** *Mycoplasma pneumoniae* in hamster tracheal organ culture studied by scanning electron microscopy, *Infec. Immun.,* 13, 229, 1976.

223. **Nachbar, M. S. and Salton, M. R. J.,** Characteristics of a lipid-rich NADH dehydrogenase-containing particulate fraction obtained from *Micrococcus lysodeikticus* membranes, *Biochim. Biophys. Acta,* 223, 309, 1970.
224. **Nakamura, K. and Watanabe, S.,** Myosin-like protein and actin-like protein from *Escherichia coli* K12 C600, *J. Biochem. Tokyo,* 83, 1459, 1978.
225. **Naot, Y., Siman-Tov, R., and Ginsburg, H.,** Mitogenic activity of *Mycoplasma pulmonis.* II. Studies on the biochemical nature of the mitogenic factor, *Eur. J. Immunol.,* 9, 149, 1979.
226. **Ne'eman, Z., Kahane, I., Kovartovsky, J., and Razin, S.,** Characterization of the mycoplasma membrane proteins. III. Gel filtration and immunological characterization of *Acholeplasma laidlawii* membrane proteins, *Biochim. Biophys. Acta,* 266, 255, 1972.
227. **Ne'eman, Z., Kahane, I., and Razin, S.,** Characterization of the mycoplasma membrane proteins. II. Solubilization and enzymic activities of *Acholeplasma laidlawii* membrane proteins, *Biochim. Biophys. Acta,* 249, 169, 1971.
228. **Ne'eman, Z. and Razin, S.,** Characterization of the mycoplasma membrane proteins. V. Release and localization of membrane-bound enzymes in *Acholeplasma laidlawii, Biochim. Biophys. Acta,* 375, 54, 1975.
229. **Neimark, H. C.,** Extraction of an actin-like protein from the prokaryote *Mycoplasma pneumoniae, Proc. Nat. Acad. Sci. U.S.A.,* 74, 4041, 1977.
230. **Newnham, A. G., and Chu, H. P.,** An *in vitro* comparison of the effect of some antibacterial, antifungal and antiprotozoal agents on various strains of *Mycoplasma* (pleuropneumonia-like organisms: P.P.L.O.), *J. Hyg. Camb.,* 63, 1, 1965.
231. **Norman, A. W., Demel, R. A., de Kruyff, B., Geurts van Kessel, W. S. M., and van Deenen, L. L. M.,** Studies on the biological properties of polyene antibiotics: comparison of other polyenes with filipin in their ability to interact specifically with sterol, *Biochim. Biophys. Acta,* 290, 1, 1972.
232. **Odriozola, J. M., Waitzkin, E., Smith, T. L., and Bloch, K.,** Sterol requirement of *Mycoplasma capricolum, Proc. Nat. Acad. Sci. U.S.A.,* 75, 4107, 1978.
233. **Oldfield, E., Chapman, D., and Derbyshire, W.,** Lipid mobility in acholeplasma membranes using deuteron magnetic resonance, *Chem. Phys. Lipids,* 9, 69, 1972.
234. **Op den Kamp, J. A. F.,** Lipid asymmetry in membranes, *Annu. Rev. Biochem.,* 48, 47, 1979.
235. **Overath, P. and Thilo, L.,** Structural and functional aspects of biological membranes revealed by lipid phase transitions, in *Biochemistry of Cell Walls and Membranes II,* Vol. 19, Metcalfe, J. C., Ed., University Park Press, Baltimore, 1978, 1.
236. **Owen, P. and Kaback, H. R.,** Antigenic architecture of membrane vesicles from *Escherichia coli, Biochemistry,* 18, 1422, 1979.
237. **Owen, P. and Salton, M. R. J.,** Membrane asymmetry and expression of cell surface antigens of *Micrococcus lysodeikticus* established by crossed immunoelectrophoresis, *J. Bacteriol.,* 132, 974, 1977.
238. **Pagano, R. E., Sandra, A., and Takeichi, M.,** Interactions of phospholipid vesicles with mammalian cells, *Ann. N. Y. Acad. Sci.,* 308, 185, 1978.
239. **Panchenko, L. F., Fedotov, N. S., and Tarshis, M. A.,** Transport properties of membrane vesicles from *Acholeplasma laidlawii.* II. Kinetic characteristics and specificity of glucose transport system, *Folia Microbiol. (Prague),* 20, 480, 1975.
240. **Panos, C.,** Chemical and physiological aspects of the bacterial L-phase variant, in *The Mycoplasmatales and the L-Phase of Bacteria,* Hayflick, L., Ed., Appleton-Century-Crofts, New York, 1969, 503.
241. **Panos, C. and Leon, O.,** Replacement of the octadecenoic acid growth requirement for *Acholeplasma laidlawii* A by *cis*-9,10-methylene-hexadecanoic acid, a cyclopropane fatty acid, *J. Gen. Microbiol.,* 80, 93, 1974.
242. **Panos, C. and Rottem, S.,** Incorporation and elongation of fatty acid isomers by *Mycoplasma laidlawii* A, *Biochemistry,* 9, 407, 1970.
243. **Paroz, P. and Nicolet, J.,** Glycoproteins in mycoplasmas. Contamination with serum aggregates from growth media, *Experientia,* 34, 1668, 1978.
244. **Pask-Hughes, R. A., Mozaffary, H., and Shaw, N.,** Glycolipids in prokaryotic cells, *Biochem. Soc. Trans.,* 5, 1675, 1977.
245. **Patel, K. R., Smith, P. F., and Mayberry, W. R.,** Comparison of lipids from *Spiroplasma citri* and corn stunt spiroplasma, *J. Bacteriol.,* 136, 829, 1978.
246. **Patriarca, P., Beckerdite, S., Pettis, P., and Elsbach, P.,** Phospholipid metabolism by phagocytic cells. VII. The degradation and utilization of phospholipids by various microbial species by rabbit granulocytes, *Biochim. Biophys. Acta,* 280, 45, 1972.
247. **Pecht, M., Giberman, E., Keysary, A., Yariv, J., and Katchalski, E.,** Hydrolysis of alanine oligopeptides by an enzyme located in the membrane of *Mycoplasma laidlawii, Biochim. Biophys. Acta,* 290, 267, 1972.

248. **Pisetsky, D. and Terry, T. M.,** Are mycoplasma membrane proteins affected by variations in membrane fatty acid composition?, *Biochim. Biophys. Acta,* 274, 95, 1972.
249. **Plackett, P.,** On the probable absence of "mucocomplex" from *Mycoplasma mycoides, Biochim. Biophys. Acta,* 35, 260, 1959.
250. **Plackett, P., Marmion, B. P., Shaw, E. J., and Lemcke, R. M.,** Immunochemical analysis of *Mycoplasma pneumoniae.* 3. Separation and chemical identification of serologically active lipids, *Aust. J. Exp. Biol. Med. Sci.,* 47, 171, 1969.
251. **Plackett, P., Smith, P. F., and Mayberry, W. R.,** Lipids of a sterol-nonrequiring mycoplasma, *J. Bacteriol.,* 104, 798, 1970.
252. **Pollack, J. D.,** Localization of reduced nicotinamide adenine dinucleotide oxidase activity in *Acholeplasma* and *Mycoplasma* species, *Int. J. Syst. Bacteriol.,* 25, 108, 1975.
253. **Pollack, J. D.,** Differentiation of *Mycoplasma* and *Acholeplasma, Int. J. Syst. Bacteriol.* 28, 425, 1978.
254. **Pollack, J. D.,** Respiratory pathways and energy-yielding mechanisms, in *The Mycoplasmas,* Barile, M. F. and Razin, S., Eds., Academic Press, New York, 1979, 188.
255. **Pollack, J. D., Razin, S., and Cleverdon, R. C.,** Localization of enzymes in *Mycoplasma, J. Bacteriol.,* 90, 617, 1965.
256. **Pollack, J. D., Razin, S., Pollack, M. E., and Cleverdon, R. C.,** Fractionation of *Mycoplasma* cells for enzyme localization, *Life Sci.,* 4, 973, 1965.
257. **Pollack, J. D., Somerson, N. L., and Senterfit, L. B.,** Isolation, characterization, and immunogenicity of *Mycoplasma pneumoniae* membranes, *Infect. Immun.,* 2, 326, 1970.
258. **Pollack, J. D. and Tourtellotte, M. E.,** Synthesis of saturated long-chain fatty acids from sodium acetate-1-C^{14} by mycoplasma, *J. Bacteriol.,* 93, 636, 1967.
259. **Powell, D. A., Hu, P. C., Wilson, M., Collier, A. M., and Baseman, J. B.,** Attachment of *Mycoplasma pneumoniae* to respiratory epithelium, *Infect. Immun.,* 13, 959, 1976.
260. **Razin, S.,** Structure, composition and properties of the PPLO cell envelope, in *Recent Progress in Microbiology,* Vol. 8, Gibbons, N. E., Ed., University of Toronto Press, Toronto, 1962, 526.
261. **Razin, S.,** Binding of nystatin by *Mycoplasma* (pleuropneumonia-like organisms), *Biochim. Biophys. Acta,* 78, 771, 1963.
262. **Razin, S.,** Osmotic lysis of mycoplasma, *J. Gen. Microbiol.,* 33, 471, 1963.
263. **Razin, S.,** Factors influencing osmotic fragility of mycoplasma, *J. Gen. Microbiol.,* 36, 451, 1964.
264. **Razin, S.,** The cell membrane of mycoplasma, *Ann. N. Y. Acad. Sci.,* 143, 115, 1967.
265. **Razin, S.,** The mycoplasma membrane, in *The Mycoplasmatales and the L-Phase of Bacteria,* Hayflick, L., Ed., Appleton-Century-Crofts, New York, 1969, 317.
266. **Razin, S.,** Structure and function in mycoplasma, *Annu. Rev. Microbiol.,* 23, 317, 1969.
267. **Razin, S.,** Reconstitution of biological membranes, *Biochim. Biophys. Acta,* 265, 241, 1972.
268. **Razin, S.,** Physiology of Mycoplasmas, in *Adv. Microb. Physiol.,* Vol. 10, Rose, A. H. and Tempest, D. W., Eds., Academic Press, London, 1973, 1.
269. **Razin, S.,** Correlation of cholesterol to phospholipid content in membranes of growing mycoplasmas, *FEBS Lett.,* 47, 81, 1974.
270. **Razin, S.,** Cholesterol incorporation into bacterial membranes, *J. Bacteriol.,* 124, 570, 1975.
271. **Razin, S.,** The mycoplasma membrane, in *Progress in Surface and Membrane Science,* Vol. 9, Cadenhead, D. A., Danielli, J. F., and Rosenberg, M. D., Eds., Academic Press, New York, 1975, 257.
272. **Razin, S.,** Cholesterol uptake is dependent on membrane fluidity in mycoplasmas, *Biochim. Biophys. Acta,* 513, 401, 1978.
273. **Razin, S.,** The mycoplasmas, *Microbiol. Rev.,* 42, 414, 1978.
274. **Razin, S.,** Isolation and characterization of mycoplasma membranes, in *The Mycoplasmas,* Vol. I, Barile, M. F. and Razin, S., Eds., Academic Press, New York, 1979, 213.
275. **Razin, S.,** Membrane proteins, in *The Mycoplasmas,* Vol. I, Barile, M. F. and Razin, S., Eds., Academic Press, New York, 1979, 289.
276. **Razin, S. and Argaman, M.,** Susceptibility of *Mycoplasma* (pleuropneumonia-like organisms) and bacterial protoplasts to lysis by various agents, *Nature (London),* 193, 502, 1962.
277. **Razin, S. and Argaman, M.,** Lysis of mycoplasma, bacterial protoplasts, spheroplasts and L-forms by various agents, *J. Gen. Microbiol.,* 30, 155, 1963.
278. **Razin, S., Argaman, M., and Avigan, J.,** Chemical composition of mycoplasma cells and membranes, *J. Gen. Microbiol.,* 33, 477, 1963.
279. **Razin, S. and Cleverdon, R. C.,** Carotenoids and cholesterol in membranes of *Mycoplasma laidlawii, J. Gen. Microbiol.,* 41, 409, 1965.
280. **Razin, S. and Cosenza, B. J.,** Growth phases of *Mycoplasma* in liquid media observed with phase-contrast microscope, *J. Bacteriol.,* 91, 858, 1966.

281. **Razin, S., Hasin, M., Ne'eman, Z., and Rottem, S.,** Isolation, chemical composition, and ultrastructural features of the cell membrane of the mycoplasma-like organism *Spiroplasma citri, J. Bacteriol.,* 116, 1421, 1973.
282. **Razin, S., Kahane, I., Banai, M., and Bredt, W.,** Adhesion of mycoplasmas to eukaryotic cells, in *Adhesion and Micro-organism Pathogenicity,* Ciba Found. Symp., Pitman Medical, London, 1980, 98.
283. **Razin, S., Kahane, I., and Kovartovsky, J.,** Immunochemistry of mycoplasma membranes, in *Pathogenic Mycoplasmas,* Ciba Foundation Symposium, Elsevier, Amsterdam, 1972, 93.
284. **Razin, S., Knyszynski, A., and Lifshitz, Y.,** Nucleases of mycoplasma, *J. Gen. Microbiol.,* 36, 323, 1964.
285. **Razin, S., Kutner, S., Ephrati, H., and Rottem, S.,** Phospholipid and cholesterol uptake by mycoplasma cells and membranes, *Biochim. Biophys. Acta,* 598, 628, 1980.
286. **Razin, S., Morowitz, H. J., and Terry, T. M.,** Membrane subunits of *Mycoplasma laidlawii* and their assembly to membranelike structures, *Proc. Nat. Acad. Sci. U.S.A.,* 54, 219, 1965.
287. **Razin, S., Ne'eman, Z., and Ohad, I.,** Selective reaggregation of solubilized mycoplasma membrane proteins and the kinetics of membrane reformation, *Biochim. Biophys. Acta,* 193, 277, 1969.
288. **Razin, S., Prescott, B., Caldes, G., James, W. D., and Chanock, R. M.,** Role of glycolipids and phosphatidylglycerol in the serological activity of *Mycoplasma pneumoniae, Infect. Immun.,* 1, 408, 1970.
289. **Razin, S., Prescott, B., and Chanock, R. M.,** Immunogenicity of *Mycoplasma pneumoniae* glycolipids: a novel approach to the production of antisera to membrane lipids, *Proc. Nat. Acad. Sci. U.S.A.,* 67, 590, 1970.
290. **Razin, S., and Rottem, S.,** Fatty acid requirements of *Mycoplasma laidlawii, J. Gen. Microbiol.,* 33, 459, 1963.
291. **Razin, S. and Rottem, S.,** Role of carotenoids and cholesterol in the growth of *Mycoplasma laidlawii, J. Bacteriol.,* 93, 1181, 1967.
292. **Razin, S. and Rottem, S.,** Techniques for the manipulation of mycoplasma membranes, in *Biochemical Analysis of Membranes,* Maddy, A. H., Ed., Chapman and Hall, London, 1976, 3.
293. **Razin, S. and Rottem, S.,** Cholesterol in membranes: studies with mycoplasmas, *Trends Biochem. Sci.,* 3, 5, 1978.
294. **Razin, S. and Shafer, Z.,** Incorporation of cholesterol by membranes of bacterial L-phase variants, with an appendix on the determination of the L-phase parentage by the electrophoretic patterns of cell proteins, *J. Gen. Microbiol.,* 58, 327, 1969.
295. **Razin, S., Tourtellotte, M. E., McElhaney, R. N., and Pollack, J. D.,** Influence of lipid components of *Mycoplasma laidlawii* membranes on osmotic fragility of cells, *J. Bacteriol.,* 91, 609, 1966.
296. **Razin, S. and Tully, J. G.,** Cholesterol requirement of mycoplasmas, *J. Bacteriol.,* 102, 306, 1970.
297. **Razin, S., Wormser, M., and Gershfeld, N. L.,** Cholesterol in mycoplasma membranes. II. Components of *Acholeplasma laidlawii* cell membranes responsible for cholesterol binding, *Biochim. Biophys. Acta,* 352, 385, 1974.
298. **Read, B. D. and McElhaney, R. N.,** Glucose transport in *Acholeplasma laidlawii* B: dependence on the fluidity and physical state of membrane lipids, *J. Bacteriol.,* 123, 47, 1975.
299. **Reinert, J. C. and Steim, J. M.,** Calorimetric detection of a membrane-lipid phase transition in living cells, *Science,* 168, 1580, 1970.
300. **Reusch, V. M. and Panos, C.,** Defective synthesis of lipid intermediates for peptidoglycan formation in a stabilized L-form of *Streptococcus pyogenes, J. Bacteriol.,* 126, 300, 1976.
301. **Robertson, J. and Smook, E.,** Cytochemical evidence of extramembranous carbohydrates on *Ureaplasma urealyticum, J. Bacteriol.,* 128, 658, 1976.
302. **Rodwell, A. W.,** The steroid growth requirement of *Mycoplasma mycoides, J. Gen. Microbiol.,* 32, 91, 1963.
303. **Rodwell, A. W.,** The stability of *Mycoplasma mycoides, J. Gen. Microbiol.,* 40, 227, 1965.
304. **Rodwell, A. W.,** The nutrition and metabolism of mycoplasma: progress and problems, *Ann. N. Y. Acad. Sci.,* 143, 88, 1967.
305. **Rodwell, A. W.,** Fatty acid composition of mycoplasma lipids: a biomembrane with only one fatty acid, *Science,* 160, 1350, 1968.
306. **Rodwell, A. W. and Peterson, J. E.,** The effect of straight-chain saturated, monoenoic and branched-chain fatty acids on growth and fatty acid composition of mycoplasma strain Y, *J. Gen. Microbiol.,* 68, 173, 1971.
307. **Rodwell, A. W., Peterson, J. E., and Rodwell, E. S.,** Macromolecular synthesis and growth of mycoplasmas, in *Pathogenic Mycoplasmas,* Elliott, K., and Birch, J., Eds., Ciba Found. Symp., 1972, 123.
308. **Rodwell, A. W. and Rodwell, E. S.,** Relationships between strains of *Mycoplasma mycoides* subsp. *mycoides* and *capri* studied by two-dimensional gel electrophoresis of cell proteins, *J. Gen. Microbiol.,* 109, 259, 1978.

309. **Rodwell, A. W., Rodwell, E. S., and Archer, D. B.,** Mycoplasmas lack a protein which closely resembles α-actin, *FEMS Lett.,* 5, 235, 1979.
310. **Rohmer, M., Bouvier, P., and Ourisson, G.,** Molecular evolution of biomembranes — structural equivalents and phylogenetic precursors of sterols, *Proc. Nat. Acad. Sci. U.S.A.,* 76, 847, 1979.
311. **Romano, N. and LaLicata, R.,** Cell fractions and enzymatic activities of *Ureaplasma urealyticum, J. Bacteriol.,* 136, 833, 1978.
312. **Romano, N., Rottem, S., and Razin, S.,** Biosynthesis of saturated and unsaturated fatty acids by a T-strain mycoplasma *(Ureaplasma), J. Bacteriol.,* 128, 170, 1976.
313. **Romano, N., Smith, P. F., and Mayberry, W. R.,** The lipids of a T-strain of *Mycoplasma, J. Bacteriol.,* 109, 565, 1972.
314. **Romijn, J. C., van Golde, L. M. G., McElhaney, R. N., and van Deenen, L. L. M.,** Some studies on the fatty acid composition of total lipids and phosphatidylglycerol from *Acholeplasma laidlawii* B and their relation to the permeability of intact cells of this organism, *Biochim. Biophys. Acta,* 280, 22, 1972.
315. **Rothman, J. E. and Engelman, D. M.,** Molecular mechanism for the interaction of phospholipid with cholesterol, *Nature (London),* 237, 42, 1972.
316. **Rothman, J. E. and Lenard, J.,** Membrane asymmetry, *Science,* 195, 743, 1977.
317. **Rottem, S.,** Heterogeneity in the physical state of the exterior and interior regions of mycoplasma membrane lipids, *Biochem. Biophys. Res. Commun.,* 64, 7, 1975.
318. **Rottem, S.,** Molecular organization of membrane lipids, in *The Mycoplasmas,* Vol. 1., Barile, M. F. and Razin, S., Eds., Academic Press, New York, 1979, 260.
319. **Rottem, S.,** Membrane lipids of mycoplasmas, *Biochim. Biophys. Acta,* 604, 65, 1980.
320. **Rottem, S. and Barile, M. F.,** Effect of cerulenin on growth and lipid metabolism of mycoplasmas, *Antimicrob. Agents Chemother.,* 9, 301, 1976.
321. **Rottem, S., Cirillo, V. P., de Kruyff, B., Shinitzky, M., and Razin, S.,** Cholesterol in mycoplasma membranes. Correlation of enzymic and transport activities with physical state of lipids in membranes of *Mycoplasma mycoides* var. *capri* adapted to growth with low cholesterol concentrations, *Biochim. Biophys. Acta,* 323, 509, 1973.
322. **Rottem, S., Gottfried, L., and Razin, S.,** Carotenoids as protectors against photodynamic inactivation of the adenosine triphosphatase of *Mycoplasma laidlawii* membranes, *Biochem. J.,* 109, 707, 1968.
323. **Rottem, S. and Greenberg, A.,** Changes in composition, biosynthesis, and physical state of membrane lipids occurring upon aging of *Mycoplasma hominis* cultures, *J. Bacteriol.,* 121, 631, 1975.
324. **Rottem, S., Hardegree, M. C., Grabowski, M. W., Fornwald, R., and Barile, M. F.,** Interaction between tetanolysin and mycoplasma cell membrane, *Biochim. Biophys. Acta,* 455, 876, 1976.
325. **Rottem, S., Hasin, M., and Razin, S.,** Binding of proteins to mycoplasma membranes, *Biochim. Biophys. Acta,* 298, 876, 1973.
326. **Rottem, S., Hasin, M., and Razin, S.,** Differences in susceptibility to phospholipase C of free and membrane-bound phospholipids of *Mycoplasma hominis, Biochim. Biophys. Acta,* 323, 520, 1973.
327. **Rottem, S., Hasin, M., and Razin, S.,** The outer membrane of *Proteus mirabilis.* II. The extractable lipid fraction and electron paramagnetic resonance analysis of the outer and cytoplasmic membranes, *Biochim. Biophys. Acta,* 375, 395, 1975.
328. **Rottem, S., Hubbell, W. L., Hayflick, L., and McConnell, H. M.,** Motion of fatty acid spin labels in the plasma membrane of mycoplasma, *Biochim. Biophys. Acta,* 219, 104, 1970.
329. **Rottem, S. and Markowitz, O.,** Membrane lipids of *Mycoplasma gallisepticum:* a disaturated phosphatidylcholine and a phosphatidylglycerol with an unusual positional distribution of fatty acids, *Biochemistry,* 18, 2930, 1979.
330. **Rottem, S. and Markowitz, O.,** Unusual positional distribution of fatty acids in phosphatidylglycerol of sterol-requiring mycoplasmas, *FEBS Lett.,* 107, 379, 1979.
331. **Rottem, S., and Markowitz, O.,** Carotenoids as reinforcers of the *Acholeplasma laidlawii* lipid bilayer, *J. Bacteriol.,* 140, 944, 1980.
332. **Rottem, S., and Panos, C.,** The synthesis of long-chain fatty acids by a cell-free system from *Mycoplasma laidlawii* A, *Biochemistry,* 9, 57, 1970.
333. **Rottem, S., Pfendt, E. A., and Hayflick, L.,** Sterol requirements of T-strain mycoplasmas, *J. Bacteriol.,* 105, 323, 1971.
334. **Rottem, S. and Razin, S.,** Adenosine triphosphatase activity of mycoplasma membranes, *J. Bacteriol.,* 92, 714, 1966.
335. **Rottem, S. and Razin, S.,** Uptake and utilization of acetate by mycoplasma, *J. Gen. Microbiol.,* 48, 53, 1967.
336. **Rottem, S. and Razin, S.,** Electrophoretic patterns of membrane proteins of mycoplasma, *J. Bacteriol.,* 94, 359, 1967.
337. **Rottem, S. and Razin, S.,** Isolation of mycoplasma membranes by digitonin, *J. Bacteriol.,* 110, 699, 1972.
338. **Rottem, S. and Razin, S.,** Membrane lipids of *Mycoplasma hominis, J. Bacteriol.,* 113, 565, 1973.

339. **Rottem, S. and Samuni, A.,** Effect of proteins on the motion of spin-labeled fatty acids in mycoplasma membranes, *Biochim. Biophys. Acta,* 298, 32, 1973.
340. **Rottem, S. Slutzky, G. M. and Bittman, R.,** Cholesterol distribution and movement in the *Mycoplasma gallisepticum* cell membrane, *Biochemistry,* 17, 2723, 1978.
341. **Rottem, S., Stein, O., and Razin, S.,** Reassembly of mycoplasma membranes disaggregated by detergents, *Arch. Biochem. Biophys.,* 125, 46, 1968.
342. **Rottem, S., Trotter, S. L., and Barile, M. F.,** Membrane-bound thioesterase activity in mycoplasmas, *J. Bacteriol.,* 129, 707, 1977.
343. **Rottem, S., Yashouv, J., Ne'eman, Z., and Razin, S.,** Composition, ultrastructure and biological properties of membranes from *Mycoplasma mycoides* var. *capri* cells adapted to grow with low cholesterol concentrations, *Biochim. Biophys. Acta,* 323, 495, 1973.
344. **Rosenbusch, J. P., Jacobson, G. R., and Jaton, J.-C.,** Does a bacterial elongation factor share a common evolutionary ancestor with actin? *J. Supramol. Struct.,* 5, 391, 1976.
345. **Ruwart, M. J. and Haug, A.,** Membrane properties of *Thermoplasma acidophila, Biochemistry,* 14, 860, 1975.
346. **Saglio, P. H. M. and Whitcomb, R. F.,** Diversity of wall-less prokaryotes in plant vascular tissue, fungi, and invertebrate animals, in *The Mycoplasmas,* Vol. 3, Whitcomb, R. F. and Tully, J. G., Eds., Academic Press, New York, 1979, 1.
347. **Saier, M. H.,** Bacterial phosphoenolpyruvate: sugar phosphotransferase systems: structural, functional, and evolutionary interrelationships, *Bacteriol. Rev.,* 41, 856, 1977.
348. **Saito, Y. and McElhaney, R. N.,** Membrane lipid biosynthesis in *Acholeplasma laidlawii* B. Incorporation of exogenous fatty acids into membrane glyco- and phospholipids by growing cells, *J. Bacteriol.* 132, 485, 1977.
349. **Saito, Y. and McElhaney, R. N.,** The positional distribution of a series of positional isomers of *cis*-octadecenoic acid in phosphatidylglycerol from *Acholeplasma laidlawii* B, *Biochim. Biophys. Acta,* 529, 224, 1978.
350. **Saito, Y., Silvius, J. R., and McElhaney, R. N.,** Membrane lipid biosynthesis in *Acholeplasma laidlawii* B. Relationship between fatty acid structure and the positional distribution of esterified fatty acids in phospho- and glycolipids from growing cells, *Arch. Biochem. Biophys.,* 182, 443, 1977.
351. **Saito, Y., Silvius, J. R., and McElhaney, R. N.,** Membrane lipid biosynthesis in *Acholeplasma laidlawii* B. The *de novo* biosynthesis of saturated fatty acids by growing cells, *J. Bacteriol.,* 132, 497, 1977.
352. **Saito, Y., Silvius, J. R., and McElhaney, R. N.,** Membrane lipid biosynthesis in *Acholeplasma laidlawii* B: elongation of medium- and long-chain exogenous fatty acids in growing cells, *J. Bacteriol.,* 133, 66, 1978.
353. **Salton, M. R. J.,** Membrane associated enzymes in bacteria, in *Advances in Microbial Physiology,* Vol. 2, Rose, A. H. and Tempest, D. W., Eds., Academic Press, London, 1974, 213.
354. **Salton, M. R. J.,** Structure and function of bacterial plasma membranes, in *Relations Between Structure and Function in Prokaryotic Cells,* Stanier, R. Y. and Rogers, H. J., Eds., Cambridge University Press, Cambridge, England, 1978, 201.
355. **Schiefer, H.-G., Gerhardt, U., and Brunner, H.,** Immunological studies on the localization of phosphatidylglycerol in the membranes of *Mycoplasma hominis, Hoppe-Seyler's Z. Physiol. Chem.,* 356, 559, 1975.
356. **Schiefer, H.-G. Gerhardt, U., and Brunner, H.,** Localization of a phosphoglycolipid in mycoplasma membranes using specific anti-lipid-antibodies, *Zbl. Bakt. Parasitenk.-1 Abt.-A,* 239, 262, 1977.
357. **Schiefer, H.-G., Gerhardt, U., Brunner, H., and Krüpe, M.,** Studies with lectins on the surface carbohydrate structures of mycoplasma membranes, *J. Bacteriol.,* 120, 81, 1974.
358. **Schiefer, H.-G., Krauss, H., Brunner, H., and Gerhardt, U.,** Ultrastructural visualization of anionic sites on mycoplasma membranes by polycationic ferritin, *J. Bacteriol.,* 127, 461, 1976.
359. **Schiefer, H.-G., Krauss, H., Brunner, H., and Gerhardt, U.,** Ultrastructural and cytochemical studies on anionic surface sites of mycoplasma membranes, *Zbl. Bakt. Parasitenk.-1 Abt.-A,* 237, 104, 1977.
360. **Schiefer, H.-G., Krauss, H., Schummer, U., Brunner, H., and Gerhardt, U.,** Cytochemical localization of surface carbohydrates on mycoplasma membranes, *Experientia,* 34, 1011, 1978.
361. **Schiefer, H.-G., Krauss, H., Schummer, U., Brunner, H., and Gerhardt, U.,** Studies with ferritin-conjugated concanavalin A on carbohydrate structures of mycoplasma membranes, *FEMS Lett.,* 3, 183, 1978.
362. **Schuldiner, S. and Kaback, H. R.,** Fluorescent galactosides as probes for the *lac* carrier protein, *Biochim. Biophys. Acta,* 472, 399, 1977.
363. **Schummer, U., Schiefer, H.-G. and Gerhardt, U.,** Mycoplasma membrane potential determined by a fluorescent probe, *Hoppe-Seylers Z. Physiol. Chem.,* 359, 1023, 1978.
364. **Schummer, U., Schiefer, H.-G., and Gerhardt, U.,** Mycoplasma membrane potentials determined by potential-sensitive fluorescent dyes, *Curr. Microbiol.,* 2, 191, 1979.

365. **Searcy, D. G., Stein, D. B., and Green, G. R.,** Phylogenetic affinities between eukaryotic cells and a thermophilic mycoplasma, *BioSystems,* 10, 19, 1978.
365a. **Seid, R. C., Smith, P. F., Guevarra, G., Hochstein, D., and Barile, M. F.,** Proc. 3rd Conf. Int. Org. for Mycoplasmology, Custer, So. Dak., September 1980, 135.
366. **Shaw, N., Smith, P. F., and Koostra, W. L.,** The lipid composition of *Mycoplasma laidlawii* strain B, *Biochem. J.,* 107, 329, 1968.
367. **Shimshick, E. J. and McConnell, H. M.,** Lateral phase separation in phospholipid membranes, *Biochemistry,* 2351, 1973.
368. **Silvius, J. R., Jinks, D. C., and McElhaney, R. N.,** Effect of membrane lipid fluidity and phase state on the kinetic properties of the membrane-bound adenosine triphosphatase of *Acholeplasma laidlawii* B, in *Microbiology-1979,* Schlesinger, D., Ed., American Society for Microbiology, Washington, D. C., 1979, 10.
369. **Silvius, J. R., Mak, N., and McElhaney, R. N.,** Lipid and protein composition and thermotropic lipid phase transitions in fatty acid-homogenous membranes of *Acholeplasma laidlawii* B, *Biochim. Biophys. Acta,* 597, 199, 1980.
370. **Silvius, J. R. and McElhaney, R. N.,** Growth and membrane lipid properties of *Acholeplasma laidlawii* B lacking fatty acid heterogeneity, *Nature (London),* 272, 645, 1978.
371. **Silvius, J. R. and McElhaney, R. N.,** Lipid compositional manipulation in *Acholeplasma laidlawii* B. Effect of exogenous fatty acids on fatty acid composition and cell growth when endogenous fatty acid production is inhibited, *Can. J. Biochem.,* 56, 462, 1978.
372. **Silvius, J. R. and McElhaney, R. N.,** Membrane lipid physical state and modulation of the (Na^+, Mg^{2+})-adenosine-triphosphatase activity in *Acholeplasma laidlawii* B, *Proc. Natl. Acad. Sci. U.S.A.,* 77, 1255, 1980.
373. **Silvius, J. R., Saito, Y., and McElhaney, R. N.,** Membrane lipid biosynthesis in *Acholeplasma laidlawii* B. Investigations into the *in vivo* regulation of the quantity and hydrocarbon chain lengths of *de novo* biosynthesized fatty acids in response to exogenously supplied fatty acids, *Arch. Biochem. Biophys.,* 182, 455, 1977.
374. **Singer, S. J. and Nicolson, G. L.,** The fluid mosaic model of the structure of cell membranes, *Science,* 175, 720, 1972.
375. **Slutzky, G. M., Razin, S., Kahane, I., and Eisenberg, S.,** Serum lipoproteins as cholesterol donors to mycoplasma membranes, *Biochem. Biophys. Res. Commun.,* 68, 529, 1976.
376. **Slutzky, G. M., Razin, S., Kahane, I., and Eisenberg, S.,** Cholesterol transfer from serum lipoproteins to mycoplasma membranes, *Biochemistry,* 16, 5158, 1977.
377. **Slutzky, G. M., Razin, S., Kahane, I., and Eisenberg, S.,** Inhibition of mycoplasma growth by human very-low-density lipoproteins, *FEMS Lett.,* 2, 185, 1977.
378. **Smith, I. C. P., Butler, K. W., Tulloch, A. P., Davis J. H. and Bloom, M.,** The properties of gel state lipid in membranes of *Acholeplasma laidlawii* as observed by ^2H NMR, *FEBS Lett.,* 100, 57, 1979.
379. **Smith, G. G., Ruwart, M. J., and Haug, A.,** Lipid phase transitions in membrane vesicles from *Thermoplasma acidophila, FEBS Lett.,* 45, 96, 1974.
380. **Smith, P. F.,** The carotenoid pigments of mycoplasma, *J. Gen. Microbiol.,* 32, 307, 1963.
381. **Smith, P. F.,** Nature of unsaponifiable lipids of a mycoplasma strain grown with isopentenyl pyrophosphate as a substitute for sterol, *J. Bacteriol.,* 95, 1718, 1968.
382. **Smith, P. F.,** The role of lipids in membrane transport in *Mycoplasma laidlawii, Lipids,* 4, 331, 1969.
383. **Smith, P. F.,** *The Biology of Mycoplasmas,* Academic Press, New York, 1971.
384. **Smith, P. F.,** A phosphatidyl diglucosyl diglyceride from *Acholeplasma laidlawii* B, *Biochim. Biophys. Acta,* 280, 375, 1972.
385. **Smith, P. F.,** Homogeneity of lipopolysaccharides from *Acholeplasma, J. Bacteriol.,* 130, 393, 1977.
386. **Smith, P. F.,** The composition of membrane lipids and lipopolysaccharides, in *The Mycoplasmas,* Vol. 1, Barile, M.F. and Razin, S., Eds., Academic Press, New York, 1979, 231.
387. **Smith P. F. and Henrikson, C. U.,** Growth inhibition of mycoplasma by inhibitors of polyterpene biosynthesis and its reversal by cholesterol, *J. Bacteriol.,* 91, 1854, 1966.
388. **Smith, P. F. and Langworthy, T. A.,** Existence of carotenoids in *Acholeplasma axanthum, J. Bacteriol.,* 137, 185, 1979.
389. **Smith, P. F., Langworthy, T. A., and Mayberry, W. R.,** Distribution and composition of lipopolysaccharides from mycoplasmas, *J. Bacteriol.,* 125, 916, 1976.
390. **Smith, P. F., Langworthy, T. A., Mayberry, W. R., and Hougland, A. E.,** Characterization of the membranes of *Thermoplasma acidophilum, J. Bacteriol.,* 116, 1019, 1973.
391. **Smith, P. F. and Lynn, R. J.,** Biological effects of acholeplasmal lipopolysaccharides, *Zbl. Bakt. Parasitenk.-1 Abt.-A,* 241, 180, 1978.
392. **Smith, P. F. and Rothblatt, G. H.,** Incorporation of cholesterol by pleuropneumonia-like organisms, *J. Bacteriol.,* 80, 842, 1960.

393. **Smith, P. F. and Sasaki, S.,** Stability of pleuropneumonia-like organisms to some physical factors, *Appl. Microbiol.,* 6, 184, 1958.
393a. **Smith, P. F.,** in Pro. 3rd Conf. Int. Org. for Mycoplasmology, Custer, So. Dak., September 1980, 150.
394. **Sobeslavsky, O., Prescott, B., and Chanock, R. M.,** Adsorption of *Mycoplasma pneumoniae* to neuraminic acid receptors of various cells and possible role in virulence, *J. Bacteriol.,* 96, 695, 1968.
395. **Somerson, N. L., James, W. D., Walls, B. E., and Chanock, R. M.,** Growth of *Mycoplasma pneumoniae* on a glass surface, *Ann. N.Y. Acad. Sci.,* 143, 384, 1967.
396. **Stalheim, O. H. V., Ritchie, A. E., and Whitcomb, R. F.,** Cultivation, serology, ultrastructure, and virus-like particles of spiroplasma 277F, *Curr. Microbiol.,* 1, 365, 1978.
397. **Stanbridge, E. J.,** A reevaluation of the role of mycoplasmas in human disease, *Annu. Rev. Microbiol., 30, 169, 1976.*
398. **Stanbridge, E. J. and Reff, M. E.,** The molecular biology of mycoplasmas, in *The Mycoplasmas,* Vol. 1, Barile, M. F., and Razin, S., Eds., Academic Press, New York, 1979, 157.
399. **Stanbrige, E. J. and Weiss, R. L.,** Mycoplasma capping on lymphocytes, *Nature,* 276, 583, 1978.
400. **Steim, J. M., Tourtellotte, M. E., Reinert, J. C., McElhaney, R. N., and Rader, R. L.,** Calorimetric evidence for the liquid-crystalline state of lipids in a biomembrane, *Proc. Nat. Acad. Sci. U.S.A.,* 63, 104,1969.
401. **Stockton, G. W., Johnson, K. G., Bulter, K. W., Polnaszek, C. F., Cyr, R., and Smith, I. C. P.,** Molecular order in *Acholeplasma laidlawii* membranes as determined by deuterium magnetic resonance of biosynthetically-incorporated specifically-labelled lipids, *Biochim. Biophys. Acta,* 401, 535, 1975.
402. **Stockton, G. W., Johnson, K. G., Butler, K., Tulloch, A. P., Boulanger, Y., and Smith, I. C. P.,** Deuterium NMR study of lipid organisation in *Acholeplasma laidlawii* membranes, *Nature (London),* 269, 267, 1977.
403. Subcommittee on the Taxonomy of Mollicutes, Proposal of minimal standards for description of new species of the class *Mollicutes, Int. J. Syst. Bacteriol.,* 29, 172, 1979.
404. **Sugiyama, T., Smith, P. F., Langworthy, T. A., and Mayberry, W. R.,** Immunological analysis of glycolipids and lipopolysaccharides derived from various mycoplasmas, *Infect. Immun.,* 10, 1273, 1974.
405. **Tarshis, M. A., Berkouzjin, A. G., and Ladygina, V. G.,** On the possible role of respiratory activity of *Acholeplasma laidlawii* cells in sugar transport, *Arch. Microbiol.,* 109, 295, 1976.
406. **Tarshis, M. A., Berkouzjin, A. G., Ladygina, V. G., and Panchenko, L. F.,** Properties of the 3-*0*-methyl-D-glucose transport system in *Acholeplasma laidlawii,* J. Bacteriol., 125, 1, 1976.
407. **Tarshis, M. A. and Kapitanov, A. B.,** Symport H^+/carbohydrate transport into *Acholeplasma laidlawii* cells, *FEBS Lett.,* 89, 73, 1978.
408. **Tarshis, M. A., Migoushina, V. L., and Panchenko, L. F.,** On the phosphorylation of sugars in *Acholeplasma laidlawii, FEBS Lett.,* 31, 111, 1973.
409. **Tarshis, M. A., Migoushina, V. L., Panchenko, L. F., Fedotov, N. S., and Bourd, H. I.,** Studies of sugar transport in *Acholeplasma laidlawii, Eur. J. Biochem.,* 40, 171, 1973.
410. **Taylor-Robinson, D. and Manchee, R. J.,** Adherence of mycoplasmas to glass and plastic, *J. Bacteriol.,* 94, 1781, 1967.
411. **Taylor-Robinson, D. and Manchee, R. J.,** Spermadsorption and spermagglutination by mycoplasmas, *Nature (London),* 215, 484, 1967.
412. **Teuber, M. and Bader, J.,** Action of polymyxin B on bacterial membranes: phosphatidylglycerol- and cardiolipin-induced susceptibility to polymyxin B in *Acholeplasma laidlawii* B, *Antimicrob. Agents Chemother.,* 9, 26, 1976.
413. **Tillack, T. W., Carter, R., and Razin, S.,** Native and reformed *Mycoplasma laidlawii* membranes compared by freeze-etching, *Biochim. Biophys. Acta,* 219, 123, 1970.
414. **Tornabene, T. G. and Langworthy, T. A.,** Diphytanyl and dibiphytanyl glycerol ether lipids of methanogenic archaebacteria, *Science,* 203, 51, 1979.
415. **Tourtellotte, M. E., Branton, D., and Keith, A.,** Membrane structure: spin labeling and freeze etching of *Mycoplasma laidlawii, Proc. Nat. Acad. Sci. U.S.A.,* 66, 909, 1970.
416. **Tourtellotte, M. E. and Zupnick, J. S.,** Freeze-fractured *Acholeplasma laidlawii* membranes: nature of particles observed, *Science,* 179, 84, 1973.
417. **Townsend, R., Archer, D. B., and Plaskitt, K. A.,** Purification and preliminary characterization of spiroplasma fibrils, *J. Bacteriol.,* 142, 694, 1980.
418. **Townsend, R., Markham, P. G., Plaskitt, K. A., and Daniels, M. J.,** Isolation and characterization of a non-helical strain of *Spiroplasma citri, J. Gen. Microbiol.,* 100, 15, 1977.
419. **Tully, J. G.,** Special features of the acholeplasmas, in *The Mycoplasmas,* Vol. I. Barile, M. F. and Razin, S., Eds., Academic Press, New York, 1979, 431.

420. **Turner, A. W.,** A study of the morphology and life cycles of the organism of *Pleuropneumonia contagiosa bovum (Borrelomyces peripneumoniae* nov. gen.) by observation in the living state under dark-ground illumination, *J. Pathol. Bacteriol.,* 41, 1, 1935.

421. **van Deenen, L. L. M.,** Phospholipids and biomembranes, in *Progress in the Chemistry of Fats and Other Lipids,* Vol. III, Part I, Holman, R.T., Ed., Pergamon Press, London, 1965.

422. **van Demark, P. J., and Plackett, P.,** Evidence for a phosphoenolpyruvate-dependent sugar phosphotransferase in *Mycoplasma* strain Y, *J. Bacteriol.,* 111, 454, 1972.

423. **Vanderkooi, G.,** Organization of proteins in membranes with special references to the cytochrome oxidase system, *Biochim. Biophys. Acta,* 344, 307, 1974.

424. **van der Neut-Kok, E. C. M., de Gier, J., Middelbeek, E. J., and van Deenen, L. L. M.,** Valinomycin-induced potassium and rubidium permeability of intact cells of *Acholeplasma laidlawii* B, *Biochim. Biophys. Acta,* 332, 97, 1974.

425. **van Golde, L. M. G., McElhaney, R. N., and van Deenen. L. L. M.,** A membrane-bound lysophospholipase from *Mycoplasma laidlawii* strain B, *Biochim. Biophys. Acta,* 231, 245, 1971.

426. **van Iterson, W. and Ruys, A. C.,** The fine structure of the mycoplasmataceae (microorganisms of the pleuropneumonia group = PPLO). 1. *Mycoplasma hominis, M. fermentans* and *M. salivarium, J. Ultrastruct. Res.,* 3, 282, 1960.

427. **van Zoelen, E. J. J., van der Neut-Kok, E. C. M., de Gier, J., and van Deenen, L. L. M.,** Osmotic behavior of *Acholeplasma laidlawii* B cells with membrane lipids in liquid-crystalline and gel state, *Biochim, Biophys. Acta,* 394, 463, 1975.

428. **Verkleij, A. J., Ververgaert, P. H. J., van Deenen. L. L. M., and Elbers, P. F.,** Phase transitions of phospholipid bilayers and membranes of *Acholeplasma laidlawii* B visualized by freeze fracturing electron microscopy, *Biochim. Biophys. Acta,* 288, 326, 1972.

429. **Verkleij, A. J. and Ververgaert, P. H. J. Th.,** Freeze-fracture morphology of biological membranes, *Biochim. Biophys. Acta,* 515, 303, 1978.

430. **Virkola, P.,** Acid-soluble nucleotides of *Acholeplasma (Mycoplasma) laidlawii A, Acta Pathol, Microbiol. Scand. B,* 86, 179, 1978.

431. **Wallace, B. A. and Engelman, D. M.,** The planar distributions of surface proteins and intramembrane particles in *Acholeplasma laidlawii* are differentially affected by the physical state of membrane lipids, *Biochim. Biophys. Acta,* 508, 431, 1978.

432. **Wallace, B. A., Richards, F. M., and Engelman, D. M.,** The influence of lipid state on the planar distribution of membrane proteins in *Acholeplasma laidlawii, J. Mol. Biol.,* 107, 255, 1976.

433. **Weber, M. M. and Kinsky, S. C.,** Effect of cholesterol on the sensitivity of *Mycoplasma laidlawii* to the polyene antibiotic filipin, *J. Bacteriol.,* 89, 306, 1965.

434. **Wieslander, Å., Christiansson, A., Walter, H., and Weibull, C.,** Fractionation of membranes from *Acholeplasma laidlawii* A on the basis of their surface properties by partition in two-polymer aqueous phase systems, *Biochim. Biophys. Acta,* 550, 1, 1979.

435. **Wieslander, Å. and Rilfors, L.,** Qualitative and quantitative variations of membrane lipid species in *Acholeplasma laidlawii* A, *Biochim. Biophys. Acta,* 466, 336, 1977.

436. **Wieslander, Å., Ulmius, J., Lindblom, G., and Fontele, K.,** Water binding and phase structures for different *Acholeplasma laidlawii* membrane lipids studied by deuteron nuclear magnetic resonance and x-ray diffraction, *Biochim. Biophys. Acta,* 512, 241, 1978.

437. **Williamson, D. L.,** Unusual fibrils from the spirochete-like sex ratio organism, *J. Bacteriol,* 117, 904, 1974.

438. **Williamson, D. L., Blaustein, D. I., Levine, R. J. C., and Elfin, M. J.,** Anti-actin peroxidase staining of the helical wall-free prokaryote *Spiroplasma citri, Curr. Microbiol.,* 2, 143, 1979.

439. **Wilson, M. H. and Collier, A. M.,** Ultrastructural study of *Mycoplasma pneumoniae* in organ culture, *J. Bacteriol.,* 125, 332, 1976.

440. **Wise, K. S., Cassell, G. H., and Acton, R. T.,** Selective association of murine T lymphoblastoid cell surface alloantigens with *Mycoplasma hyorhinis, Proc. Natl. Acad. Sci. U.S.A.,* 75, 4479, 1978.

441. **Woese, C. R., Magrum, L. J., and Fox, G. E.,** Archaebacteria, *J. Mol. Evol.,* 11, 245, 1978.

442. **Wròblewski, H.,** Spiralin: its topomolecular anatomy and its possible function in the *Spiroplasma citri* cell membrane, *Zbl. Bakt. Parasitenk.-1 Abt.-A,* 241, 179, 1978.

443. **Wròblewski, H.,** Cross-linking of spiralin in the *Spiroplasma citri* cell membrane, *J. Bacteriol.,* in press.

444. **Wròblewski, H., Burlot, R., and Johansson, K.-E.,** Solubilization of *Spiroplasma citri* cell membrane proteins with the anionic detergent sodium lauryl sarcosinate (Sarkosyl), *Biochimie,* 60, 389, 1978.

445. **Wròblewski, H., Johannson, K.-E., and Burlot, R.,** Crossed immunoelectrophoresis of membrane proteins from *Acholeplasma laidlawii* and *Spiroplasma citri, Int. J. Syst. Bacteriol.,* 27, 97, 1977.

446. **Wròblewski, H., Johansson, K.-E. Hjérten, S.,** Purification and characterization of spiralin, the main protein of the *Spiroplasma citri* membrane, *Biochim. Biophys. Acta,* 465, 275, 1977.

447. **Wurtz, M., Jacobson, G. R., Steven, A. C., and Rosenbusch, J. P.,** Paracrystalline arrays of protein-synthesis elongation factor Tu, *Eur. J. Biochem.,* 88, 593, 1978.
448. **Yaguzhinskaya, O. E.,** Detection of serum proteins in the electrophoretic patterns of total proteins of mycoplasma cells, *J. Hyg. Camb.,* 77, 189, 1976.
449. **Yamakawa, T. and Nagai, Y.,** Glycolipids at the cell surface and their biological functions, *Trends Biochem. Sci.,* 3, 128, 1978.
450. **Yang, L. L. and Haug, A.,** Purification and partial characterization of a procaryotic glycoprotein from the plasma membrane of *Thermoplasma acidophilum, Biochim., biophys. Acta,* 556, 265, 1979.
451. **Yang, L. L. and Haug, A.,** Structure of membrane lipids and physicobiochemical properties of the plasma membrane from *Thermoplasma acidophilum,* adapted to growth at 37°C, *Biochim. Biophys. Acta,* 573, 308, 1979.
452. **Zucker-Franklin, D., Davidson, M., and Thomas, L.,** The interaction of mycoplasmas with mammalian cells. I. HeLa cells, neutrophils, and eosinophils, *J. Exp. Med.,* 124, 521, 1966.
453. **Kahane, I. and Gilliam, J. M., personal communication.**
454. **Gross, Z., Markowitz, O., and Rottem, S.,** unpublished data.
455. **Ephrati, H.,** unpublished data.
456. **Razin, S. and Greenberg, A.,** unpublished data.
457. **Shinar, D. and Rottem, S.,** unpublished data.
458. **Johansson, K.-E.,** personal communication.
459. **Mouches, C. and Bové, J.,** personal communication.
460. **Townsend, R.,** personal communication.
461. **Cirillo. V. P.,** unpublished data.
462. **Banai, M.,** unpublished data.
463. **Kubicki, J.,** personal communication.
464. **Efrati, H., Razin, S., and Rottem,** unpublished data.

INDEX

A

B

C

D

E

F

G

H

I

K

L

M

N

O

P

R

S

T

U

V

W

X

Z